PROTON PASSAGE ACROSS CELL MEMBRANES

The Ciba Foundation is an international scientific and educational charity. It was established in 1947 by the Swiss chemical and pharmaceutical company of CIBA Limited—now CIBA-GEIGY Limited. The Foundation operates independently in London under English trust law.

The Ciba Foundation exists to promote international cooperation in biological, medical and chemical research. It organizes about eight international multidisciplinary symposia each year on topics that seem ready for discussion by a small group of research workers. The papers and discussions are published in the Ciba Foundation symposium series. The Foundation also holds many shorter meetings (not published), organized by the Foundation itself or by outside scientific organizations. The staff always welcome suggestions for future meetings.

The Foundation's house at 41 Portland Place, London, W1N 4BN, provides facilities for meetings of all kinds. Its Media Resource Service supplies information to journalists on all scientific and technological topics. The library, open seven days a week to any graduate in science or medicine, also provides information on scientific meetings throughout the world and answers general enquiries on biomedical and chemical subjects. Scientists from any part of the world may stay in the house during working visits to London.

Ciba Foundation Symposium 139

PROTON PASSAGE ACROSS CELL MEMBRANES

A Wiley – Interscience Publication

1988

JOHN WILEY & SONS

Chichester · New York · Brisbane · Toronto · Singapore

Published in 1988 by John Wiley & Sons Ltd, Baffins Lane,
Chichester, Sussex PO19 1UD, UK.

Suggested series entry for library catalogues:
Ciba Foundation Symposia

Ciba Foundation Symposium 139
x + 268 pages. 67 figures, 13 tables

Library of Congress Cataloging in Publication Data

Proton passage across cell membranes.
p. cm. — (Ciba Foundation symposium; 139)
Proceedings of the Symposium on Proton Passage across Cell Membranes, held at the Ciba Foundation, London, 9–11 Feb. 1988.
Edited by Gregory Bock and Joan Marsh.
'A Wiley Interscience publication.'
Includes index.
ISBN 0 471 91903 9
1. Biological transport — Congresses. 2. Protons — Congresses.
3. Cell membranes — Congresses. I. Bock, Gregory. II. Marsh, Joan.
III. Ciba Foundation. IV. Symposium on Proton Passage Across Cell Membranes (1988: Ciba Foundation) V. Series.
QH509.P77 1988
574.87′5—dc19 88-20526
CIP

British Library Cataloguing in Publication Data

Proton passage across cell membranes.
1. Organisms. Cells. Membranes. Protons.
Transport phenomena.
I. Bock, Gregory II. Marsh, Joan.
III. Series
574.87′5

ISBN 0 471 91903 9

Typeset by Inforum Ltd, Portsmouth
Printed and bound in Great Britain by Bath Press,
Bath, Avon

Contents

Symposium on Proton passage across cell membranes, held at the Ciba Foundation, London 9–11 February 1988

This symposium is based on a proposal made by Professor Roger Thomas and Dr Brian Harvey

Editors: Gregory Bock (Organizer) and Joan Marsh

Participants

C.C. Aickin Department of Pharmacology, University of Oxford, South Parks Road, Oxford OX1 3QT, UK

Q. Al-awqati Department of Medicine & Physiology, College of Physicians & Surgeons of Columbia University, 630 West 168th Street, New York, NY 10032, USA

W.F. Boron Department of Physiology, Yale University School of Medicine, B-106 SHM, PO Box 333, 333 Cedar Street, New Haven CT 06510-8026, USA

N.A. Curtin Department of Physiology, Charing Cross & Westminster Medical School, Fulham Palace Road, London W6 8RF, UK

A. de Hemptinne Laboratorium voor Normale & Pathologische Fysiologie, Rijksuniversiteit Gent, A.Z. Blok (B), De Pintelaan 185, B-900 Ghent, Belgium

E. Frömter Zentrum der Physiologie Klinikum der J W Goethe-Universität, Theodor-Stern-Kai 7, D-6000 Frankfurt 70, Federal Republic of Germany

J.I. Gillespie Department of Physiological Sciences, University of Newcastle upon Tyne, School of Medicine, Framlington Place, Newcastle upon Tyne NE2 4HH, UK

J.R. Green Rosenthal R-1056.P.11, CIBA-GEIGY AG, CH-4002 Basle, Switzerland

S. Grinstein Department of Cell Biology, Hospital for Sick Children, 555 University Avenue, Toronto, Canada M5G 1X8

B.J. Harvey Laboratoire Jean Maetz, Department de Biologie, Commissariat à l'Energie Atomique, Station Marine, BP 68, F-06230 Villefranche-sur-Mer, France

K. Kaila Department of Zoology, Division of Physiology, University of Helsinki, Arkadiankatu 7, SF-00100 Helsinki 10, Finland

T.E. Machen Department of Physiology-Anatomy, University of California, Berkeley, CA 94720, USA

G. Malnic Department of Physiology & Biophysics, Instituto de Ciencias Biomedicas, Universidade de Sao Paulo, Cidade Universitaria – CEP 05508, Caixa Postal 4365 – CEP 01051, Sao Paulo, SP, Brazil

W.H. Moolenaar Int Embryologie Institut, Hubrecht Laboratorium, Uppsalalaan 8, 3584 CT Utrecht, The Netherlands

M. Morad Department of Physiology, University of Pennsylvania School of Medicine, Philadelphia, PA 10104-6084, USA

H. Oberleithner Institut für Physiologie, Universität Würzburg, Röntgenring 9, D-8700 Würzburg, Federal Republic of Germany

J.M. Russell Department of Physiology & Biophysics, University of Texas, Medical Branch, Galveston, TX 77550-2781, USA

I. Sabolic Department of Physiology, University of Zagreb, Faculty of Medicine, Box 978, Salata 3, Yu-41001 Zagreb, Yugoslavia

W.-R. Schlue Institut für Zoologie, Lehrstuhl Neurobiologie, Universität Düsseldorf, Universitätsstrasse 1, D-4000 Düsseldorf 1, Federal Republic of Germany

P. Steels Department of Mathematics, Physics, Computer Sciences & Physiology, Limburgs Universitair Centrum, Universitaire Campus, B-3610 Diepenbeek, Belgium

P.R. Steinmetz Department of Medicine, University of Connecticut School of Medicine, Health Center, Farmington, CT 06032-9984, USA

D.K. Stone Department of Internal Medicine, University of Texas, Health Sciences Center, 5323 Harry Hines Boulevard, Dallas, TX 75235-9030, USA

E. Syková Institute of Physiological Regulations, Czechoslovak Academy of Science, Bulovka, Pav 11, CS-18085 Prague 8, Czechoslovakia

R.C. Thomas (*Chairman*) Department of Physiology, School of Medical Sciences, University of Bristol, Bristol BS8 1TD, UK

G. Vairo (*Bursar*) Department of Medicine, University of Melbourne, Royal Melbourne Hospital, Parkville, Melbourne 3050, Australia

R.D. Vaughan-Jones Department of Physiology, University of Oxford, Parks Road, Oxford OX2 3PT, UK

Introduction

It is generally thought that the first cellular organisms were formed by the enclosure of certain components of the 'primeval soup' within a membrane that had to be impervious to macromolecules such as proteins and nucleic acids. Since the ionic composition of these first cells was, by definition, similar to that of the surrounding medium, the permeability of the membrane to ions was, initially, of little consequence. The adaptation of cells to survival in a variable environment with the accompanying need to regulate osmolarity meant that ionic concentrations within cells had to be maintained independently of external conditions. This required a membrane that formed a barrier to passive ion movement but contained specific mechanisms for the uptake or extrusion of particular ion species. In most animals, the major extracellular ions are sodium, chloride and bicarbonate; the chief intracellular ions are potassium, magnesium, phosphates, sulphate, bicarbonate and charged groups on proteins. Transport of many of these ions is linked to the movement of protons.

Intracellular pH can vary considerably for distinct cell types, from 4.5 in cells of the prostate to 8.5 in osteoblasts. Within the cell, individual organelles often have a different pH to the bulk cytoplasm, which is important for their particular functions. Thus, the ion transporters within the membranes of these organelles may have properties distinct from transport mechanisms in the plasma membrane. When the organelle membrane is derived from the plasma membrane by endocytosis, this raises the interesting question of the origin of the specialized transporter.

Cellular pH is affected by the activity of the cell; protons produced by metabolism in the form of lactic acid, for example, have to be excreted from the cell. Such acid products of metabolism acidify the blood plasma, and this has to be corrected in the kidney, where salt excretion and plasma pH are regulated by specific ion transporters. A recent discovery is that the passage of protons across an intact epithelium involves the asymmetric distribution of different transporters on the cell membranes. In nerve cells, propagation of the signal occurs via rapid ion fluxes down electrochemical gradients. Proton movements may affect this in various ways: proton pumping may be driven by the same ion gradient as used for transmission of nerve impulses; protons can induce transformation of ion channels, altering the affinity for their respective ion; and, in snail neurons, depolarization opens proton channels directly.

Proton transport can occur either by exchange for another cation or by association with a weak acid anion, particularly bicarbonate, to form an uncharged molecule which can diffuse across the membrane. Studies of the properties of the different types of transporters are becoming more sophisticated as techniques are refined: the availability of accurate and sensitive microelectrodes facilitates measurements in small cells, electron microscopy is revealing fine structural details, and the tools of molecular biology are now being used to probe the physical composition of the channels.

Protons move across the membranes of all types of cells. In this symposium the characteristics of the range of transport processes mentioned above are described, together with the details of their similarities and differences, particularly with respect to their pharmacological properties. The role of the individual ion transporters in cellular metabolism, and the significance of results obtained from studies *in vitro* for the physiological state, are discussed.

Joan Marsh

Movement of acid equivalents across the mammalian smooth muscle cell membrane

C. Claire Aickin

University Department of Pharmacology, South Parks Road, Oxford OX1 3QT

Abstract. Factors affecting intracellular pH (pH_i) in the smooth muscle of guinea pig ureter have been investigated using pH-sensitive microelectrodes. Associated acids and bases appear to have free passage across the cell membrane but results suggest very low permeability to charged acid equivalents, thus implicating carrier-mediated movements in many of the observed pH_i transients. Recovery from acidosis in the nominal absence of CO_2 was inhibited by removal of Na^+ and by the presence of amiloride, indicating that Na^+/H^+ exchange was responsible. The presence of CO_2 resulted in a faster recovery from acidosis but, since intracellular buffering power was not increased, not a substantially faster effective extrusion of protons. Surprisingly, amiloride no longer caused discernable inhibition. Recovery from moderate acidosis remained Na^+ dependent but was not inhibited by DIDS or acetazolamide or by the absence of Cl^-, suggesting a dominant Na^+-, and HCO_3^--dependent mechanism unlike any hitherto described. Recovery from alkalosis was inhibited by DIDS and Cl^--free conditions, indicating that Cl^-/HCO_3^- exchange was involved. Results suggest reversal of this mechanism on extreme acidosis. Experiments in vascular smooth muscle with fluorescent indicators confirm the presence of Na^+/H^+ exchange but provide conflicting evidence about the presence and properties of the HCO_3^--dependent mechanism.

1988 Proton passage across cell membranes. Wiley, Chichester (Ciba Foundation Symposium 139) p 3–22

Smooth muscle is a latecomer to the field of transmembrane proton movements. Microelectrodes are difficult to use in the small cells (about 3 μm in diameter), nuclear magnetic resonance appears to require a degree of metabolic poisoning in order to obtain a large enough signal (Spurway & Wray 1987), but fluorescent pH indicators seem ideally suited and are being used with increasing incidence. Results obtained using this technique, together with my current research using microelectrodes, have been responsible for a recent rapid advance.

The first indications of the transport mechanisms involved in effective proton extrusion from smooth muscle cells came indirectly from observation of an amiloride-sensitive fraction of Na^+ uptake, assuming the action of amiloride to be inhibition of Na^+/H^+ exchange (Smith & Brock 1983, Owen 1984), and of an amiloride-sensitive recovery from increased tension following intracellular acidosis (Ighoroje & Spurway 1985). At the same time, I made the first direct measurements of intracellular pH (pH_i) in smooth muscle with a trustworthy pH-sensitive microelectrode (Aickin 1983). My initial work gave evidence for a Cl^-/HCO_3^- exchange mechanism which affected transport of acid equivalents when extracellular Cl (Cl_o) was altered, but which was not involved in the extrusion of acid equivalents after moderate intracellular acidosis (Aickin & Brading 1984). Later, I showed that recovery from acidosis occurred via HCO_3^--dependent and HCO_3^--independent mechanisms, both of which were dependent upon Na^+ (Aickin 1985). This work was reviewed in 1986 (Aickin 1986) and forms the background to the recent research presented here.

My objective is to characterize the mechanisms responsible for control of pH_i in smooth muscle. Currently, there is no guarantee that the changes in pH_i observed in response to acid or alkaline loads are brought about by movement of protons rather than other acid equivalents or even that the movements are across the cell membrane. Nevertheless, since I have recorded an unchanged rate of recovery from acidosis over an eight hour period in a single cell with repeated intracellular acid loadings, it seems unlikely that the recovery can be explained by internal sequestration. Therefore, until it is proved otherwise, I am happy to assume that the majority of the changes observed in pH_i are caused by trans-sarcolemmal movements of acid equivalents, but not necessarily of protons.

Methods

My experiments have been done in the smooth muscle of isolated guinea pig ureter with double-barrelled microelectrodes using the proton cocktail of Ammann et al (1981) and the reference liquid ion exchanger of Thomas & Cohen (1981) to measure pH_i and membrane potential (E_m), respectively. Full details of the electrode and other experimental procedures have been published (Aickin 1984, 1987). All experiments were performed at 35 °C.

Results and discussion

Application of an acid load

Fig. 1 illustrates the, now standard, method used to induce an intracellular acidosis (see Thomas 1984). In brief, application of an ammonium salt causes a rapid alkalinization due to NH_3 entry and combination with protons to form NH_4^+, followed by an assumed, slower entry of NH_4^+ and consequent slow

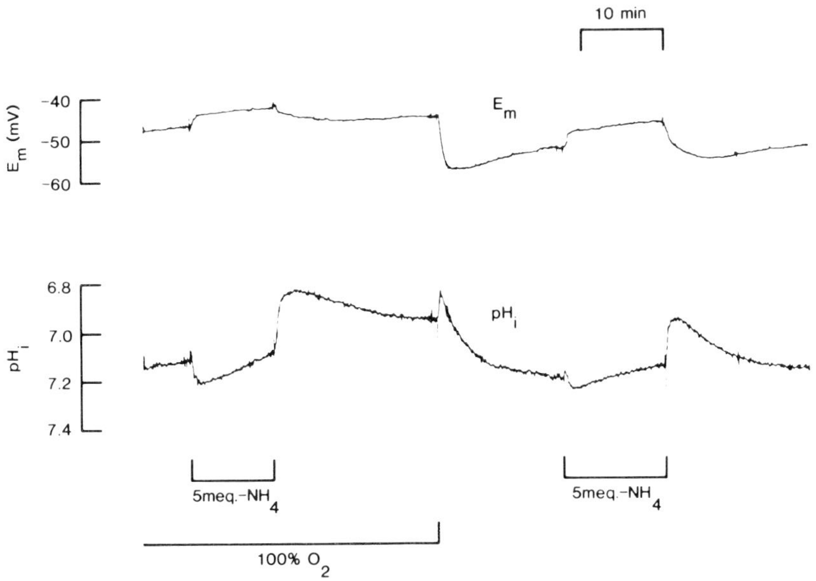

FIG. 1. Pen recordings of membrane potential (E_m) and intracellular pH (pH_i) in the smooth muscle of guinea pig ureter obtained using a double-barrelled pH-sensitive microelectrode. An intracellular acid load was applied by addition and removal of 2.5 mM $(NH_4)_2SO_4$ in the nominal absence of CO_2 (solution buffered with 21 mM NaHepes and equilibrated with 100% O_2). The external buffer was then changed to 5% CO_2, 21 mM HCO_3^- at the same pH (7.47). This caused a second acidosis and, after stabilization of pH_i, 2.5 mM $(NH_4)_2SO_4$ was again added and removed. Note that recovery from acidosis was faster and the steady-state pH_i more alkaline in the presence than in the nominal absence of CO_2.

acidification. Removal of the ammonium salt then causes a large acidification which overshoots the steady-state pH_i as virtually all of the accumulated NH_4^+ leaves as NH_3. Addition of CO_2 similarly causes a rapid acidification due to entry of CO_2 and production of protons from its subsequent hydration. Recovery from acidification must reflect effective extrusion of acid equivalents against the electrochemical gradient. Of course, all these changes in pH_i are brought about by effective transfer of protons. Some are via movements of freely permeable uncharged weak acids or bases, or via the supposed passive entry of a charged weak acid. But it is the physiological response to attempt to return pH_i to normal that is of prime interest.

Dependence of acid extrusion on HCO_3^-

Fig. 1 shows that recovery from acidosis is faster in the presence than in the nominal absence of CO_2 and HCO_3^- (Aickin 1985). The same result has also been obtained using fluorescence to monitor pH_i in isolated segments of small arteries from the rat (C. Aalkjær & E.J. Cragoe, personal communication) but not in cultured vascular smooth muscle (Korbmacher et al 1988). Since it is generally assumed that the intracellular buffering power is greater in the presence than nominal absence of CO_2, even an unchanged rate of pH_i recovery should represent significant slowing of the rate of effective acid extrusion.

Fig. 1 also illustrates the unusual finding of a more acidic steady-state pH_i in the nominal absence, rather than presence, of CO_2 and HCO_3^-. This has also been observed in guinea pig vas deferens (Aickin 1984), rabbit aorta (using distribution of the weak acid, 5,5-dimethyloxazolidine-2,4-dione, Gerstheimer et al 1987), and cultured vascular smooth muscle cells (using fluorescence; Korbmacher et al 1988), but otherwise only in glial cells (Deitmer & Schlue 1987) and some cardiac ventricular cells. (R.D. Vaughan-Jones, personal communication). Such behaviour is consistent with the presence of a dominant, HCO_3^--dependent mechanism for acid extrusion although other explanations are not excluded.

Dependence of acid extrusion on Na^+

Recovery from acidosis requires Na^+_o (see Fig. 2), both in the presence and nominal absence of CO_2 (Aickin 1985), but the apparent affinity for Na^+ is

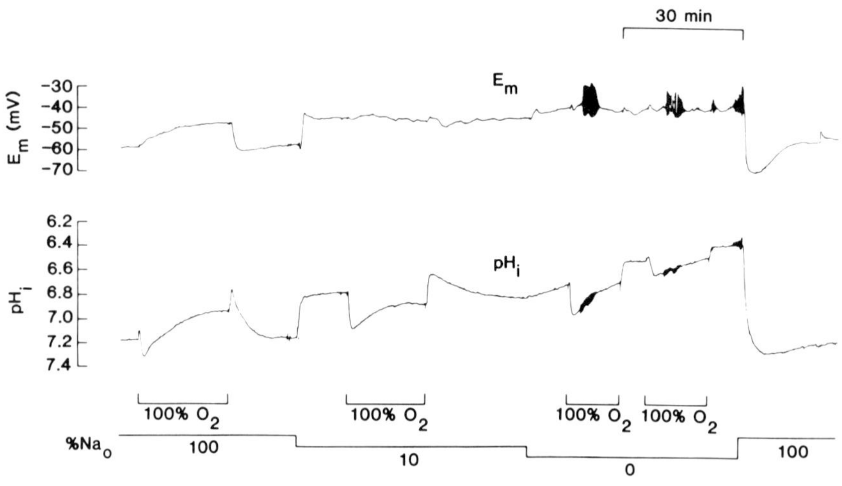

FIG. 2. Pen recordings of an experiment showing the effect of external Na^+ on pH_i and on the recovery of pH_i from acidosis induced by addition of 5% CO_2. Reduction of Na^+_o (substituted by Tris) inhibited recovery from acidosis. In the absence of Na^+_o and nominal absence of CO_2, pH_i continued to fall well below the level predicted by a passive distribution: at the E_m of −40 mV, protons would be in equilibrium across the sarcolemma at a pH_i of 6.8. Readdition of Na^+_o caused a dramatic recovery of pH_i and marked hyperpolarization.

greater in the presence of CO_2. $Na^+{}_o$ had to be reduced to 1% to prevent recovery from moderate acidosis in the presence of CO_2 but only to 10% (13.6 mM) in its nominal absence. Inhibition of acid extrusion by removal of $Na^+{}_o$ has also been found in vascular smooth muscle using fluorescence measurements (Weissberg et al 1987, Korbmacher et al 1988, C. Aalkjær & E.J. Cragoe, personal communication) and a K_m of 14.6 mM in the nominal absence of CO_2 has been reported (Weissberg et al 1987). Readdition of $Na^+{}_o$ causes a rapid recovery of pH_i (see Fig. 2 and Weissberg et al 1987): pH_i can increase by one unit in four minutes. Perhaps significantly, this is accompanied by a pronounced hyperpolarization.

Such findings are not unlike those reported in other preparations, but the behaviour of smooth muscle pH_i in the absence of $Na^+{}_o$ is unusual. First, pH_i falls well below the level predicted for a passive distribution (see Fig. 2). Therefore this fall must be because of either internal production of acid or inward transport of acid equivalents. This result also suggests that the membrane permeability to acid equivalents is very low (see Aickin 1984). Second, when pH_i has fallen so low, the presence of CO_2 and HCO_3^- prevents any further fall, as shown in Fig. 2, or even causes some recovery. I believe that this is due to activation of $Cl^-{}_i/HCO_3^-{}_o$ exchange by extreme intracellular acidosis (see Vanheel et al 1984, Vaughan-Jones 1986).

Effect of amiloride

It seemed likely that the Na^+-dependent, HCO_3^--independent acid extrusion would be via Na^+/H^+ exchange (e.g. Smith & Brock 1983, Owen 1984). This

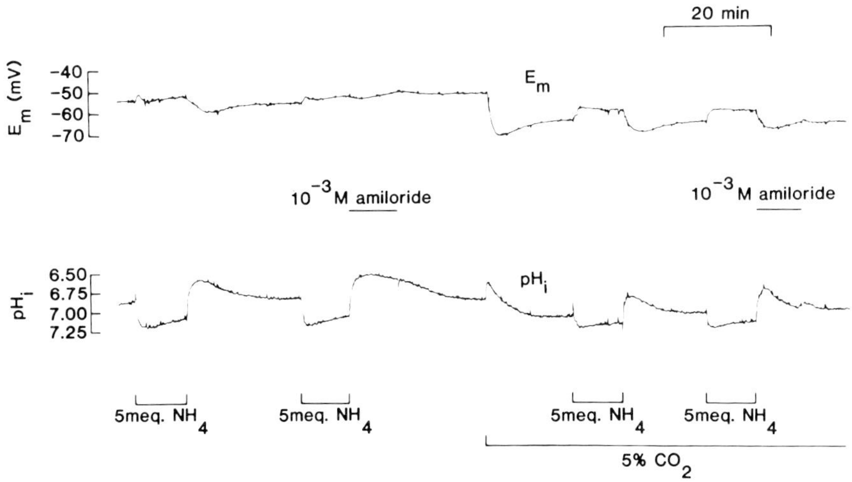

FIG. 3. Pen recordings of an experiment illustrating inhibition of pH_i recovery from acidosis by amiloride in the nominal absence of CO_2 but not in the presence of 5% CO_2.

was confirmed by inhibition of pH_i recovery after application of amiloride in the nominal absence of CO_2, as shown in Fig. 3. The same result has been reported in vascular smooth muscle (Weissberg et al 1987, C. Aalkjær & E.J. Cragoe, personal communication). Rather surprisingly, I have been unable to find any effect of amiloride in the presence of CO_2 (Fig. 3), even when Na^+_o was reduced to 2%. This is at variance with results in vascular smooth muscle where Korbmacher et al (1988) find equal effectiveness in the presence and nominal absence of CO_2; C. Aalkjær & E.J. Cragoe (personal communication) report a much smaller effect in the presence than nominal absence of CO_2.

Buffering power in the presence and nominal absence of CO_2 and HCO_3^-

A possible explanation of my failure to observe an effect of amiloride in the presence of CO_2 is that in this condition Na^+/H^+ exchange only contributes a small fraction to the recovery, which is then dominated by the HCO_3^--dependent mechanism. This would not be unreasonable, if the expected large increase in intracellular buffering power due to the addition of CO_2 and HCO_3^- actually occurred. My previous best estimate of intrinsic (non-CO_2)

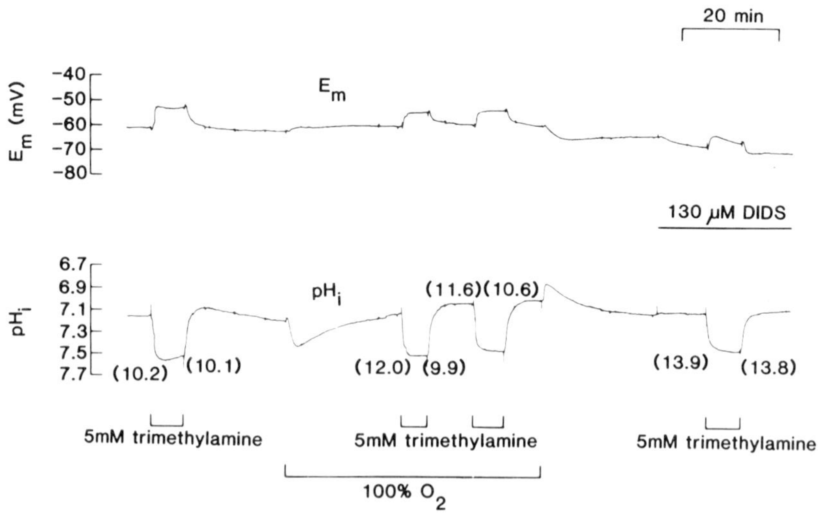

FIG. 4. Estimation of intracellular buffering power from the changes in pH_i recorded on application and removal of 5 mM trimethylamine in the presence (5%) and nominal absence of CO_2. Calculated values, in meq H^+/pH unit/litre are given in parentheses. Note that in the presence of CO_2 there was a slow recovery of pH_i after the initial alkalinization on addition of trimethylamine (first exposure), which then resulted in an overshoot of pH_i on removal of trimethylamine. This was not observed in the nominal absence of CO_2 (second and third exposures) and was inhibited by the presence of DIDS (fourth exposure).

buffering power in vas deferens was 8.6 meq H^+/pH unit/litre (Aickin 1984). In the presence of 5% CO_2, the intracellular buffering power would be expected to be at least doubled at the peak acidification recorded in most experiments and increased by up to fivefold at normal pH_i. Thus the rate of effective proton extrusion could be 10–20 times faster in the presence than nominal absence of CO_2. However, buffering power calculated from the acidification on removal of NH_4^+, in experiments like that shown in Fig. 3, was very similar in the presence and nominal absence of CO_2. Equally, determination of buffering power from application and removal of trimethylamine (Fig. 4) gave the same values in the presence and nominal absence of CO_2. C. Aalkjær & E.J. Cragoe (personal communication) also find very similar values under both conditions in vascular smooth muscle. It has been reported that smooth muscle contains no carbonic anhydrase (Maren 1967, Mühleisen & Kreye 1985). Thus equilibration of CO_2 and HCO_3^- may be too slow for this system to contribute to intracellular buffering. In confirmation, I have found that acetazolamide (10^{-4} M, up to two hours' exposure) has no effect on any pH_i transient. Although a novel concept, this unfortunately does not help explain the differing sensitivities to amiloride!

Effect of DIDS and the absence of Cl^-_o

Na^+-dependent, HCO_3^--dependent mechanisms involved in the effective transport of protons have been characterized as Na^+-dependent Cl^-/HCO_3^- exchange (for review see Roos & Boron 1981) and Na^+-HCO_3^- co-transport (Boron & Boulpaep 1983). Both mechanisms are inhibited by DIDS (4,4′-diisothiocyanatostilbene-2,2′-disulphonic acid). However, Fig. 5 shows no effect of DIDS on the recovery from acidosis in response to removal of NH_4^+. This result has also been found by Korbmacher et al (1988) in cultured vascular smooth muscle but C. Aalkjær & E.J. Cragoe (personal communication) report a significant inhibition in arteriole segments. This difference is intriguing.

Pharmacological identification of transport processes can be dangerous; non-specific inhibition is common but lack of inhibition may be equally misleading. Therefore, to characterize the Na^+-dependent, HCO_3^--dependent mechanism further I looked for a sensitivity to Cl^-. Removal of Cl^-_o causes an alkalinization due to loss of intracellular Cl^- via Cl^-/HCO_3^- exchange (Aickin & Brading 1984, Gerstheimer et al 1987, Korbmacher et al 1988) but recovery from intracellular acidosis, after equilibration in Cl^--free solution, was unaffected (see Aickin & Brading 1984). Thus the mechanism is not Na^+-dependent Cl^-/HCO_3^- exchange. The question remains whether the mechanism is similar to the Na^+/HCO_3^- co-transport described in proximal tubule cells (Boron & Boulpaep 1983). This normally operates to acid load the cell, i.e. in the opposite direction. It is therefore possible that DIDS needs

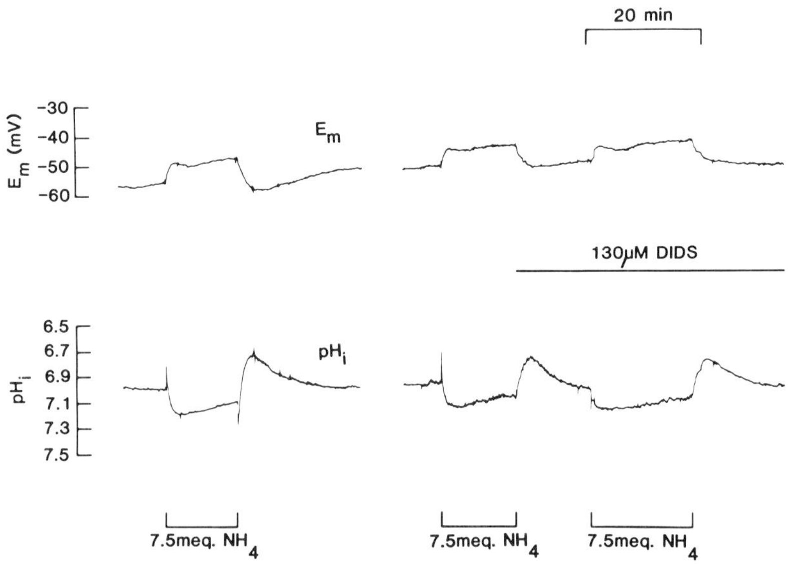

FIG. 5. Pen recordings of an experiment to investigate the effect of DIDS on the regulation of pH_i in the presence of 5% CO_2. DIDS had no effect on the recovery of pH_i from acidosis, induced by removal of 3.75 mM $(NH_4)_2SO_4$, but slowed the recovery of pH_i from alkalosis, induced by addition of 3.75 mM $(NH_4)_2SO_4$. The gap in the record was for a period of 1h during which 3.75 mM $(NH_4)_2SO_4$ was added and removed in the nominal absence of CO_2.

to be at an internal rather than external site in order to inhibit the mechanism in the ureter. Certainly the hyperpolarization observed on recovery from acidosis, particularly marked on readdition of Na^+_o (Fig. 2), is consistent with the electrogenic nature of the co-transport described in proximal tubule.

Recovery from alkalosis

Fig. 5 shows that although DIDS did not inhibit recovery from acidosis, it did inhibit the slow acidification observed in the presence of NH_4^+. Similarly, its presence inhibited the slow acidification seen during prolonged application of trimethylamine in CO_2/HCO_3^- buffered solution (Fig. 4). The slow acidification in the presence of NH_4^+ was also almost completely inhibited in Cl^--free conditions. These results strongly suggest that Cl^-/HCO_3^- exchange is stimulated by intracellular alkalosis and causes the effective *inward* transport of protons. Such a role for Cl^-/HCO_3^- exchange has previously been described in cardiac Purkinje fibres (Vaughan-Jones 1982). It should be noted that these results also indicate that the permeability to NH_4^+ is very low.

The finding of carrier-mediated recovery from alkalosis raises the question of whether the rapid acidification following removal of CO_2 (Fig. 2) may be explained in the same way. Clearly, in the nominal absence of CO_2 and HCO_3^-, Cl^-/HCO_3^- exchange can operate only slowly (Aickin & Brading 1984). Therefore it was little surprise to find that this acidification was unaffected by the complete absence of Cl^- (see Fig. 9 of Aickin & Brading 1984). In six out of 10 cells, DIDS had no effect either, but in the other four application of DIDS greatly inhibited the acidification. Such variation is difficult to explain but it seems unlikely that Cl^-/HCO_3^- exchange contributes to this acidification. Neither was the acidification dependent on Na^+, as suggested for leech glial cells (Deitmer & Schlue 1987). Since smooth muscle cells rapidly lose intracellular Na^+ on removal of Na^+_o (Aickin 1987), completely Na^+-free conditions are readily attained, yet the acidification on removal of CO_2 was unaffected (see Fig. 2). Interestingly, repeated exposure to nominally CO_2-free solution over a period of several hours decreased the degree of acidification. This has also been observed in papillary muscle (R.D. Vaughan-Jones, personal communication). Time appears to be a relevant factor, since 24 hours' cold storage of the ureter reduced the acidification and resulted in the same steady-state pH_i in the presence and nominal absence of CO_2. I would therefore suggest that the acidification on removal of CO_2 results from metabolic production of acid equivalents and inhibition of the dominant HCO_3^--dependent mechanism for acid extrusion. Smooth muscle has a notably high metabolic rate and it is probable that this declines with time after isolation from the animal.

Conclusions

These results suggest that there are virtually no passive movements of protons or charged acid equivalents through membrane channels in smooth muscle. Therefore almost all movements of acid equivalents must result from some form of carriage: from movement of associated weak acids and bases, or from turn over of co-transport or counter-transport mechanisms. Na^+/H^+ exchange, stimulated by intracellular acidosis to extrude protons against their electrochemical gradient, is definitely present. In ureter, at least, there is another Na^+-dependent mechanism which effectively extrudes protons. This is HCO_3^- dependent but Cl^- independent and is not inhibited by DIDS, amiloride or acetazolamide. It therefore has characteristics unlike any mechanism previously reported. It is unclear whether this mechanism is present in vascular smooth muscle, since conflicting reports suggest that it is sensitive to both HCO_3^- and DIDS (C. Aalkjær & E.J. Cragoe, personal communication) or sensitive to neither (Korbmacher et al 1988). Cl^-/HCO_3^- exchange is also present and causes an effective inward transport of protons on intracellular alkalosis. It may also operate in the opposite direction during extreme intracellular acidosis.

Acknowledgements

I am grateful to Drs C. Aalkjær and C. Korbmacher for sending me copies of submitted manuscripts, to Dr A.F. Brading for helpful comments, and to the MRC for support through a senior research fellowship.

References

Aickin CC 1983 Intracellular pH of the smooth muscle cells of the guinea-pig vas deferens. J Physiol (Lond) 334:112–113P

Aickin CC 1984 Direct measurement of intracellular pH and buffering power in smooth muscle cells of guinea-pig vas deferens. J Physiol (Lond) 349:571–585

Aickin CC 1985 The effect of Na^+ and HCO_3^- ions on recovery from an acid load in the smooth muscle of guinea-pig ureter. J Physiol (Lond) 369:80P

Aickin CC 1986 Intracellular pH regulation by vertebrate muscle. Annu Rev Physiol 48:349–361

Aickin CC 1987 Investigation of factors affecting the intracellular sodium activity in the smooth muscle of guinea-pig ureter. J Physiol (Lond) 385:483–505

Aickin CC, Brading AF 1984 The role of chloride-bicarbonate exchange in the regulation of intracellular chloride in guinea-pig vas deferens. J Physiol (Lond) 349:587–606

Ammann D, Lanter F, Steiner RA, Shulthess P, Shijo Y, Simon W 1981 Neutral carrier based hydrogen ion selective microelectrode for extra- and intracellular studies. Anal Chem 53:2267–2269

Boron WJ, Boulpaep EL 1983 Intracellular pH regulation in the renal proximal tubule of the salamander. Basolateral HCO_3^--transport. J Gen Physiol 81:53–94

Deitmer JW, Schlue W-R 1987 The regulation of intracellular pH by identified glial cells and neurones in the central nervous system of the leech. J Physiol (Lond) 388:261–283

Gerstheimer FP, Mühleisen M, Nehring D, Kreye VAW 1987 A chloride-bicarbonate exchanging anion carrier in vascular smooth muscle of the rabbit. Pflügers Arch Eur J Physiol 409:60–66

Ighoroje AD, Spurway NC 1985 How does vascular muscle in the isolated rabbit ear artery adapt its tone after alkaline or acid loads? J Physiol (Lond) 367:46P

Korbmacher C, Helbig H, Stahl F, Wiederholt M 1988 Evidence for Na/H exchange and Cl/HCO_3 exchange in A10 vascular smooth muscle cells. Pflügers Arch Eur J Physiol 412:29–37

Maren TH 1967 Carbonic anhydrase: chemistry, physiology and inhibition. Physiol Rev 47:618–635

Mühleisen M, Kreye VAW 1985 Lack of soluble carbonic anhydrase in aortic smooth muscle of the rabbit. Pflügers Arch Eur J Physiol 405:234–236

Owen NE 1984 Platelet-derived growth factor stimulates Na^+ influx in vascular smooth muscle cells. Am J Physiol 247:C501–C505

Roos A, Boron WJ 1981 Intracellular pH. Physiol Rev 61:296–434

Smith JB, Brock TA 1983 Analysis of angiotensin-stimulated sodium transport in cultured smooth muscle from rat aorta. J Cell Physiol 114:284–290

Spurway NC, Wray S 1987 A phosphorus nuclear magnetic resonance study of metabolites and intracellular pH in rabbit vascular smooth muscle. J Physiol (Lond) 393:57–71

Thomas RC 1984 Experimental displacement of intracellular pH and the mechanism of its subsequent recovery. J Physiol (Lond) 354:3–22P

Thomas RC, Cohen CJ 1981 A liquid ion-exchanger alternative to KC1 for filling intracellular reference microelectrodes. Pflügers Arch Eur J Physiol 390:96–98
Vanheel B, de Hemptinne A, Leusen I 1984 Analysis of Cl^--HCO_3^- exchange during recovery from intracellular acidosis in cardiac Purkinje strands. Am J Physiol 246:C391–C400
Vaughan-Jones RD 1982 Chloride activity and its control in skeletal and cardiac muscle. Philos Trans R Soc Lond B Biol Sci 299:537–548
Vaughan-Jones RD 1986 An investigation of chloride-bicarbonate exchange in the sheep cardiac Purkinje fibre. J Physiol (Lond) 379:377–406
Weissberg PL, Little PJ, Cragoe EJ, Bobik A 1987 Na-H antiport in cultured rat aortic smooth muscle: its role in cytoplasmic pH regulation. Am J Physiol 253:C193–C198

DISCUSSION

Thomas: Claire, you say that there is a classic Na^+/H^+ exchanger in smooth muscle which is inhibited by amiloride, but it's very feeble, worth may be 10% of the normal pH_i regulating activity?

Aickin: No; if I'm right about the buffering power, then Na^+/H^+ exchange forms about 50% of the recovery from acidosis.

Thomas: There is a second mechanism which might be rheogenic Na^+-HCO_3^- co-transport. These membrane potential effects are an important area where the sort of experiments that you are doing can provide more information than work being done using fluorescent dyes.

Then there is also some sort of Cl^-/HCO_3^- exchange, transporting Cl^- into the cell. DIDS inhibits this exchanger but doesn't inhibit the Na^+-HCO_3^- co-transporter in smooth muscle of the guinea pig ureter. People working with epithelial cells find that the Na^+-HCO_3^- co-transporter in those cells is inhibited by DIDS.

Aickin: It may be relevant that the epithelial mechanism is working in the opposite direction, carrying HCO_3^- into the cell.

Thomas: So you think that each of these two transport systems contributes as much as 50% of the normal recovery of intracellular pH in the presence of bicarbonate?

Aickin: Yes, if it's reasonable to calculate percentages in that way. I don't know of any precedent for it, but when there is bicarbonate present, maybe Na^+-HCO_3^- co-transport takes over and the Na^+/H^+ exchanger shuts down.

Oberleithner: If we acid load an amphibian distal tubule cell, where we know that there is a Na^+/H^+ exchanger, in the absence of the CO_2/HCO_3^- buffer system we can inhibit the recovery of pH_i almost completely by amiloride. If we acid load the cell in the presence of HCO_3^-, the pH_i recovery occurs even in the presence of 1mM amiloride. If we acid load in HCO_3^- buffered solution, inhibit with amiloride and then add DIDS, we again have an inhibition of the pH_i recovery. Obviously there are two transport systems in the renal cells located on opposite sides i.e. the Na^+/H^+ exchanger on the luminal cell membrane and

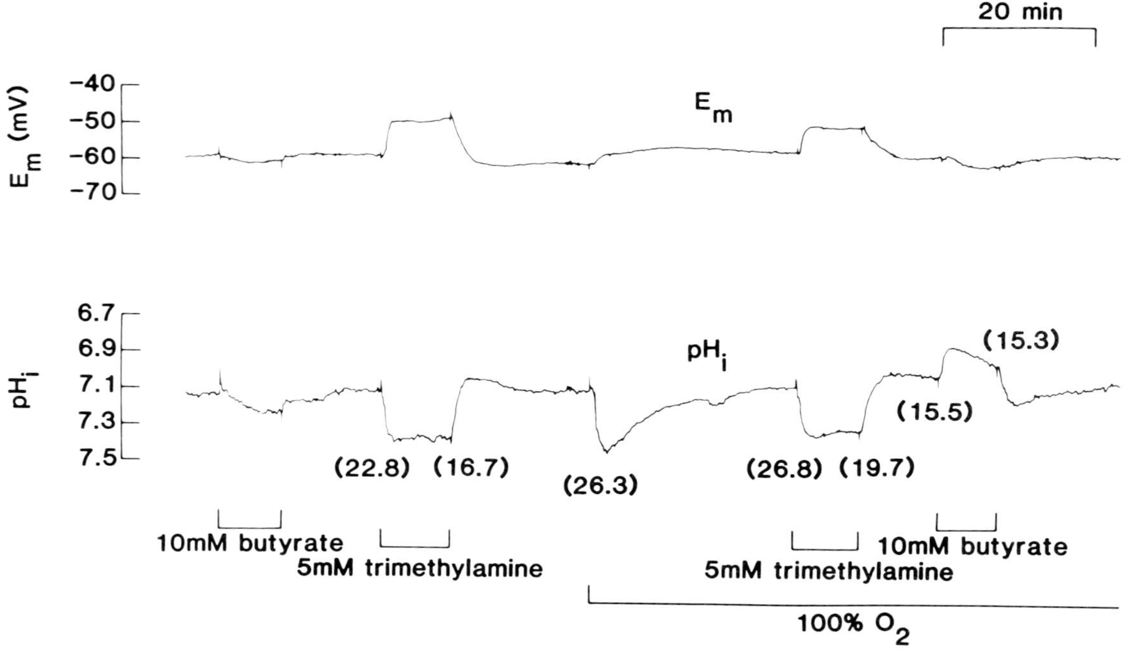

FIG. 1. Pen recordings of an experiment showing the effect of a weak acid, butyrate, and a weak base, trimethylamine, on pH_i in the presence of (5%) and nominal absence of CO_2. Calculated values of intracellular buffering power, in meq H^+/pH unit/litre, are given in parentheses. Application of butyrate caused an intracellular acidosis in the nominal absence of CO_2 but an immediate alkalosis in the presence of CO_2. Buffering power was unaffected by the presence of CO_2.

the Na^+-HCO_3^- co-transport system on the basolateral membrane. If we acidify the cell, for example by the ammonia pre-pulse method, then block the Na^+/H^+ exchanger with amiloride in the presence of HCO_3^-, the Na^+-HCO_3^- co-transporter restores pH_i. If we inhibit the co-transporter, then we see the inhibitory effect of amiloride on pH_i recovery.

Thomas: So amiloride plus SITS (or DIDS) together completely wipe out the pH_i recovery.

Aickin: Are you saying that in the presence of bicarbonate you can't see an amiloride block?

Oberleithner: Either no block or a very reduced one.

Aickin: Do you know what happens to the buffering power? That's an important point.

Oberleithner: No, I haven't considered that yet.

Curtin: Claire, have you ever tried to measure buffering power using a different permeant acid, one other than carbon dioxide?

Aickin: Yes. Using CO_2, particularly when switching between CO_2-free and CO_2-containing solutions, there are problems with the rapid recovery processes, from both acidosis and alkalosis. In one set of experiments, I could take any value between five and infinity! (Aickin 1984). Obviously, the recovery processes must be inhibited before the intrinsic buffering power can be measured. Adding permeant acids, for example butyrate, reveals another quaint response of smooth muscle (Fig.1).

On addition of butyrate, the cell ought to go acid, and in the nominal absence of CO_2 this is what happens. But in the presence of CO_2 it doesn't, it becomes alkaline. You take butyrate away and the pH_i recovers. Acetate and propionate also cause an alkalinization in the presence of CO_2. I began to think that these weak acids were modifying metabolism because all of the anions are substrates of the Krebs cycle. So I tried benzoate, which is not metabolized, and it did exactly the same as butyrate. So these permeant acids would give a value for the buffering power of infinity!

Curtin: It looks like the ionized form is permeant.

Thomas: Why would taking away bicarbonate make something permeant?

Aickin: Exactly.

Boron: The calculated buffering power of such an open-system buffer ought to be unaffected by inhibitors of carbonic anhydrase, provided sufficient time is allowed for the uncatalysed CO_2 hydration reaction. What was the effect of acetazolamide?

Aickin: It didn't have any effect.

Boron: I would think that if you added trimethylamine to the cell, the alkaline load should be partially buffered by the open-system CO_2/HCO_3^- buffer, provided you wait for an 'infinite' time. The evidence that the system is 'open' for CO_2 is that application of CO_2 produces a decrease in pH_i.

Aickin: The pH_i transients are a little more complicated though because in the presence of CO_2, Cl^-/HCO_3^- exchange acts as the alkaline regulator. Therefore you can't wait for a new steady state to estimate the buffering power, either on addition or removal of trimethylamine, because this will incorporate the effect of transport processes. You can wait an infinite time when Cl^-/HCO_3^- exchange is inhibited, but even then the calculated buffering power is essentially the same in the presence and absence of CO_2 (Fig.4 and Discussion Fig.1).

Boron: The experiment could be done in the presence of DIDS to block HCO_3^- transport. You should be able to compare the effect of trimethylamine in the presence of CO_2/HCO_3^- plus DIDS with that of trimethylamine in the absence of HCO_3^-. I would expect the total buffering power to increase substantially in the presence of CO_2/HCO_3^-.

Aickin: I quite agree. This is exactly the experiment that was done in Fig. 4 but not the answer that it produced!

Boron: I don't like the idea of adding 5% CO_2 and not seeing that cause a substantial increase in buffering power. I don't have any problem with the open-system CO_2 buffering power being virtually zero over a period of a couple of seconds—because the CO_2 doesn't have time to equilibrate across the cell membrane. After five minutes in the steady state that CO_2 should have equilibrated.

Grinstein: The half time for spontaneous equilibration of the CO_2 hydration reaction is about 40 secs.

Aickin: I can only say that I am puzzled by the lack of effect.

Thomas: I once did a simple experiment to measure buffering power by injecting HCl into a snail neuron (Thomas 1984). In a CO_2-free environment, the pH_i became acid, and the calculated value of the buffering power was about 12. I added 4% CO_2 which caused the usual transient change, then I injected the same amount of acid and the subsequent pH change was much smaller than in the first part of the experiment. That to me suggests that the *functional* buffering power had greatly increased. Obviously the physico-chemical buffering power at infinite time cannot have been changed. I then added acetazolamide, which blocks carbonic anhydrase and the response of pH_i to another injection of HCl was larger. Acetazolamide did not affect pH regulation, therefore blocking CA had decreased the functional buffering power.

Claire's experiment is vaguely similar (Fig. 4). She is measuring buffering power with 5 mM trimethylamine at intervals. She takes the cells from 5% CO_2 to 0% CO_2 and there is a pH change which Claire calculates to give a buffering power of 44. When CO_2 is added back the buffering power is 28. How certain are you that the amounts of CO_2 are correct?

Aickin: Not very, BOC's mixtures are not very accurate. The important thing is that I know how much HCO_3^- is present at any pH from titration of solutions equilibrated with the supposed 5% CO_2.

Vaughan-Jones: The high values of 44 and 28 for the buffering power are consistent with what we see in multicellular cardiac tissue. It is well known that rapid regulation blunts changes of pH_i so if you block pH regulation, the changes of pH_i induced by adding and removing CO_2 will be larger. This would result in a much lower estimate of buffering power.

Thomas: Perhaps CO_2 entry from the outside is not fast enough to beat the HCO_3^- entry on a transporter. That could explain the values of 44 and 28 but does not explain the lack of any effect of CO_2.

Frömter: Dr Aickin, what was the HCO_3^- concentration in your experiment and did it change when the CO_2 pressure was altered?

Aickin: The 5% CO_2 is buffered with 21 mM HCO_3^-. 100% oxygen is buffered with 21 mM Hepes and zero HCO_3^-.

Boron: How reliable is trimethylamine as a means of measuring buffering power? Do you get similar values with ammonia?

Aickin: Yes, if Cl^-/HCO_3^- exchange is first inhibited with DIDS to prevent recovery from alkalosis on application of NH_4^+. Otherwise, of course, you would get much higher values, particularly in the presence of CO_2.

Boron: We recommend calculating buffering power from the decrease in pH_i produced by withdrawing ammonia, and not worrying about the plateau phase acidification. The ammonia pulse should be carried out under conditions that block acid extrusion, for example, zero Na^+, or in the presence of DIDS or amiloride. I think that this is the most reliable method.

Aickin: I agree, if you can be certain that there is no transport left to cause pH_i recovery. I could only use this method by removing external Na^+ and waiting for pH_i to stabilize. The cells are then extremely acidic and unstable. I would also have to add DIDS to inhibit Cl^-/HCO_3^- exchange which then appears to operate in the opposite direction to normal. Why should trimethylamine give a different answer to NH_4^+?

Boron: I am suspicious of trimethylamine. What if it were affecting metabolism in some way or failing to equilibrate? When there is some doubt about what is happening, I recommend returning to use of the simplest weak base.

Vaughan-Jones: Claire, you are assuming that it is at equilibrium across the membrane, i.e. that the undissociated form is at equal concentrations on both sides. This may not be the case.

Boron: That's why I think Claire should use something simple like ammonia.

Thomas: Ammonia is not simple because it enters as NH_4^+ and even when you wash it out some may leave as NH_4^+.

Boron: Very little.

Thomas: Ammonia may also be metabolized, there are as many problems with ammonia as with trimethylamine.

Boron: Several labs have considerable experience of measuring buffering power by using ammonia; there is much less experience of the use of trimethylamine.

Thomas: But ammonia is not applicable in all situations, I can't use ammonia in snail neurons because it is too slow—the recovery after acid loading takes too long. It takes about half an hour for all the ammonium to leave the cell, during which time all sorts of things have happened.

Boron: I still think that ammonia should be tried in this system. You could use a lot of weak bases, e.g. nicotine, 4-amino pyridine, but I wouldn't trust any of them, if the results are suspicious.

Thomas: Do you not like the value for the buffering power of 10 or 11?

Boron: I love the buffering power of 10-11.

Thomas: Is it possible that this result is perfectly valid and that CO_2 is not affecting the buffering power in these cells?

Vaughan-Jones: I would simply like to see these experiments done on an isolated single cell of smooth muscle. I predict that you won't get the same answer. A lot of your anomalous values for buffering power may be due to inadequate equilibration of weak acids or weak bases in multicellular tissue.

Thomas: I believe in these results I just think that they are difficult to explain! Your experiment with butyrate (Discussion Fig. 1) shows that there is very little effect on the membrane potential when butyrate is added in the presence of CO_2, so I don't think that it is an electrogenic transport process. Could butyrate be affecting surface pH in some way so as to reset the pH_i regulating system which then pumps out protons? This is in 21 mM HCO_3^-. The surface pH is much more likely to be disturbed in the presence of CO_2 than in 100% O_2, when you have 21 mM Hepes. When you add butyrate the surface pH goes alkaline, which would help to explain this. If you could eliminate that effect by having 21 mM Hepes in the HCO_3^- solution, that would prove that it was due to surface pH. That is one possibility, any other ideas?

Aickin: It could be a Na^+-dependent butyrate co-transporter, which also requires HCO_3^-.

Thomas: So butyrate substitutes for HCO_3^-.

Oberleithner: But shouldn't you see a depolarization when you apply butyrate, assuming that there is Na^+-butyrate co-transport?

Aickin: No, assuming the same electrogenicity as the Na^+- HCO_3^- co-transporter of epithelial cells, it should hyperpolarize. As Roger (Thomas) said, there is little effect on membrane potential. I was thinking more of Walter Boron's new Na^+-monocarboxylate co-transporter, but it would have to be functional only in the presence of CO_2.

Boron: Claire, is it known whether the alkalinization that you see in the presence of HCO_3^- when you add butyrate is Na^+ dependent?

Aickin: No, nothing is known about it at all. Benzoate, acetate and propionate have the same effect.

de Hemptinne: Have you tried the effect of lactate or pyruvate?

Aickin: No, I haven't.

Boron: There are a number of enzymes in cells whose activities are critically affected by the presence or absence of CO_2. So, it is conceivable that the effect of the HCO_3^- is not direct.

It is possible that the butyrate acts as it does in the renal proximal tubules. Thus, a Na^+-monocarboxylate co-transporter would allow Na^+-butyrate to enter the cell. This would be followed by butyric acid exit by non-ionic diffusion. The net effect of butyrate recycling would be an increase in pH_i.

Thomas: There would have to be massive inward transport of butyrate for this explanation to be feasible.

Grinstein: Your Na^+-HCO_3^- co-transport system is electrogenic. If the stoichiometry is 3:1 HCO_3^-:Na^+, like it is in the epithelial system, what makes it go inward rather than outward? Is the membrane potential very depolarized?

Aickin: I would neither like to say that it is electrogenic nor ascribe a coupling ratio to it. A coupling ratio of $3HCO_3^-$:$1Na^+$ would drive the co-transport outwards at normal pH_i and membrane potential. A coupling ratio of 3:2 would turn it inwards and would still be adequate to produce a hyperpolarization. These cells have a fairly healthy membrane potential of about -60 mV.

Thomas: So your hyperpolarization may be nothing to do with the rheogenic transport?

Aickin: It's possible it's nothing to do with it—what would be nice would be to have a specific inhibitor of this acid regulator. If the hyperpolarization was then also abolished, perhaps we would be nearer to saying it really is electrogenic. I would only say that the Na^+/HCO_3^- co-transporters that have been characterized are electrogenic and there is a membrane potential change in my system which would fit with it being electrogenic.

Al-Awqati: Do you have any explanation for the hyperpolarization with alkalinization, since the charge doesn't look like it is going through the Na^+-HCO_3^- co-transporter?

Aickin: Rheogenic Na^+-HCO_3^- co-transport working to extrude acid equivalents would cause a hyperpolarization.

Al-Awqati: The hyperpolarization could be caused by the alkalinization opening K^+ channels.

Aickin: It could be, all I know is that whenever I see a recovery from an acid load there is an accompanying hyperpolarization and it is much more marked in the presence of CO_2 than in the absence. But you could say it's a conductance change and bigger in the presence of HCO_3^- than the absence because the membrane resistance is greater. I can't argue against that.

Al-Awqati: Have you done this in the presence of Ba^{2+} or Cs^{2+}?

Aickin: No.

Boron: The hyperpolarization seen on the addition of Na^+ is characteristic of the electrogenic Na^+-HCO_3^- co-transport system. Thus, it seems that this transporter could be present in your cells.

Schlue: You have shown that the HCO_3^- dependent mechanism in your preparation is unaffected by DIDS; on the other hand, you have compared this with results of Wiederholt and his group on epithelial cells who were able to block the HCO_3^- dependent mechanism using the stilbene derivatives, SITS or

DIDS. They preincubated with 1mM DIDS for 30 minutes (personal communication, Jentsch et al 1985); in your recordings the preincubation time was much less. Have you ever tried a longer preincubation and/or higher concentration of DIDS?

Aickin: In the experiment shown in Fig.5, the preparation had been exposed to DIDS for nearly 40 minutes by the time the recovery from the third acid load was recorded. It is usually more difficult to acid load a cell in the presence if DIDS than shown in Fig.5. More often, almost all the slow acidification in the presence of NH_4^+ is inhibited. In 4/10 cells there is then no acidification after removal of CO_2, only the alkalinization. So I am often left with no method of acid loading the cell in the presence of DIDS, which makes it rather difficult to study the effect of a long preincubation.

Harvey: Molecular oxygen is thought to affect the Na^+/H^+ exchanger. After an NH_4^+ pulse do you get recovery in a pure oxygen atmosphere?

Aickin: Oh yes.

Thomas: You go from pure oxygen to 95% oxygen, so it's not exactly a major change in oxygen.

de Hemptinne: What about surface effects? When you suddenly increase the CO_2 pressure or add an organic acid to the superfusate, the intracellular pH becomes acid but the pH at the surface of the cell increases transiently. The magnitude of the surface pH change is larger, the lower the buffer concentration in the medium. You are using 21 mM Hepes, which is a reasonable buffer concentration.

Aickin: I also tried taking away and putting back CO_2 using only 5 mM Hepes in the nominal absence of CO_2 and the effect was exactly the same.

de Hemptinne: In your preparation there is no carbonic anhydrase present, this presumably diminishes the capacity of CO_2/HCO_3^- to buffer rapid and transient pH changes either 'in' or at the surface of cells. In many other muscle preparations (cardiac and skeletal muscle), the presence of carbonic anhydrase has been shown. I am wondering whether the relatively large buffer value that you calculate after applying a CO_2 pulse could result from an undisturbed layer or surface effects which were neglected in your calculation.

Aickin: I think that's possible, yes.

de Hemptinne: Contrary to what can be seen in other muscle preparations, you don't measure an acid layer at the surface. When you apply acids such as carbonic acid or butyric acid, do you notice any transient pH change?

Aickin: I have never managed to measure a surface pH so I can't answer the question.

Vaughan-Jones: I would like to make three comments. Firstly, I agree with you about the possibility that the buffering power inside multicellular tissue may be over-estimated due to slow movements of CO_2 and ammonia across the membrane.

Secondly, I am confused, like Walter (Boron), about the non-contribution of

CO_2/HCO_3^- to the buffering power. I can't accept that it's due to lack of carbonic anhydrase. There is said to be very little carbonic anhydrase in heart, but in the sheep Purkinje fibre we have been able to demonstrate an increase in the intracellular buffering power in the presence of CO_2/HCO_3^-. The estimation of intracellular buffering power is critically dependent on whether or not you are blocking acid or alkaline regulation mechanisms. If these are in operation, even if they are fairly slow, the 'back titration' method used for the estimation of buffering power following an ammonia pre-pulse or after addition of CO_2 can produce major errors (see Vaughan-Jones, this volume).

The third point is that the electrode that you are using is what's typically called the LIE (liquid ion-exchanger) electrode. This is an electrode whose tip is filled with a non-specific cation resin; it records only membrane potential because the sum of the cations on both sides of the membrane is roughly equal. My graduate student, Mei-Lin Wu, has tested the LIE electrode in Purkinje fibres. Under a number of conditions, most notably in sodium-free solutions, it gives very spurious readings of membrane potential which when subtracted from the pH recording electrode lead to spurious readings of intracellular pH. What do you think about the ability of the LIE electrode to measure membrane potential in smooth muscle?

Aickin: Obviously it's something one always should be bothered about when you start changing Na^+. Any change in intracellular Na^+ without an equal change in K^+ (or another cation to which the electrode is equally sensitive) would mean that the LIE would be lying. Membrane potential measured with a LIE electrode in cells exposed to Na^+- or K^+ -free solution for a long time may well be a few mV out. But as far as I can tell from repeating the same experiments using a conventional KCl-filled electrode, the LIE does not lie by much. Changes in intracellular Na^+ and K^+ must be very much smaller during my experiments to determine buffering power than during prolonged exposure to K^+-free solution, so I don't think the LIE electrode can be held responsible for my problems with buffering power.

The problem is that addition of CO_2 should increase buffering power – it does in every other preparation and it's a natural assumption. All the discussion here has centred on why the buffering power is so low in the presence of CO_2. Perhaps the other side of the question is why it's so big in the nominal absence of CO_2—if we agree that 10 is big.

Thomas: 10 is fine.

Russell: When you did the Cl^--free experiments, what did you use to replace Cl^-?

Aickin: Gluconate and glucuronate.

Russell: And you did not see intracellular alkalinization?

Aickin: No. I do see considerable alkalinization, of more than 0.7 pH units in *vas deferens* (Aickin & Brading 1984). Ureter is a little different, for some reason it does not maintain the alkalosis in the continued absence of external Cl^-.

Russell: You would attribute this to the Cl^-/HCO_3^- exchanger running backwards?

Aickin: Right.

Russell: Have you done this in SITS and DIDS?

Aickin: Yes, in the presence of DIDS the changes in intracellular Cl^- caused by changing external Cl^- are slowed about tenfold and the changes in pH_i are abolished. DIDS works instantly in this case.

Russell: Were you saying that the reason Cl^- is highest in smooth muscle is because of the Cl^-/HCO_3^- exchanger? I was under the impression there was an Na-K-Cl co-transporter in smooth muscle (Owen 1984). Do you know what physiological purpose is served by a high intracellular Cl^- concentration in smooth muscle?

Aickin: Cl^- is high partly because the HCO_3^- gradient drives it there via the Cl^-/HCO_3^- exchanger and partly because, as you say, there is a Na^+-K^+-Cl^- co-transport mechanism (Aickin 1987). The rather slow changes in intracellular Cl^- observed in the presence of DIDS are mediated by this mechanism. It is important to remember that the Cl^- permeability of these cells is very low. Because the Cl^- leak is so low, there is very little net inward transport of Cl^- in the steady state—so the acid regulating system is not shunted. In my experiments, application of DIDS has no effect on the steady-state pH_i, but this may not be true for all smooth muscles.

References

Aickin CC 1984 Direct measurement of intracellular pH and buffering power in smooth muscle cells of guinea-pig vas deferens. J Physiol (Lond) 349:571–585

Aickin CC 1987 Na,K,Cl co-transport is involved in Cl accumulation in the smooth muscle of isolated guinea-pig vas deferens. J Physiol (Lond) 394:87P

Aickin CC, Brading AF 1984 The role of chloride-bicarbonate exchange in the regulation of intracellular chloride in guinea-pig vas deferens. J Physiol (Lond) 349:587–606

Jentsch TJ, Schill BS, Schwartz P, Matthes H, Keller SK, Wiederholt M 1985 Kidney epithelial cells of monkey origin (BSC-1) express a sodium bicarbonate cotransport. J Biol Chem 260:15554–15560

Owen NE 1984 Regulation of Na/K/Cl cotransport in vascular smooth muscle cells. Biochem Biophys Res Comm 125:500–508

Thomas RC 1984 Experimental displacement of intracellular pH and the mechanism of its subsequent recovery. J Physiol (Lond) 354:3P–22P

Vaughan-Jones RD 1988 Regulation of intracellular pH in cardiac muscle. In: Proton passage across cell membranes (Ciba Found Symp 139) Wiley, Chichester p 23–46

Regulation of intracellular pH in cardiac muscle

R.D. Vaughan-Jones

University Laboratory of Physiology, Parks Road, Oxford OX1 3PT, UK

Abstract. Intracellular pH (pH_i) in sheep cardiac Purkinje fibres is controlled by sarcolemmal Na^+/H^+ and Cl^-/HCO_3^- exchange. At normal pH_o (7.4), Na^+/H^+ exchange mediates an acid efflux whenever pH_i falls and Cl^-/HCO_3^- exchange mediates an equivalent acid influx in response to a *rise* in pH_i. Intracellular pH is also influenced by $Ca^{2+}{}_i$, which can activate force development leading to the anaerobic production of lactic acid. This is evident after an increase in stimulation rate which reversibly reduces both pH_i and extracellular surface pH (pH_s). Rate-dependent pH_i changes are inhibited following inhibition of glycolysis, indicating that they are caused by accumulation of lactic acid. In some cases, the efflux of lactic acid may provide a faster method for recovery of pH_i from a metabolic acidosis than that provided by Na^+/H^+ exchange. Finally, direct pH_i measurement in isolated mammalian ventricular myocytes suggests that the intrinsic intracellular buffering power (β) of ventricular tissue may be considerably lower than previously believed. An accurate knowledge of β is essential for calculating net membrane fluxes of acid equivalents from changes in pH_i.

1988 Proton passage across cell membranes. Wiley, Chichester (Ciba Foundation Symposium 139) p 23–46

Contraction in the heart is sensitive to changes of intracellular pH (pH_i). For example, in skinned ventricular fibres (Fabiato & Fabiato 1978) and intact sheep cardiac Purkinje fibres (Vaughan-Jones et al 1987), tension can decline with a fall in pH_i roughly in accordance with the equation: tension α $[H^+]^{-2}$. The reasons for this inhibitory influence of acidosis are discussed elsewhere (Fabiato & Fabiato 1978, Vaughan-Jones et al 1987). The important point to note is that because of the pH_i sensitivity of force development the maintenance of normal cardiac contractility depends, in part, on the tissue's ability to control its intracellular pH. In this chapter we consider the mechanisms that regulate pH_i, such as sarcolemmal Na^+/H^+ or Cl^-/HCO_3^- exchange and lactic acid transport. We also consider some other factors that exert an important influence upon pH_i, most notably changes of intracellular Ca^{2+} ($Ca^{2+}{}_i$) and cellular metabolism. Finally, we examine the ability of intracellular buffers in cardiac muscle to minimize excursions in pH_i. In particular, we consider the possibility that the intracellular buffering power (β) of ventricular tissue may, hitherto, have been greatly overestimated.

Methods

In most of the present work, intracellular pH was recorded in isolated pieces of cardiac muscle *in vitro* (sheep heart Purkinje fibres, guinea pig papillary muscle and guinea pig ventricular strips). In some other experiments, pH_i was recorded in single isolated ventricular myocytes of guinea pig (see Mitchell et al 1984 for method of cell isolation). In all experiments, pH_i was measured with an intracellular pH-selective microelectrode. This was fabricated by incorporating a short column (<100 μM) of liquid pH sensor (Ammann 1986) in the tip of a silanized glass micropipette; the remainder of the electrode being filled with a convenient, well-buffered salt solution. Methods of electrode construction are described fully elsewhere (Vaughan-Jones & Aickin 1987, Vaughan-Jones 1988). For experiments with sheep Purkinje fibres, separate single-barrelled pH and conventional (3M KCl) voltage-recording microelectrodes were inserted (<100 μm apart) and pH_i recorded as the difference between the two electrode signals. For experiments with papillary muscle and isolated myoctes, a double-barrelled microelectrode (one barrel is pH selective, the other contains 3M KCl) was used (see Vaughan-Jones 1988 for details of double-barrelled electrode). In some experiments, intracellular Na^+ or Cl^- were also recorded using liquid-sensor ion-selective microelectrodes. In all experiments with multicellular tissue, the preparation was attached to a tension transducer so that twitch and/or tonic tension could be observed.

The composition of basic Tyrode solution was: NaCl 140 mM; KCl 4.5 mM; $CaCl_2$ 2.5 mM; $MgCl_2$ 1.0 mM. Solutions were buffered to pH 7.4 (37 °C) using either 10 mM Hepes buffer (equilibrated with 100% O_2 or air) or 22 mM $NaHCO_3$/5% CO_2 + 95% O_2.

Results and discussion

Intracellular pH in quiescent cardiac muscle, like that in other tissues, is usually maintained in the range 7.0–7.2 (for a pH_o 7.4). There appear to be two sarcolemmal mechanisms that mediate the basic control of pH_i, Na^+/H^+ and Cl^-/HCO_3^-exchange.

Na^+/H^+ exchange

The first part of the experiment shown in Fig. 1 demonstrates that if an intracellular acid load is induced in a sheep Purkinje fibre while maintaining a constant bulk pH_o, then the intracellular acidosis is only transient, pH_i recovers to more alkaline levels within about 30 minutes (note that much faster pH_i recoveries are seen in ventricular tissue; see e.g. Fig. 5). The intracellular acid load was induced by adding and removing 10 mM NH_4Cl in the external perfusate (the ammonia pre-pulse technique, Roos & Boron 1981).

The recovery of pH_i from intracellular acidosis appears to occur mainly via acid extrusion by sarcolemmal Na^+/H^+ exchange (Deitmer & Ellis 1980, Ellis & MacLeod 1985, Piwnica-Worms et al 1985). Perhaps the best evidence for this is that, as shown in Fig. 1, removal of extracellular Na^+ (Na^+ replaced isosmotically by choline) slows greatly this recovery of pH_i, while recovery can be restored by re-admitting Na^+_o. The experiment shown in Fig. 1 was performed in the nominal absence of CO_2/HCO_3 buffers; furthermore, the pH_i recovery has been shown to be insensitive to the application of the stilbene anion-exchange inhibitor, DIDS (4,4′-diisothiocyanoto-stilbene-2,2′-disulphonic acid) (Vaughan-Jones 1982, Vanheel et al 1984). Consequently, there is unlikely to be an influence on pH_i recovery from Cl^-/HCO_3^- exchange or from a Na^+-dependent Cl^-/HCO_3^- exchanger. Further evidence for the operation of Na^+/H^+ exchange has been provided by the observation that pH_i recovery is accompanied by a transient rise (1–2 mM) in intracellular Na^+ activity, a^i_{Na}, (Deitmer & Ellis 1980) consistent with Na^+ influx in exchange for

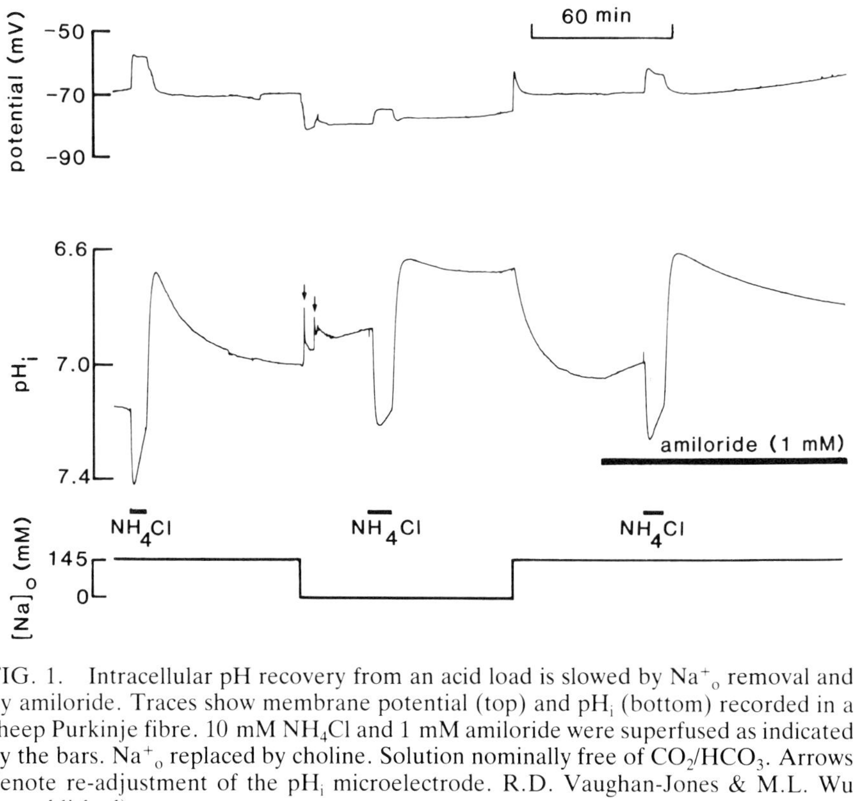

FIG. 1. Intracellular pH recovery from an acid load is slowed by Na^+_o removal and by amiloride. Traces show membrane potential (top) and pH_i (bottom) recorded in a sheep Purkinje fibre. 10 mM NH_4Cl and 1 mM amiloride were superfused as indicated by the bars. Na^+_o replaced by choline. Solution nominally free of CO_2/HCO_3. Arrows denote re-adjustment of the pH_i microelectrode. R.D. Vaughan-Jones & M.L. Wu (unpublished).

acid efflux. The rise in a^i_{Na} is only transient because it stimulates the extrusion of Na^+ by the Na^+/K^+ pump (Piwnica-Worms et al 1985, Kaila & Vaughan-Jones 1987).

The final part of Fig. 1 shows that the diuretic drug amiloride, an inhibitor of Na^+/H^+ exchange in other tissues, slows the recovery of pH_i. It is notable, however, that pH_i recovery is not inhibited completely by amiloride. Assuming that Na^+/H^+ exchange is 100% inhibited in Na^+-free (choline) solution, then it is possible to estimate from Fig. 1 that amiloride has inhibited the initial rate of Na^+_o-sensitive pH_i recovery by 90%. Since Na^+_o and amiloride compete for the same external binding site (Mahnensmith & Aronson 1985), this degree of inhibition would indicate a K_i for amiloride of about 7.0 μM (assuming a K_m for Na^+_o of 10 mM; Ellis & MacLeod 1985), which is similar to the K_i estimated for amiloride in other tissues (see Mahnensmith & Aronson 1985).

Other features in Fig. 1 merit comment. Firstly, removal of Na^+_o or addition of amiloride both produce a slow fall of pH_i at similar rates, consistent with inhibition of a small resting H^+ efflux via Na^+/H^+ exchange. This argues against the suggestion of Ellis & MacLeod (1985) that Na^+/H^+ exchange fails to contribute to the maintenance of normal resting pH_i. Fig. 2 shows that in quiescent tissue amiloride produces a small fall of both pH_i and a^i_{Na}, providing further evidence for a small activation of Na^+/H^+ exchange at rest.

Fig. 1 also shows that pH_i recovery is not inhibited entirely in Na^+-free solution: pH_i recovers from the acid load by 0.05 units over the first 25 minutes after NH_4Cl removal. This has been observed consistently in many fibres and it occurs in solutions where Na^+ has been replaced isosmotically by choline, N-methylglucamine, Ca^{2+} or K^+. The recovery is also insensitive to amiloride or DIDS. As yet, its cause is unidentified but it could possibly represent a slow intracellular sequestration of H^+ ions or a sarcolemmal efflux of acid equivalents via some novel mechanism. It means, however, that pH_i recovery in normal Na^+_o may not proceed exclusively via Na^+/H^+ exchange. Hence, in order to examine the pH_i dependence of the exchanger, it is necessary to perform a pharmacological dissection using amiloride. One method of achieving this is shown in Fig. 2A.

In this experiment, pH_i and a^i_{Na} were recorded simultaneously and amiloride was added before and after inhibiting the Na^+/K^+ pump with a cardiotonic steroid, strophanthidin. This procedure is another way, other than by using NH_4Cl removal, of inducing an intracellular acid load. Inhibition of the Na^+/K^+ pump leads to an elevated a^i_{Na} and, via sarcolemmal Na^+/Ca^{2+} exchange, an elevated Ca^{2+}_i; a rise of Ca^{2+}_i in heart has been demonstrated to induce an intracellular acid load (Vaughan-Jones et al 1983). Consequently, 11 minutes after adding strophanthidin, pH_i started to fall. This acidosis stimulates Na^+/H^+ exchange, as shown by the fact that addition of amiloride

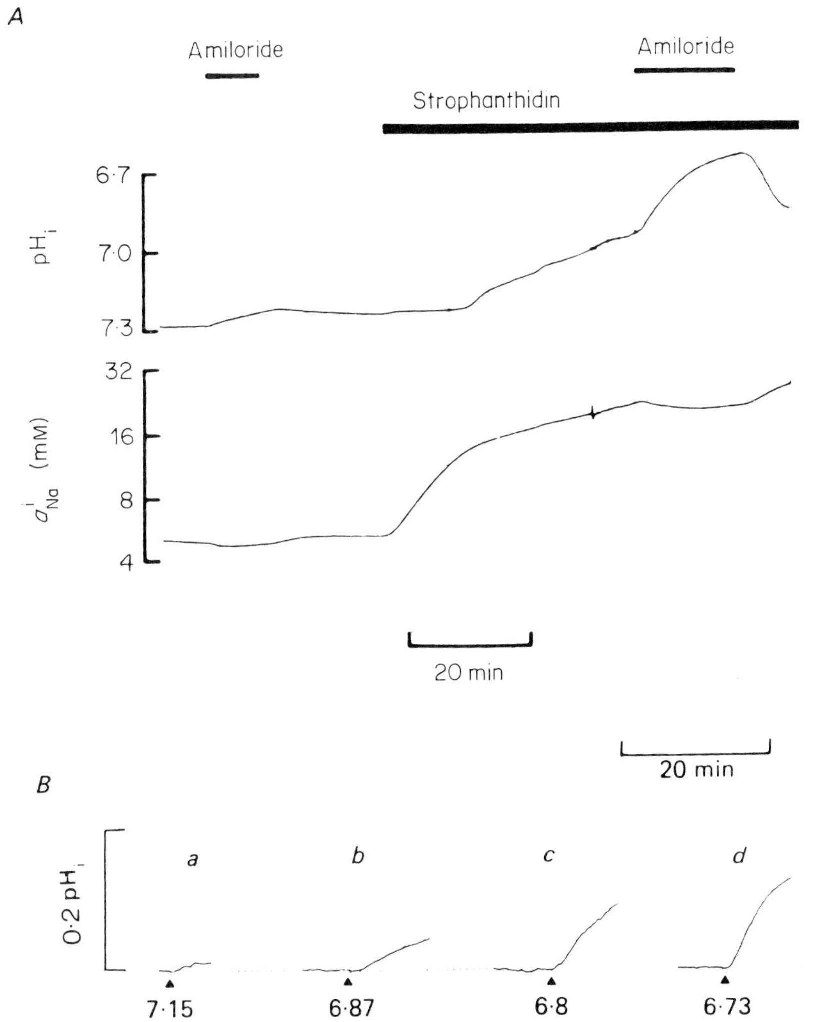

FIG. 2. Amiloride-sensitive Na^+/H^+ exchange is stimulated by a fall in pH_i. (A) pH_i and a^i_{Na} were recorded simultaneously in a sheep Purkinje fibre. The membrane was voltage-clamped (2 microelectrode voltage-clamp) at a holding potential of −70 mV (not shown). 1 mM amiloride and 10^{-5} M strophanthidin were added as indicated by the bars. Note the large effect of amiloride on pH_i and a^i_{Na} when applied in the presence of strophanthidin. B) Data from another fibre. Traces show the acidifying effect of amiloride (1 mM) when applied (triangles) at successively lower values of pH_i (pH_i indicated below the triangles in the records). pH_i was reduced by adding 10^{-5} M strophanthidin (as in A). a was strophanthidin free, b-d were recorded successively at roughly 20–30 min intervals after adding strophanthidin. In between each record, amiloride was removed for about 15 min. The base-line of the recordings has been adjusted such that the initial rate of intracellular acidification equals zero before amiloride application. Note that the influence of amiloride on pH_i increases steeply with a fall in pH_i from 7.15–6.73. Modified from Kaila & Vaughan-Jones, 1987.

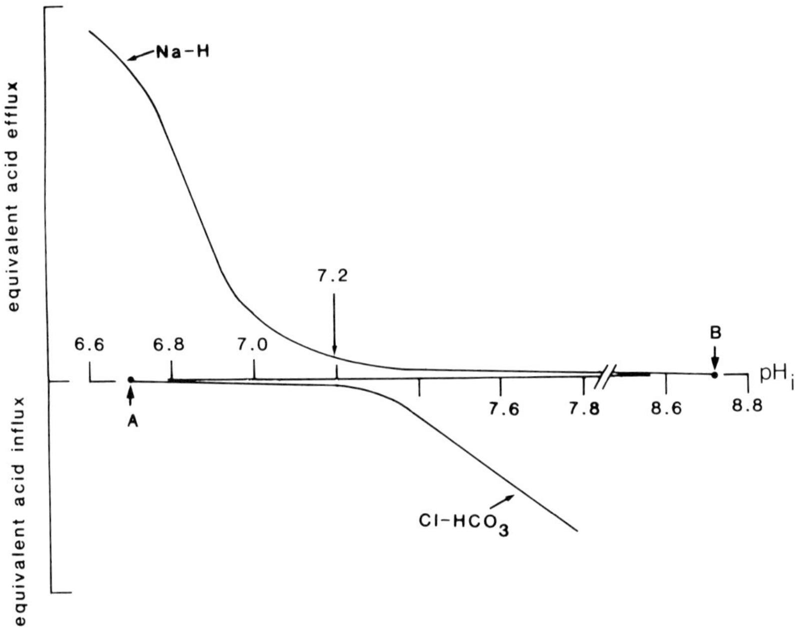

FIG. 3. Contribution of Na^+/H^+ and Cl^-/HCO_3^- exchange to pH_i regulation. The curve labelled Na-H shows the dependence of acid efflux on pH_i determined for Na^+/H^+ exchange from experiments like that shown in Fig. 2. The curve has been re-drawn from Fig. 9 of Kaila & Vaughan-Jones 1987. The curve labelled Cl-HCO_3 is drawn tentatively for a constant p_{CO2} of 38 mmHg and a pH_o of 7.4 (see text for explanation of curve) and indicates that the system will normally mediate an equivalent acid influx. The arrows A & B indicate, respectively, the calculated pH_i at equilibrium for Cl^-/HCO_3^- and Na^+/H^+ exchange, assuming no chemical input of energy and assuming a^i_{Cl} 20 mM; a^o_{Cl} 100 mM; a^i_{Na} 7 mM; a^o_{Na} 108 mM; a^o_{HCO3} 22 mM. Arrow indicates 'normal' pH_i of 7.2. The ordinate axes for Na^+/H^+ and Cl^-/HCO_3^- exchange are not drawn to scale.

at pH_i 6.95 induced a further fall of pH_i and a decrease in a^i_{Na}. These amiloride-induced changes in pH_i and a^i_{Na} are tenfold larger than those observed under control conditions (strophanthidin-free) when pH_i was 7.28, thus indicating clearly that Na^+/H^+ exchange was stimulated at the lower pH_i. Fig. 2B investigates, in another fibre, the influence of amiloride on pH_i at various vaues of pH_i (the acid load was again induced by adding strophanthidin). Since amiloride inhibits acid efflux via Na^+/H^+ exchange, it is clear from Fig. 2B that the exchanger is steeply activated (about tenfold) over the range of pH_i, 7.15–6.73. This result has been approximated graphically in Fig. 3 (curve labelled Na-H). Also shown (arrow B) is the pH_i at which Na^+/H^+ exchange is thermodynamically at equilibrium (see legend for details). The steep activation of acid efflux over the physiological range (6.7–7.2) is qualitatively similar to that described by Frelin et al (1985) in cultured chick ventricular cells and it also matches that described in other tissues (Mahnensmith & Aronson 1985). The curve illustrated in Fig. 3 emphasizes that Na^+/H^+ exchange will act physiologically as an acid 'outward rectifier' in response to a fall in pH_i. It should be noted that, following an intracellular *alkalosis* in the Purkinje fibre, the exchanger does not contribute to net acid fluxes (see Fig. 8 of Vaughan-Jones 1981).

Many factors are believed to alter the position of the Na^+/H^+ activation curve (Fig. 3) relative to the pH_i axis and also to alter the steepness of this curve (Moolenaar 1986). These include extracellular pH (Jean et al 1985) and compounds that stimulate protein kinase C (e.g. phorbol esters) and analogues of diacylglycerol, which may be a natural regulator of protein kinase C and hence of Na^+/H^+ exchange. In our own work (R.D. Vaughan-Jones & M.L. Wu, unpublished), we have observed that certain metabolic inhibitors, most notably inhibitors of glycolysis (e.g. 10 mM 2-deoxyglucose added to glucose-free Tyrode) can dramatically impede pH_i recovery from an acid load (ammonia pre-pulse) while also inhibiting the transient rise in a^i_{Na} that normally accompanies the pH_i recovery. The reason for this inhibition is not yet established but it would, for example, be consistent with a requirement of Na^+/H^+ exchange for phosphorylation by protein kinase C.

Cl^-/HCO_3^- exchange

The possible existence of Cl^-/HCO_3^- exchange in non-epithelial tissues has been obscured by the recent realization that some anion-exchange systems also transport Na^+ ions (Thomas 1984). These systems are not, therefore, examples of simple Cl^-/HCO_3^- exchange. At present, an anion exchanger in non-epithelial cells that is independent of Na^+ has been identified in cardiac muscle (Vaughan-Jones 1981, 1986), smooth muscle (Aickin & Brading 1984) and neutrophils (Simchowitz & Roos 1985).

In cardiac muscle, it is notable that the thermodynamic gradients for Cl and HCO_3 ions predict that a simple anion exchange system, if it operated kinetically, would mediate Cl^- influx-HCO_3^- efflux, which is equivalent to a net acid *influx* (intracellular Cl^- activity is about 20 mM, so that $Cl_o/Cl_i >> HCO_{3o}/HCO_{3i}$). Indeed the system would not reverse until pH_i was about 6.7 (see arrow A in Fig. 3). It has been shown previously (Vaughan-Jones 1981, 1982) that an intracellular alkalosis (at a constant p_{CO2} of 38 mmHg) can activate a DIDS-sensitive, HCO_3^- efflux and Cl^- influx via anion exchange. This tends to restore normal pH_i following alkalosis. Equally clearly, it has been shown (Vaughan-Jones 1982, Vanheel et al 1984) that anion-exchange inhibitors do not affect pH_i recovery from an intracellular *acid* load (at a normal bulk pH_o of 7.4). One must therefore conclude that at constant p_{CO2} the activation of anion exchange at pH_i 6.7–7.2 is small or negligible, whereas it becomes increasingly large in the range of pH_i 7.2–7.6. For this reason, the pH_i dependence of anion-exchange activation must be highly non-linear, as drawn in the lower half of Fig. 3. A similar, non-linear activation by pH_i of acid influx via Cl^-/HCO_3^- exchange has recently been postulated by Olsnes et al (1986) in cultured *VERO* cells. An acid-equivalent influx via anion exchange has also been observed in human neutrophils (Simchowitz & Roos 1985).

The role of sarcolemmal ion-exchange transport in the control of cardiac pH_i can now be summarized by the combined curves drawn for Na^+/H^+ and Cl^-/HCO_3^- exchange in Fig. 3. The net effect is that a symmetrical system for pH_i control exists following intracellular acid–base disturbances. This can be divided into two rectifying systems: i) Na^+/H^+ exchange, which is an outward acid-transporting rectifier, and ii) Cl^-/HCO_3^- exchange, which is an inward acid equivalent-transporting rectifier.

Other factors influencing pH_i

One factor influencing pH_i in heart is intracellular Ca^{2+} ($Ca^{2+}{}_i$). A rise of $Ca^{2+}{}_i$, if sufficiently large, can reduce pH_i (e.g. Fig. 2A and Vaughan-Jones et al 1983). An important physiological example of this in heart is shown in Fig. 4. This shows the effect of stimulating a Purkinje fibre at rates varying from 0.1 to 4 Hz. This encompasses the physiological range of heart rate (~1–3 Hz). A rate increase produces a small and reversible fall in pH_i which usually stabilizes within 3–10 min. On average, pH_i falls by 0.02 units Hz^{-1}. Although this acidosis is small, the high sensitivity of contraction to changes in pH_i means that the acidosis exerts an important down-regulatory influence on contraction (Bountra et al 1988a).

The rate-dependent acidosis can be inhibited by applying the Ca^{2+} antagonist D600 (20 μM), implying an involvement of Ca^{2+} (Bountra et al 1988b). This influence of Ca^{2+} on pH_i, however, is likely to be indirect, being mediated largely by a stimulation of glycolysis following an increase in contraction. This can be demonstrated by stimulating a Purkinje fibre in the presence of glycolytic inhibitors (10 mM 2,deoxyglucose or 0.5 mM iodoacetate in glucose-free Tyrode). Under these conditions, the rate-dependent acidosis is reduced by up to 80%, even when contraction is identical to that observed in the absence of the inhibitors (Bountra et al 1988b). This suggests that at higher stimulation frequencies the increased demand for ATP following an increased work load increases the anaerobic resynthesis of ATP via glycolysis. Such a situation will generate intracellular lactic acid leading to a fall in pH_i. The hypothesis gains further credence from the fact that, in unbuffered solutions, extracellular surface pH (pH_s) of a Purkinje fibre (measured either using a pH microelectrode pushed against the extracellular surface or by using the extracellular pH dye, phenol red) can decrease by up to 0.15 units following a 2 Hz train of contractions (Bountra et al 1988c). The rate of development of this acidosis at the external membrane surface matches the rate of fall in pH_i and is reduced by 50% or more in the presence of 4 mM

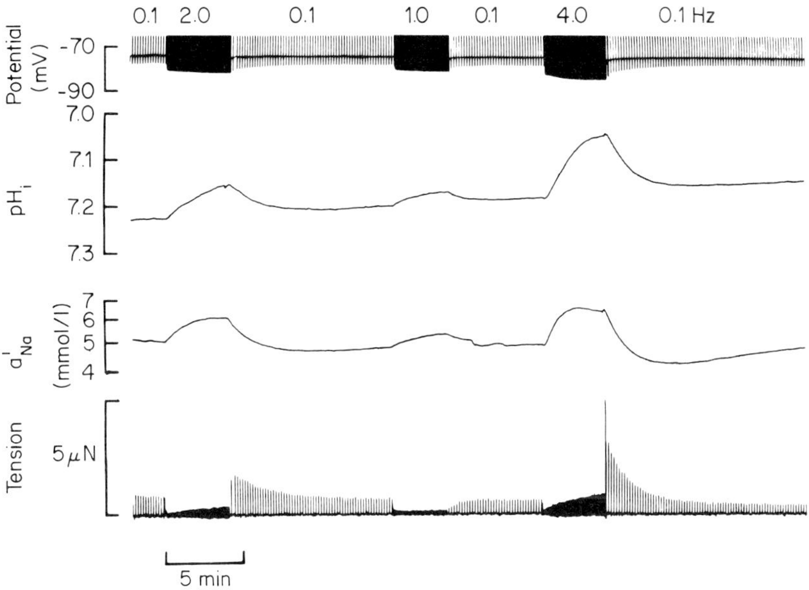

FIG. 4. pH$_i$ in the sheep Purkinje fibre is rate dependent. Traces show (top to bottom) membrane potential (only the diastolic membrane potential is shown, the peaks of action potentials are off scale); pH$_i$; a^i_{Na} and tension (tension trace is band-pass filtered 0.1–10 Hz in order to facilitate comparison of successive twitches). The fibre was stimulated (field stimulation) at 0.1, 1, 2, and 4 Hz (indicated at top of figure). The pen recorder cannot resolve action potentials at the higher frequencies, so that, under these conditions, the potential trace becomes a continuous black bar. Note that pH$_i$ falls reversibly with increasing rates of stimulation. From Bountra et al 1988a.

α-cyano-4-hydroxy cinnamate, an inhibitor of sarcolemmal lactic acid transport (deHemptinne et al 1983). A similar rate-dependent decrease of pH$_s$ has been reported in mammalian papillary muscle (Vanheel et al 1985). A rate-dependent fall of pH$_i$ attributable to the generation of lactic acid has also been identified recently in whole, perfused ferret hearts (Allen et al 1986).

These experiments indicate that changes of pH$_i$ in cardiac muscle are likely to be influenced greatly by the metabolic status of the tissue. For example, under hypoxic conditions and more particularly during periods of cardiac ischaemia, changes in intracellular lactic acid concentration play an important role in determining the control of pH$_i$. It is notable that in Fig. 4 recovery of pH$_i$ following a *decrease* in rate is extremely rapid (half-time 1–2 min; Bountra et al 1988b) compared with the recovery rate attributed previously to Na^+/H^+ exchange (cf Fig. 1). This suggests that the sarcolemmal efflux of metabolically generated lactic acid may often be the fastest and most efficient way of restoring normal pH$_i$ following a metabolic acidosis.

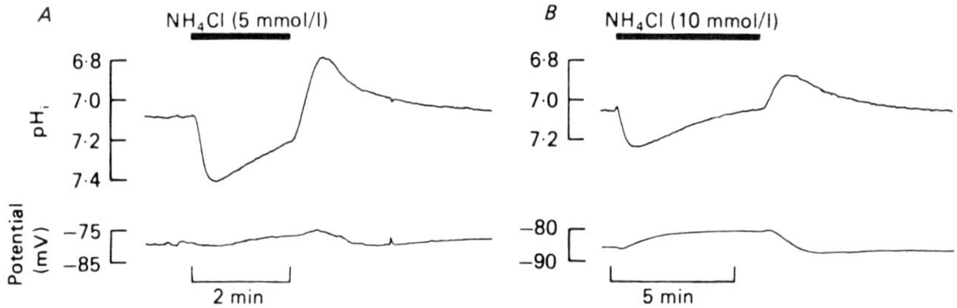

FIG. 5. Comparison in guinea pig of pH_i in an isolated ventricular myocyte (A) and an isolated strip of ventricular muscle (200–300 μm wide) (B). pH_i recorded with a double-barrelled pH-selective electrode. The other barrel (3M KCl) records membrane potential as shown. Traces show effects of adding and removing NH_4Cl as indicated by bars above the pH_i traces. CO_2/HCO_3-free Tyrode buffered with 10 mM Hepes (pH_o 7.4). Note that pH_i changes are smaller and slower in the ventricular strip. From Bountra et al 1987.

Intracellular buffering power

The first line of defence of a cardiac cell to an intracellular acid–base disturbance is its intracellular, cytoplasmic buffer system. Various methods have been devised for estimating intracellular buffering power (β) in intact cells (see Roos & Boron 1981). Interestingly, estimated values of β in cardiac ventricular tissue are among the highest yet reported in any cell other than the erythrocyte (~70 mM; see Ellis & Thomas 1976). A high intracellular buffering power has been suggested to be of functional advantage in stabilizing pH_i, since cardiac muscle is so active metabolically. Recent measurements of pH_i, however, in isolated mammalian ventricular myocytes question these apparently high values of buffering power (Bountra et al 1987). Fig. 5 compares the addition and removal of NH_4Cl to either a single isolated ventricular myocyte (A)or to a multicellular strip of ventricular muscle (B)(both preparations taken from a guinea pig heart). It is clear that the magnitude of the pH_i changes following NH_4Cl removal is much larger in the isolated myocyte than in the strip of ventricular muscle (note also the lower concentration of NH_4Cl used in 5A). In addition, the speed of recovery in pH_i is about threefold faster in the isolated myocyte. The larger changes of pH_i in the single cell might therefore imply a much lower intracellular buffering power.

Intracellular buffering power can be quantified from these sorts of experiment. This is done by i) measuring the rapid fall of pH_i following NH_4Cl removal, and ii) estimating the concentration of intracellular NH_4^+ ions using the Henderson–Hasselbalch equation and the value of pH_i prior to removal of external NH_4Cl (a further assumption is that the pK_a of NH_4Cl is the same inside and outside the cell). The value of β is estimated as: intracellular acid load (mmol/l) (equal to the concentration of $[NH_4^+]_i$ at time zero) divided by the subsequent pH_i change upon NH_4Cl removal. Adjustment is traditionally made to allow for the effects of subsequent pH_i regulation by extrapolation of the pH_i recovery back to the moment of NH_4Cl removal. The estimated values of β are 3–4 times larger for the ventricular muscle strip (β = 71.4 ± 4.1 mmol/l; ± SEM, n = 12 for multicellular ventricular muscle and 18.2 ± 2, n = 10 for isolated ventricular myocytes).

One, albeit unlikely, possibility is that these major differences in β are real and are caused by the enzymatic method of cell dispersion used in the isolation of ventricular myocytes. An alternative explanation, however, is that the differences are spurious and caused by inadequacies in the 'back extrapolation' method used to allow for the effects of pH_i regulation. We have tested for this latter possibility by slowing pH_i recovery in the presence of amiloride (1 mM). Under these conditions, the initial rapid fall in pH_i observed in isolated myocytes upon NH_4Cl removal is similar to that observed under control conditions. The initial rapid acidosis in ventricular muscle, however, is consistently over twice as large as that in the absence of amiloride and this results in a dramatic *decrease* in the estimated value of apparent intracellular buffering power. We tentatively conclude that in multicellular tissue, because of access delays in the complex extracellular spaces, external NH_4Cl removal does not result in the 'instantaneous' disappearance of $[NH_4^+]_i$ and that this is because molecular NH_3 takes several minutes to leave this sort of preparation completely. A slow departure of NH_3 will result in an underestimation of the initial acid load, if sarcolemmal acid extrusion mechanisms are functional and can restore pH_i before the full intracellular acidosis develops. This, in turn, will lead to an overestimation in the apparent value of β. Such problems would appear to be minimal in experiments with single myocytes. It seems possible, therefore, that the *real* intracellular buffering power of ventricular tissue is closer to that estimated in myocytes, i.e. 18 mmol/l, a value which is closer to that estimated previously in other types of single cell. Such downward revision of the value of β in ventricular tissue will obviously have important implications for the calculation of net membrane acid fluxes based upon measurements of changes in pH_i.

Conclusion

Intracellular pH in cardiac muscle is maintained at 7.0–7.2 largely by means of an outward-rectifying acid transporter (Na^+/H^+ exchange) that is activated by a fall in pH_i and an inward-rectifying acid equivalent transporter (Cl^-/HCO_3^- exchange) activated by a rise in pH_i. Intracellular pH is also significantly influenced by changes of Ca^{2+}_i which activate contraction and can therefore stimulate the anaerobic generation of lactic acid. This means that the sarcolemmal transport of metabolic acid also becomes important in the control of pH_i. Finally, estimates of β in single isolated ventricular myocytes suggest that it may not be as high as previously believed, but may rather match more closely the values of β quoted for other unicellular preparations.

Acknowledgements

The author wishes to thank the British Heart Foundation for financial support and Elena Vaughan-Jones for typing the manuscript.

REFERENCES

Aickin CC, Brading AF 1984 The role of chloride-bicarbonate exchange in the regulation of intracellular chloride in guinea-pig vas deferens. J Physiol (Lond) 349:587–606

Allen DG, Elliott AC, Smith GL 1986 Intracellular acidosis on increasing the frequency of stimulation in isolated ferret hearts is partly attributable to increased lactic acid production. J Physiol (Lond) 377:113P

Ammann D 1986 Ion-selective microelectrodes. Principles, design and application. Springer Verlag, Berlin, Heidelberg, New York, Tokyo

Bountra C, Powell T, Vaughan-Jones RD 1987 Comparison of microelectrode measurement of intracellular pH in cardiac ventricular tissue and isolated ventricular cells of guinea-pig. J Physiol (Lond) 390:58P

Bountra C, Kaila K, Vaughan-Jones RD 1988a Effect of repetitive activity upon intracellular pH, sodium and contraction in sheep cardiac Purkinje fibres. J Physiol (Lond) 398:341–360

Bountra C, Kaila K, Vaughan-Jones RD 1988b Mechanism of rate-dependent pH changes in sheep cardiac Purkinje fibres. J Physiol (Lond), in press

Bountra C, Hilgemann D, Vaughan-Jones RD 1988c Intracellular pH transients measured with phenol red and pH-sensitive microelectrodes in sheep cardiac Purkinje fibres. Proceedings of Physiological Society, Paris 1988, in press

Deitmer JW, Ellis D 1980 Interaction between the regulation of the intracellular pH and sodium activity of sheep cardiac Purkinje fibres. J Physiol (Lond) 304:471–488

DeHemptinne A, Marranes R, Vanheel B 1983 Influence of organic acids on intracellular pH. Am J Physiol 245:C178–C183

Ellis D, Thomas RC 1976 Direct measurement of the intracellular pH of mammalian cardiac muscle. J Physiol (Lond) 262:755–771

Ellis D, MacLeod K 1985 Sodium-dependent control of intracellular pH in Purkinje fibres of sheep heart. J Physiol (Lond) 359:81–106

Fabiato A, Fabiato F 1978 Effects of pH on the myofilaments and the sarcoplasmic reticulum of skinned cells from cardiac and skeletal muscles. J Physiol (Lond) 276:233–255

Frelin C, Vigne P, Lazdunski M 1985 The role of the Na^+/H^+ exchange system in the regulation of the internal pH in cultured cardiac cells. Eur J Biochem 149:1–4

Jean T, Frelin C, Vigne P, Barbry P, Lazdunski M 1985 Biochemical properties of the Na^+/H^+ exchange system in rat brain synaptosomes. J Biol Chem 260:9678–9684

Kaila K, Vaughan-Jones RD 1987 Influence of sodium-hydrogen exchange on intracellular pH, sodium and tension in sheep cardiac Purkinje fibres. J Physiol (Lond), 390:93–118

Mahnensmith RL, Aronson PS 1985 The plasma membrane sodium-hydrogen exchanger and its role in physiological and pathophysiological processes. Circ Res 56:773–788
Mitchell MR, Powell T, Terrar DA, Twist VW 1984 The effect of ryanodine, EGTA and low sodium potentials in rat and guinea-pig ventricular myocytes: evidence for two inward currents during the plateau. Br J Pharmacol 81:543–550
Moolenaar WH 1986 Effects of growth factors on intracellular pH regulation. Am Rev Physiol 48:363–376
Olsnes S, Tønnessen TI, Sandvig K 1986 pH regulated anion antiport in nucleated mammalian cells. J Cell Biol 102:967–971
Piwnica-Worms D, Jacob R, Horres RC, Lieberman M 1985 Na/H exchange in cultured chick heart cells -pH_i regulation. J Gen Physiol 85:43–64
Roos A, Boron WJ 1981 Intracellular pH. Physiol Rev 61:296–434
Simchowitz L, Roos A 1985 Regulation of intracellular pH in human neutrophils. J Gen Physiol 85:443–470
Thomas RC 1984 Experimental displacement of intracellular pH and the mechanism of its subsequent recovery. J Physiol (Lond) 354:3–22P
Vanheel B, DeHemptinne A, Leusen I 1984 Analysis of Cl^--HCO_3^- exchange during recovery from intracellular acidosis in cardiac Purkinje strands. Am J Physiol 246:C391–C400
Vanheel B, DeHemptinne A, Leusen I 1985 Intracellular pH and contraction of isolated rabbit and cat papillary muscle: effect of superfusate buffering. J Mol Cell Cardiol 17:23–29
Vaughan-Jones RD 1981 Chloride-bicarbonate exchange in the sheep cardiac Purkinje fibre. In: Nuccitelli R, Deamer D (eds) Intracellular pH: its measurement, regulation and utilization in cellular functions. Alan R Liss, New York p239–252
Vaughan-Jones RD 1982 Chloride activity and its control in skeletal and cardiac muscle. Philos Trans R Soc Lond B Biol Sci 299:537–568
Vaughan-Jones RD 1986 An investigation of chloride- bicarbonate exchange in the sheep cardiac Purkinje fibre. J Physiol (Lond) 379:377–406
Vaughan-Jones RD 1988 pH-selective microelectrodes: construction and use in investigations of transmembrane sodium-hydrogen exchange. In: Grinstein S (ed) Sodium-hydrogen exchange. CRC Press, Boca Raton, Florida, in press
Vaughan-Jones RD, Aickin CC 1987 Ion selective microelectrodes. In: Gray P, Standen N, Whittaker M (eds), Microelectrode techniques for cell physiology. Cambridge, Company of Biologists p 137–167
Vaughan-Jones RD, Eisner DA, Lederer WJ 1987 Effects of changes of intracellular pH on contraction in sheep cardiac Purkinje fibres. J Gen Physiol 89:1015–1032
Vaughan-Jones RD, Lederer WJ, Eisner DA 1983 Ca^{2+} ions can affect intracellular pH in mammalian cardiac muscle. Nature (Lond) 303:522–524

DISCUSSION

de Hemptinne: The problem with using multicellular preparations is partly the result of surface effects. When one calculates the apparent buffer capacity of the cells using the NH_4^+ pulse technique, one often assumes that the concentration of NH_4^+ in the unstirred layer covering the surface of the cells is immediately and at all times equal to that of the bulk solution. This is obviously not correct, as suggested by the transient pH change measured at the surface when the NH_4^+ pulse is applied.

Vaughan-Jones: In single cells we have looked at surface pH in well-buffered media, 10-20 mM Hepes; even with 10 mM NH_4Cl we do not see very large changes in surface pH, the changes are much smaller than those one can measure in multicellular tissue. However, even in multicellular Purkinje fibres in well-perfused conditions, 10–20 mM Hepes, the surface pH changes that we measure are not always very large, often about 50% smaller than those that you have reported in Purkinje fibres (Vanheel et al 1986), but that may be due to different perfusion conditions in the experimental chamber. I agree that a change in surface pH would be a contributing factor and certainly it would contribute to the slow access of CO_2 or efflux of ammonia.

Thomas: Why do you think that Claire (Aickin), using a multicellular preparation, gets low buffering power values? What is the difference between her preparation and yours?

Vaughan-Jones: In some of her experiments using ammonia, Claire was getting a high buffering power. I can only say that in multicellular ventricular muscle, experimental estimates of intracellular buffering power (β) are high whereas intracellular buffering power estimated in ventricular myocytes seems to be lower. Interestingly, the average buffering power calculated for the sheep Purkinje fibre is more like 30, which is closer to the value for a single cell. Variation in the apparent value of β may be due to variation in the thickness of the preparation or the general architecture of multicellular strips which varies according to the person dissecting.

Syková: Do you find this acid shift only on the surface of the fibre or also in the space between the fibres?

Vaughan-Jones: In clefts? We can't say. We measure the surface acidosis with an electrode but we can also have phenol red perfused throughout the whole preparation. The latter gives a mean value of acidosis in clefts, at the surface and deeper in the tissue. We get similar changes in pH with both methods so there is probably acidosis in the clefts as well as on the surface.

Thomas: You have not poked a blunt electrode into the middle of the tissue?

Vaughan-Jones: No. One other thing is that surface acidosis measured with phenol red is usually more acidic in absolute terms than that measured with an electrode, which might indicate that in the centre of the fibre extracellular pH may be slightly more acid than at the surface (the fibres are 200–300 μm in diameter).

Syková: Siesjö et al (1985) found that lactic acid concentrations in the brain during seizures were not altered for the first 30-60 seconds, therefore lactic acid accumulation cannot explain the initial rapid acidification. The gradual fall in pH_e attributable to lactic acid production occurred later as a result of non-ionic diffusion of lactic acid from the cells into the extracellular space. What is the permeability of the cell membrane to lactate in the heart?

Vaughan-Jones: High. Recovery from metabolic acidosis occurs with a half time of about one minute and seems to be due to the efflux of lactic acid, whereas pH_i recovery from an ammonia pre-pulse is much slower and is produced by Na^+/H^+ exchange. So I would say that the lactic acid pathway is probably the most efficient way of restoring pH_i in heart after metabolic acidosis. It's a two-way system: heart normally consumes lactate in response to, for example, whole body metabolic acidosis, which would mean lactic acid entering the cell and causing movement of pH in the opposite direction. This means that there is a bi-directional influence on pH_i through the lactic acid pathway. This influence is supplementary to the control of pH_i that occurs through Na^+/H^+ and Cl^-/HCO_3^- exchange.

Thomas: Do you have lactate present in your normal superfusion solution?

Vaughan-Jones: No, in our experiments we deliberately bias the lactic acid carrier towards the outward mode by having no lactate in our perfusing solutions.

Boron: I agree that the transporter should be able to go back and forth but I am not sure that what you say is exactly correct. If lactic acid goes into the cell and then releases H^+ and lactate, and if that lactate is used to make glucose or to produce energy, then both the lactate and the protons will be consumed. Therefore, I would expect to see no change in intracellular pH.

Vaughan-Jones: Alex de Hemptinne's group have shown that in the short-term, if you increase the extracellular lactate concentration, you get an acid load inside, which is sustained for the period of lactate loading (de Hemptinne et al 1983). So I think the answer is: it can go backwards and, when it does, it can reduce intracellular pH.

Boron: But if the lactate is used for energy production that acidosis will be blunted.

Vaughan-Jones: It will be blunted, yes.

Machen: You showed (Fig. 2) that when you add stropanthidin and then add amiloride the cells acidify at increasing rates as time passes. You use that as evidence that the Na^+/H^+ exchanger is increasing its rate of turnover with acidosis. What causes the increasing rate of acidosis?

Vaughan-Jones: It's due to the fact that intracellular Ca^{2+} is rising and this is promoting acidosis at a particularly rapid rate. The acidosis is most probably due to the fact that a rise in intracellular Ca^{2+} activates contraction, leading to the generation via glycolysis of metabolic acid. Displacement by Ca^{2+} of H^+ from intracellular binding sites may also be occurring. All this means that the cell is producing acid at a particular rate and this stimulates the Na^+/H^+ exchanger—if you then block the exchanger, you uncover the real rate of production of acid inside the cell in response to Ca^{2+}.

Machen: Does the calcium come from the extracellular space?

Vaughan-Jones: Yes, in that it depends on the operation of the sarcolemmal Na^+/Ca^{2+} exchanger.

Machen: If you remove extracellular Ca^{2+}, does that block this effect?

Vaughan-Jones: Yes, you can remove the calcium-activated acidosis by reducing external Ca^{2+}. This reduces the internal Ca^{2+} concentration and then the cell becomes more alkaline.

Thomas: So a major pathway for calcium entry under those conditions of sodium pump inhibition and high internal Na^+ is reversed Na^+/Ca^{2+} exchange. Is that a normal mechanism for calcium entry in the action potential in heart muscle?

Vaughan-Jones: No, the gradients for calcium entry on a reversed Na^+/Ca^{2+} exchanger are such that you get net Ca^{2+} entry only for a very short time, just at the peak of the upstroke of the action potential.

Harvey: You think that the effect of stropanthidin is due to calcium activating metabolism. Might it not have an effect on the exchanger via changes in intracellular Na^+?

Another point: when you compare the activation curves of intracellular pH versus activity of the exchanger (Fig. 3), one of the parameters which is essential for the net forward movement of the exchanger, i.e. the intracellular Na^+ concentration is changing.

Vaughan-Jones: The first question is: do variations in Ca^{2+} concentration influence Na^+/H^+ exchange directly? If you raise the internal Ca^{2+} concentration, this produces a fall in pH_i, even in the presence of amiloride, so the influence of Ca^{2+} on pH_i is independent of the Na^+/H^+ exchanger.

The second question is: does a rise of intracellular Na^+ affect the exchanger? We can't answer that conclusively. What we can do is look at the rate of recovery of pH_i from an internal acid load (ammonia pre-pulse) plotted as a function of pH_i, under conditions where the intracellular Na^+ concentration is either high or low. When internal Na^+ is high, there is a shift of the graph along the pH_i axis to lower values of pH_i, which might suggest that the raised intracellular Na^+ is inhibiting Na^+/H^+ exchange. However, when internal Na^+ is high, internal Ca^{2+} is also high (due to Na^+/Ca^{2+} exchange) and this will *generate* intracellular acid. When we allow for this extra acid, then there often appears to be no additional effect of the raised internal Na^+ on the rate of recovery of pH_i from an intracellular acid load. Sergio Grinstein has proposed that intracellular Na^+ has an allosteric inhibitory effect on Na^+/H^+ exchange in thymic lymphocytes (Grinstein et al 1984). As yet, we have been unable to confirm that a similar effect occurs in cardiac muscle.

Boron: These are elegant experiments: you are looking at a rate of pH change, then you add amiloride and look for the change in the slope of the pH versus time trace. That should be independent of whatever acid loading is being produced by Ca^{2+}. In the presence of strophanthidin, the initial dpH_i/dt versus pHi curve had a foot on it and was accelerating going up. But in one of the last slides that you showed us, presumably derived from an NH_4^+ pulse experiment where you compared multicellular versus single cell data, the curves were almost straight lines.

Vaughan-Jones: Those latter curves are plots of dpH_i/dt as a function of pH_i for pH_i recovery from an NH_4^+ pre-pulse. I made no correction for any other possible mechanisms, they are just a simple description.

Boron: But if anything, those curves would have been shifted upward because you are underestimating Na^+/H^+ exchange.

Vaughan-Jones: By a small amount, yes. If you plot the curves obtained from an ammonia pre-pulse experiment conducted when internal Na^+ is at its normal low level, then they are roughly the same as what you get if you plot $-dpH_i/dt$ as a function of pH_i using the amiloride dissection technique (illustrated in Fig. 2b). If you raise intracellular Na^+ by adding stropanthidin, then the dpH_i/dt versus pH_i curve produced by using the ammonia pre-pulse method shifts to the left along the pH_i axis (to lower pH_i values). But if you compensate for the fact that extra acid is being generated inside the cell in the presence of stropanthidin (because Ca^{2+} has been elevated), then this is sufficient to account for much of the leftward shift (Kaila & Vaughan-Jones 1987).

The rate of recovery of pH_i is due to the balance between net acid efflux and acid generation *de novo* within the cell. Acid generation increases enormously in the presence of stropanthidin and this will slow the rate of recovery of pH_i. This accounts for a major fraction of the shift to the left.

Morad: Both the gradient for Na^+ and the acid production in the cell have changed, is that right?

Vaughan-Jones: Yes, the gradient has changed because the internal Na^+ concentration has changed.

Morad: And the acid production is because of Ca^{2+}/H^+ exchange across the mitochondria, or what?

Vaughan-Jones: It's Ca^{2+} promoting acidosis, which you can remove by reducing the concentration of Ca^{2+}.

Al-Awqati: If the sodium gradient has changed, you would expect a change in this curve.

Vaughan-Jones: You are relating thermodynamics to kinetics. The only thing that shifts when you shift the transmembrane Na^+ gradient is the equilibrium pH_i for Na^+/H^+ exchange. We are talking about non- equilibrium thermodynamics and you don't necessarily expect a change in rate simply by changing the concentration of internal substrate.

Al-Awqati: Rate is proportional to the gradient.

Vaughan-Jones: This is a fundamental misunderstanding, rate is not necessarily proportional to driving force. It is ultimately related but the relationship is not a simple one.

Boron: Did you mention that there are some data from vesicles which suggest that internal Na^+ affects Na^+/H^+ exchange?

Vaughan-Jones: What I am saying is that we can't see, as yet, any convincing data that internal Na^+ allosterically inhibits Na^+/H^+ exchange in sheep Purkinje fibres. We account for our observations in other ways.

Aickin: How do you calibrate the amount of acid? You seem to be saying that Ca^{2+} is elevated so the acid load is elevated and you can subtract that bit. Are you subtracting by going to low Ca^{2+} or by calculating how much of an acid load the elevated Ca^{2+} causes and simply taking that away from the pH_i recorded?

Vaughan-Jones: We estimate the rate of background acid production by adding amiloride (to inhibit acid extrusion) and observing the subsequent fall in pH_i. The rate of fall of pH_i with amiloride is about ten times faster in the presence of stropanthidin than in control conditions, indicating a large increase in the background rate of acid generation (see Kaila & Vaughan-Jones 1987).

Thomas: What about the differences in buffering power?

Vaughan-Jones: Intrinsic intracellular buffering power looks fairly similar with and without stropanthidin—we have measured it.

Harvey: Amiloride has been shown to inhibit metabolism and this may interfere with metabolic acid load.

Vaughan-Jones: We are getting into details. You asked whether I have evidence for inhibition of Na^+/H^+ exchange by internal Na^+. I said that we can observe a slowing of pH_i recovery from an acid load, if we raise internal Na^+. However, we can account for much of this observation in terms of an increase in the rate of background acid production within the cell. If I said the opposite, that the observation provides evidence for allosteric inhibition of the exchanger by internal Na^+, you would be much more correct to point out that the result might be due to increased intracellular acid production.

Malnic: You show that amiloride blocks this Na^+/H^+ exchanger and that this exchanger acts differently as intracellular pH changes. This also occurs in the brush border vesicles in the proximal tubule. Could what you have shown be in some part due to changes in buffering power at different internal pH?

Vaughan-Jones: If we measure intracellular buffering power using the ammonia pre-pulse technique, then it looks roughly the same at high and low pH_i—given all the errors associated with measuring buffering power.

Malnic: The last curve also suggests that the buffering power might vary with changes in internal pH.

Vaughan-Jones: There have been reports that the apparent buffering power changes slightly with pH_i.

de Hemptinne: You showed a shift to the left of the curve relating proton extrusion to pH_i when the intracellular Na^+ concentration increases. What happens after depletion of intracellular Na^+?

Vaughan-Jones: Fig. 1 shows that if you deplete a Purkinje fibre of Na^+, then acid load using an ammonia pre-pulse and then readmit external Na^+, the recovery of pH_i is much faster than in control conditions. This again raises two possibilities: 1) on readmitting external Na^+, intracellular Na^+ is initially zero. This may remove an allosteric inhibition by Na^+ of Na^+/H^+ exchange, thus producing pH_i recovery at a faster rate than in control conditions. 2) In Na^+-free conditions, where it's known that the Ca^{2+} concentration is high, the moment you put back external Na^+, intracellular Ca^{2+} will fall to low levels (via Na^+/Ca^{2+} exchange) and this will remove a Ca^{2+}-induced acid load. The pH_i recovery will thus be accelerated. This returns to the same problem we discussed previously—it needs to be looked at very carefully.

Thomas: Claire had a remarkably fast pH_i recovery when she put Na^+ back after acid loading in Na^+ free conditions.

Aickin: At that stage the internal Na^+ was shooting up as well.

Thomas: Could it be important that the ATP level may be much higher than normal? You have taken all the load off the Na^+ pump, for example.

Vaughan-Jones: Yes, but in Na^+-free solution, Ca^{2+} metabolism is disrupted because the intracellular Ca^{2+} concentration is raised. This could increase the metabolic demand for ATP by the Ca^{2+} pump at the sarcoplasmic reticulum.

Grinstein: There is a slight acidification when you add deoxyglucose—can you explain that?

Vaughan-Jones: No, one possibility is that it's due to the net depletion of ATP in the cell.

Grinstein: Is it because phosphorylation of deoxyglucose to deoxyglucose 6-phosphate is releasing protons?

Vaughan-Jones: That is possible. In addition, if you accept that the ATP is normally kept in balance by re-phosphorylation of ADP, then there is no net generation of protons. If, however, there is net depletion of ATP, that would cause acidification.

Grinstein: Could it be an ATP-dependent shift of the Na^+/H^+ antiporter?

Vaughan-Jones: Sure.

Grinstein: You showed the pH-dependent activation of the Cl^-/HCO_3^- exchanger, is that derived purely from thermodynamics or have you measured the kinetics?

Vaughan-Jones: I have measured the kinetics. It is also interesting to note that a simple model of Cl^-/HCO_3^- exchange (where rate is a hyperbolic saturating function of Cl^- and HCO_3^- levels on both sides of the membrane) predicts that the activation of acid-equivalent fluxes plotted as a function of pH_i, at constant p_{CO_2}, will be linear in the range pH_i 6.6–7.8, reversing at the equilibrium point. The fact that activation is markedly non-linear (Fig. 3) could mean that there is some allosteric modification of rate versus pH_i or, alternatively, that a more complex kinetic scheme is operating.

Grinstein: Glaser and Cassell (Rothenberg et al 1983) have shown that ouabain and stropanthidin-treated cells show a stimulation of Na^+/H^+ exchange despite the Na^+ gain. We never found this.

Vaughan-Jones: That's what we find here. We add stropanthidin, which is equivalent to ouabain, inhibit the Na^+/K^+ pump and stimulate the Na^+/H^+ exchanger. The mechanism seems to work through the fact that you are generating an acid load.

Grinstein: In your case, it's acid induced by higher concentrations of Ca^{2+}.

Thomas: The question is, do you stimulate the exchanger at exactly the same intracellular pH or was the stimulation purely because you increased the work for it to do?

Vaughan-Jones: As far as we can tell, it seems to be a function of the cell becoming acidic. Whether or not there are shifts in ATP under these conditions which play a role would be difficult to say.

Grinstein: Your argument is that it would only work in cells that have plenty of Na^+/Ca^{2+} exchangers?

Vaughan-Jones: Yes it would be most prominent. It depends on whether a change of Na^+ inside the cell can lead to a change of Ca^{2+} and whether changing Ca^{2+} can affect pH_i.

Grinstein: In our cells we never get changes in cytoplasmic Ca^{2+}, as we change internal or external Na^+. When Na^+ is increased inside the cell, there is always an inhibition of Na^+/H^+ exchange, much like your shift to the left.

Morad: If you patch-clamp these cells, basically you clamp the internal Ca^{2+}. If you then give an NH_4^+ pulse, what happens?

Vaughan-Jones: We have not done that, it would be nice to try.

Thomas: If you are controlling Ca^{2+} by exchange with the patch electrode filling solution, wouldn't you necessarily be controlling pH as well?

Vaughan-Jones: These are the sort of things we would like to do but we have only just started doing measurements in myocytes so all this remains to be done.

Al-Awqati: You showed that the Cl^-/HCO_3^- exchanger is activated at a specific pH. Is the HCO_3^- concentration constant under these conditions? Are you changing pH and HCO_3^-?

Vaughan-Jones: Intracellular HCO_3^- increases under conditions where you tend to activate the exchanger at the inside site. In other words, if pH_i increases at a constant p_{CO2}, then internal HCO_3^- levels also increase. It seems likely that Cl^-/HCO_3^- exchange is seeing this rise of intracellular HCO_3^-. However, the non-linear activation of acid-equivalent fluxes as a function of pH_i suggests that the situation is more complex than a simple exchanger obeying Michaelis-Menten kinetics.

Steinmetz: But the K_m is not known?

Vaughan-Jones: I have estimated it. The K_m for HCO_3^- in the absence of chloride is about 1 mM. Similarly, the K_m for Cl^-, with no HCO_3^- competition, is about 1mM (Vaughan-Jones 1986). These values are similar to those found in the red blood cell.

Thomas: Are you sure that there is no Na^+ requirement for that system?

Vaughan-Jones: I have bled the cell of Cl^-, then given Cl^- back: Cl^- is re-accumulated principally via Cl^-/HCO_3^- exchange. The Cl^- re-accumulation can proceed in the absence of Na^+; the initial rate of re-uptake of Cl^- is little affected by removing Na^+.

Thomas: To summarize, you think that normally the pH regulation in cardiac muscle is by Na^+/H^+ exchange primarily in an acid direction.

Vaughan-Jones: There is also lactic acid transport, metabolic acid loads are one of the major forms of acid loading that the heart is going to see.

Thomas: But many of your experiments have been done without a contraction.

Vaughan-Jones: Stimulation rate affects pH_i, which means the heart must have some way to minimize the changes of pH_i and it seems to do this through lactic acid transport.

Morad: I am interested in knowing your thoughts on cardiac contractility. Electron probe analysis shows that there is charge imbalance after release of Ca^{2+} in skeletal muscle. It has been postulated that this is partially balanced by counter movements of K^+, Mg^{2+} and H^+ moving back into the sarcoplasmic reticulum. A similar thing must happen in a heart except much more dramatically because the heart is beating all the time. How do you see this regulation, is there an alkalinization of the myoplasm, vis à vis your Na^+/H^+ exchange?

Vaughan-Jones: To give a trivial answer, empirically, as far as the electrode is concerned, you don't see it because the electrode is too slow. There have been reports that skeletal muscle becomes alkaline during contraction. But the interpretation given then was that it was due to changes in phosphocreatine rather than movement of Ca^{2+} and H^+ into the sarcoplasmic reticulum (Baylor et al 1982). The latter possibility will not be seen until you image single cardiac cells while looking at internal pH using fluorescent chemical probes. What one might expect to see then is a non-homogeneity of pH within the cell and flashes of alkalosis which give you an outline of the sarcoplasmic reticulum (assuming that the process does occur). The degree of alkalosis would presumably depend on the rate constants of the buffers present. It may be minimal, if there is a very fast buffering system with high capacity.

Oberleithner: Can you comment on the role of the Na^+/K^+-ATPase. I assume that it's major function is to maintain the Na^+ gradient. Is it pH-sensitive itself? Apart from the systems you described, is there any rheogenic HCO_3^- transport system, for example Na^+-HCO_3^- co-transport?

Vaughan-Jones: Yes, I would see the Na^+/K^+ pump as being principally a controller of Na^+. In heart that means it is also indirectly a controller of Ca^{2+} through Na^+/Ca^{2+} exchange. Its function is the control of electrical and contractile activity through setting the gradients for Na^+ and K^+.

There is evidence for pH sensitivity of the Na^+/K^+ pump, so you could get pH-sensitive changes in Na^+ via the Na^+/K^+ pump or inhibition thereof.

As regards the rheogenic transporter, I have not looked for it directly but it would seem that there is not good evidence for a rheogenic system simply because pH recovery from an acid load is little affected by adding back HCO_3^-. There is evidence that there may be an effect due to a HCO_3^- leak which is very small; but there is no evidence for any great change in the net outward fluxes of acid equivalents in response to adding back HCO_3^- or CO_2, which would suggest that the role of Na^+-HCO_3^- co-transport is minimal.

Russell: I was very interested in the effects of metabolic inhibition and I would like to suggest a possible explanation. In the barnacle muscle we found that raising magnesium inhibits a similar transport system, the Na^+-dependent Cl^-/HCO_3^- exchanger. Because one of the important buffers for Mg^{2+} is ATP, as the ATP concentration falls, the concentration of Mg^{2+} rises, and we see a shift of the pH activation curve to the left, similar to the one you described. We measured SITS-sensitive Cl^- efflux, which is a convenient marker for this Na^+-dependent Cl^-/HCO_3^- exchange system. At pH 6.6 in the dialysis fluid this transport process is almost maximally activated. However, as we raise the internal Mg^{2+} concentration, we see inhibition of Cl^- efflux and acid extrusion. When the internal pH is quite acidic, about 6.6, the K_i for intracellular Mg^{2+} is about 12–13 mM. If we then do the same thing at pH 6.9, somewhat closer to normal pH for barnacle, we find that K_i is around 7–8 mM; at pH 7.05, K_i is 2–3 mM. Therefore, as you get closer to the normal pH inside the cell (7.35), the inhibition by intracellular Mg^{2+} falls into a range that for the barnacle is near the physiological concentration of Mg^{2+}, 1–2 mM. So the problem of why does the acid extruder turn off at a pH at which thermodynamics would predict that the Na^+-dependent Cl^-/HCO_3^- exchanger should still be quite active, may be partly due to an inhibitory effect of intracellular Mg^{2+}, at least for this system.

Vaughan-Jones: At the moment we can't say whether the effects of metabolic inhibitors on Na^+/H^+ exchange are mediated through changes of ATP or inorganic phosphate, which will rise, or even cAMP. Preliminarily, we find that the glycolytic inhibitors are more effective blockers than aerobic inhibitors, even though it's reported that ATP levels fall under both circumstances.

Curtin: In the experiments where you inhibit glycolysis and see intracellular acidification, do the contractions go away with the same time course as the change in the pH?

Vaughan-Jones: It's variable, with deoxyglucose force often drops along with the initial acidification. We have some experiments with iodoacetate, which has the same effect as deoxyglucose in that it eliminates the rate- dependent acidosis, where the force has remained reasonably constant even though the cell is going acid; we don't know the explanation for that. One possibility is that the Ca^{2+} concentration might be rising even though the concentration of H^+ is rising and that the increase in Ca^{2+} is offsetting the inhibitory effect of acidosis.

When testing the inhibitory effect of deoxyglucose on the rate-dependent acidosis, we used a voltage-clamp apparatus to adjust force development during a test train in the presence of deoxyglucose, so that it matched force development during a control train in the absence of deoxyglucose. This ensured that the work placed on the preparation was roughly equal in both conditions so that we could compare the control and test acidoses (Bountra et al 1988).

Curtin: Do you have any direct evidence about what the calcium is doing?

Vaughan-Jones: No.

Russell: One of the fundamental properties of the pH regulators is their sensitivity to changes in internal pH; nearly all are activated by a fall in pH_i or inactivated by a rise in pH_i. There has not been much study of the mechanisms by which pH_i regulates the activity of its own regulator. Generally, it is considered that some sort of modifier site exists on the internal face of the membrane. It is usually assumed that this modifier site is on the transporter molecule, where it is directly titrated by protons as pH_i changes. I would like to report some results of two studies, one in internally dialysed barnacle muscle fibres performed in collaboration with Dr MS Brodwick, and one on squid axons performed in collaboration with Dr Walter Boron. I believe that the results of these studies may have an important bearing on at least some kinds of regulation of pH_i regulators. Both of these preparations rely almost exclusively on Na^+-dependent Cl^-/HCO_3^- exchange for pH_i regulation (Boron & Russell 1983, Russell et al 1983).

Recently, we have shown that in barnacle muscle raising the intracellular Mg^{2+} concentration, but not the external concentration, inhibits acid extrusion from acid-loaded fibres as well as the obligatory, SITS-sensitive Cl^- efflux. More interesting than the simple inhibition by intracellular Mg^{2+} is the relation between pH_i and the inhibitory potency of intracellular Mg^{2+}. The more closely pH_i approaches the normal set-point (pH = 7.35), the more potent intracellular Mg^{2+} becomes as an inhibitor. Thus, the K_i is 13 mM at pH_i 6.8 and only 3.8 mM at pH_i 7.1 (normal intracellular Mg^{2+} concentration is 1–4 mM). Therefore, it is possible that the set-point for pH_i regulation could be the result of this interaction between intracellular Mg^{2+} and H^+, since, on thermodynamic grounds, there is sufficient energy in the relevant ion gradients to raise pH_i to at least 7.8.

A somewhat different observation was made in squid axon. In this preparation, where pH_i regulation has an absolute requirement for ATP (Russell & Boron 1976, Boron & Russell 1983), we found that dialysis with ATP-γ-S at acidic pH_i, followed by complete nucleotide removal, resulted in persistent activation of the acid-extruding mechanism for at least two hours after washing out the ATP-γ-S. However, if the pH_i was raised to normal levels (7.35) for even relatively short periods of time, the acid extruder was inactivated. This alkaline inactivation was time dependent with a time constant of about 10 minutes. We therefore propose that alkaline pH_i promotes either dephosphorylation of the transport protein (or a necessary co-factor), or the release of ATP. Since Mg^{2+} also inhibits acid extrusion in squid axon, it is tempting to speculate that intracellular Mg^{2+} works by promoting dephosphorylation (possibly by activating a phosphatase) or by promoting the release of ATP. Further work will be directed towards elucidation of the mechanisms for both Mg^{2+} and alkaline inactivation of the pH_i regulator.

Grinstein: I had the impression that ATP-γ-S was used by kinases but was not hydrolysed by phosphatases, is that correct?

Russell: That is the standard story because for biochemists the time course for an experiment is often a couple of minutes and our results have a half-time of ten minutes. There is evidence in the literature that after 10–20 minutes there is some hydrolysis by phosphatases (Hauptman et al 1985). I think if you could do our experiment with real ATP, the half-time would be very fast, as fast as we can change the pH.

References

Baylor SM, Chandler WK, Marshall MW 1982 Optical measurements of intracellular pH and magnesium in frog skeletal muscle fibres. J Physiol 331:105–137

Boron WF, Russell JM 1983 Stoichiometry and ion dependencies of the intracellular-pH-regulating mechanism in squid giant axons. J Gen Physiol 81:373–399

Bountra C, Kaila K, Vaughan-Jones RD 1988 Mechanism of rate-dependent pH changes in sheep cardiac Purkinje fibres. J Physiol (Lond), in press

de Hemptinne A, Marranes R, Vanheel B 1983 Influence of organic acids on intracellular pH. Am J Physiol 245:178–183C

Grinstein S, Cohen S, Rothstein A 1984 Cytoplasmic pH regulation in thymic lymphocytes by an amiloride-sensitive Na^+/H^+ antiport. J Gen Physiol 83:341–369

Hauptman M, Wilson DF, Erecinska M 1985 Effect of ATPγS in isolated rat brain synaptosomes. Biochem Pharmacol 34:1247–1254

Kaila K, Vaughan-Jones RD 1987 Influence of sodium-hydrogen exchange on intracellular pH, sodium and tension in sheep cardiac Purkinje fibres. J Physiol (Lond) 390:93–118

Rothenberg P, Glaser L, Schlesinger P, Cassel D 1983 Activation of Na/H exchange by epidermal growth factor. J Biol Chem 258:12644–12653

Russell JM, Boron WF, Brodwick MS 1983 Intracellular pH and Na fluxes in barnacle muscle with evidence for reversal of the ionic mechanism of intracellular pH regulation. J Gen Physiol 82:47–78

Russell JM, Boron WF 1976 Role of chloride transport in regulation of intracellular pH. Nature (Lond) 264:73–74

Siesjö BK, von Hanwehr R, Nergelius G, Nevander G, Ingvar M 1985 Extra- and intracellular pH in the brain during seizures and in the recovery period following the arrest of seizure activity. J Cereb Blood Flow Metabol 5:47–57

Vanheel B, de Hemptinne A, Leusen I 1986 Influence of surface pH on intracellular pH regulation in cardiac and skeletal muscle. Am J Physiol 250:748–760C

Vaughan-Jones RD 1986 An investigation of chloride-bicarbonate exchange in the sheep cardiac Purkinje fibre. J Physiol (Lond) 379:377–406

Ionic mechanisms of intracellular pH regulation in the nervous system

Wolf-R. Schlue and Joachim W. Deitmer

Institut für Zoologie, Lehrstuhl für Neurobiologie, Universität Düsseldorf, Universitätsstr. 1, D-4000 Düsseldorf 1, Federal Republic of Germany

Abstract. Two separate mechanisms responsible for intracellular pH (pH_i) regulation in neuronal membranes of the nervous system have been studied so far: they are Na^+/H^+ and Na^+-H^+-HCO_3^-/Cl^- exchange. The involvement of these mechanisms in pH_i regulation of neurons and glial cells has been investigated in the leech central nervous system using ion-selective microelectrodes. The amiloride-sensitive Na^+/H^+ exchange is the predominant mechanism of pH_i regulation in nominally HCO_3^- free, Hepes-buffered saline of both neurons and glial cells of this nervous system. In the presence of CO_2-HCO_3^- buffer, however, the SITS-sensitive Na^+-H^+-HCO_3^-/Cl^- exchanger contributes to acid extrusion in neurons and probably also in glial cells. Unlike neuronal pH_i, glial pH_i increases when Hepes is replaced by CO_2-HCO_3^- as the extracellular buffer, and decreases again on return to Hepes buffer. The glial alkalinization occurs in the opposite direction, as would be expected from the CO_2 movement across the cell membrane and its hydration to form carbonic acid which dissociates into H^+ and HCO_3^- ions. The expected acidification, however, is observed in neurons, and is reduced by acetazolamide and ethoxzolamide, inhibitors of carbonic anhydrase, which catalyses the formation of carbonic acid. On the other hand, these drugs are shown to produce no change of the CO_2–HCO_3^--induced alkalinization in glial cells. The observations suggest that Na^+-HCO_3^- co-transport across the glial cell membrane, mediating the influx of HCO_3^- ions into the cell interior, could be responsible for the unusual alkalinization. Further evidence for the activation of Na^+-HCO_3^- co-transport, as a third mechanism involved in pH_i homeostasis of the nervous system, is presented.

1988 Proton passage across cell membranes. Wiley, Chichester (Ciba Foundation Symposium 139) p 47–69

In most cells of the nervous system, intracellular pH (pH_i) is close to the external pH, too alkaline to be explained by a passive distribution of H^+ or HCO_3^- ions (for references see Roos & Boron 1981, Boron 1983, Moody 1984, Thomas 1984). Net passive fluxes would tend to acidify pH_i, so that an active acid extrusion mechanism is required to keep pH_i constant.

Mechanisms of acid extrusion

The major difference between cells of the nervous system is that some have only a single pH_i-regulating mechanism involving Na^+-dependent anion exchange, whereas others have two separate systems, one of which is independent of HCO_3^- transport but still requires Na^+ influx.

Na^+-H^+-HCO_3^-/Cl^- exchanger. The Na^+-H^+-HCO_3^-/Cl^- exchanger (or Na^+-dependent HCO_3^-/Cl^- exchanger) was first proposed to explain acid extrusion by snail neurons (Thomas 1977; see diagram in Fig. 1). This acid extrusion system is inhibited by the anion exchange inhibitors SITS or DIDS (see Methods), by removing external Na^+ or HCO_3^-, or by depleting internal Cl^- (Thomas 1977). The acid extrusion is accompanied by an increase in the intracellular Na^+ activity and a decrease in internal Cl^- activity. The system is thought to be electroneutral because pH_i recovery occurs with no change in membrane potential (E_m), and changes in the latter do not affect the rate of pH_i recovery (Thomas 1978). The Na^+-H^+-HCO_3^-/Cl^- exchanger, as a single pH_i-regulating system, has subsequently been found in several other cells of invertebrate nervous systems, e.g. squid axon (Boron & Russell 1983), crayfish central neurons, (Moody 1981) and crayfish stretch receptor sensory neurons (Moser 1985).

Na^+-H^+-HCO_3^-/Cl^- and Na^+/H^+ exchanger. In contrast to the cells mentioned above, other preparations appear to use two separate mechanisms to regulate pH_i: Na^+-H^+-HCO_3^-/Cl^- exchange and Na^+/H^+ exchange (Fig. 1). Simultaneous action by these two mechanisms was first demonstrated to

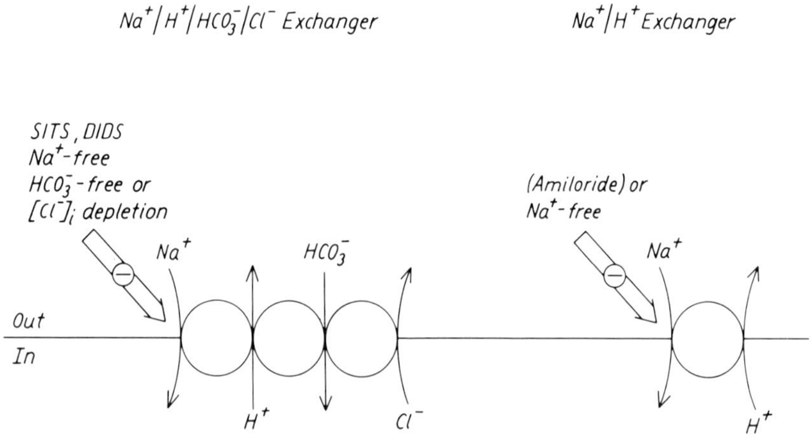

FIG. 1. Diagrams of two possible acid extrusion mechanisms in the nervous system: Na^+-H^+-HCO_3^-/Cl^- and Na^+/H^+ exchangers. The different experimental conditions for inhibition of both exchangers are indicated (arrows).

operate during pH_i recovery in crayfish neurons (Moody 1981). Recent evidence suggests that a similar pair of mechanisms may be involved in pH_i regulation by lamprey reticulospinal neurons (Chesler 1986).

The Na^+/H^+ exchanger is probably identical to that described in sea urchin eggs (Johnson et al 1976) and mouse soleus muscle fibres (Aickin & Thomas 1977). This exchange process depends on the presence of Na^+ in the external solution and involves the entry of Na^+ into the cell (Moody 1981). It is independent of the concentration of HCO_3^- in the external solution and is not blocked by SITS. Furthermore, the K^+-sparing diuretic, amiloride, has been shown to block the Na^+/H^+ exchanger in most preparations, but not in all.

The second component of pH_i regulation resembles the Na^+-dependent HCO_3^-/Cl^- exchanger. It is sensitive to the external HCO_3^- concentration, is blocked completely by SITS or DIDS and involves the efflux of Cl^- ions from the cell (Moody 1981). This second component of pH_i regulation appears also to be Na^+ sensitive, since pH_i recovery is almost totally blocked when the cells are kept in Na^+-free solution.

Leech central nervous system

The involvement of both pH_i regulating mechanisms has been investigated in neurons and glial cells of the leech central nervous system using ion-selective microelectrodes. The leech nervous system is particularly suitable for this kind of study for the following reasons: 1) individual neurons and glial cells can be identified and penetrated by microelectrodes (Schlue & Walz 1984); and 2) several types of ion-selective microelectrodes have been successfully used for intracellular recording of ion activities (Deitmer & Schlue 1981, 1983, Schlue & Wuttke 1983, Schlue & Thomas 1985).

Our present results reveal both similarities and differences concerning the maintenance of pH_i homeostasis in the two types of cells. A more detailed account of this investigation has been given elsewhere (Schlue & Thomas 1985, Schlue & Deitmer 1987, Deitmer & Schlue 1987).

Methods

The preparation and dissection procedures used to isolate single ganglia of the leech, *Hirudo medicinalis*, and the selection of cells have been described before (Schlue & Deitmer 1980). The isolated ganglia were pinned by their connectives and lateral nerve roots to the silicone base of a perspex chamber and superfused continually. The experiments were done on the neuropile glial cells and on Retzius neurons at room temperature.

Solutions

The normal leech saline (nominally HCO_3^--free) had the following compo-

TABLE 1 Intracellular pH (pH_i), membrane potential (E_m) and H^+ equilibrium potential (E_{H^+}) in glial cells and neurons of leech nervous system

Neuropile glial cell

Buffer	*pH_i*	*E_m (mV)*	*E_H^+ (mV)*
Hepes	6.85±0.06 (n = 23)	−68.4±7.5 (n = 23)	−32.2
2% CO_2/11 mM HCO_3^-	7.18±0.13 (n = 25)	−73.5±5.6 (n = 17)	−12.9

Retzius neuron

Buffer	*pH_i*	*E_m (mV)*	*E_H^+ (mV)*
Hepes	7.28±0.08 (n = 35)	−47.3±5.5 (n = 35)	−7.0
2% CO_2/11 mM HCO_3^-	7.20±0.12 (n = 15)	−43.8±3.8 (n = 15)	−11.7

sition: NaCl, 115 mM; KCl, 4 mM; $CaCl_2$, 1.8 mM; Hepes (adjusted to pH 7.4 with NaOH), 10 mM; and glucose, 11 mM. HCO_3^- solutions were equilibrated with 2% CO_2 and 98% oxygen and contained 11 mM $NaHCO_3$ instead of Hepes (pH 7.4). Ammonium-containing solutions were made by replacing 20 mM NaCl with equivalent amounts of NH_4Cl. Na^+-free solutions were made by replacing Na^+ with *N*-methyl-D-glucamine neutralized with HCl and/or CO_2 as appropriate. The Cl^--free solutions were made by equimolar replacements of the Cl^- salts with gluconate salts.

Amiloride (3,5-diamino-6-chloropyrazinoyl guanidine; Sigma), furosemide (4-chloro-*N*-furfuryl-5-sulphamoylanthranilic acid; Sigma), SITS (4-acetamido-4′-isothiocyanatostilbene-2,2′-disulphonic acid; Sigma), DIDS (4,4′-diisothiocyanatostilbene-2,2′-disulphonic acid; Sigma), acetazolamide (5-acetamido-1,3,4-thiadiazole-2-sulphonamide; Cyanamid) and ethoxzolamide (6-ethoxy-2-benzothiazole sulphonamide; gift of Dr. Siffert, Bochum) were used within a few minutes of dissolving in solutions.

Microelectrodes

The double-barrelled microelectrodes were prepared as described by Zeuthen (1980). Two glass capillaries differing in outside diameter (1.2 and 1.5 mm) were glued together. In a microelectrode puller, the thinner tube was softened and twisted 180° around the thicker tube, and then the two tubes were pulled out together to a sharp tip. While dry nitrogen was blown through the thin barrel, the thicker barrel was then silanized by exposing the back end to hexamethyldisilazane vapour.

The pH-selective microelectrodes were made as described by Schlue & Thomas (1985). The silanized barrel was filled at the tip with a proton cocktail (tridodecylamine as H^+ ionophore; developed by Ammann et al 1981) and behind that with a Cl^--containing citrate buffer, pH 6, saturated with CO_2. Finally the reference barrel was filled with 3 M KCl, and the microelectrode was tested. When used in snail neurons, such electrodes gave very similar pH_i values to those recorded simultaneously by recessed-tip glass microelectrodes (Schlue & Thomas 1985).

The Na^+-selective microelectrodes were made as described by Deitmer & Schlue (1983). The tip of the silanized barrel was filled with a column of Na^+ cocktail (ETH 227 as Na^+ ionophore; Steiner et al 1979). The Na^+-selective microelectrode barrel was filled with 100 mM NaCl + 10 mM Hepes, pH 7.0, and the reference barrel with 3 M KCl. These electrodes were calibrated in solutions with different Na^+ concentrations and a constant ionic background containing: KCl, 100 mM; $MgCl_2$, 2 mM; $CaCl_2$, 0.01 mM; Hepes, 2 mM (pH 7.0).

Results

Steady-state pH_i

The pH_i, E_m and H^+ equilibrium potential in neuropile glial cells and Retzius neurons of the leech nervous system are summarized in Table 1. In both types of cell the pH_i is maintained at a higher value than that predicted where H^+ ions are distributed across the membrane in equilibrium with E_m. In leech glial cells and neurons, active acid extrusion is required to maintain the high pH_i and to overcome passive leakage of H^+ ions into the cells.

pH_i regulation

Intracellular acidifications were induced and the active recovery of pH_i was observed. Two methods were used to cause intracellular acidification: the NH_4Cl pre-pulse technique and the application of CO_2.

NH_4Cl acid loading has been widely used in studies of pH_i regulation ever since it was introduced by Boron & De Weer (1976). The preparation was exposed to 20 mM NH_4Cl for 1–2 min (see Figs. 2 and 3) during which time there was a rapid alkalinization followed by a slower acidification. The rapid alkalinization resulted as NH_3 rapidly permeated the cell; the slower acidification was caused by the permeation of $NH_4{}^+$, probably through K^+ channels. Removal of NH_4Cl from the external solution produced a rapid acidification, due to the efflux of NH_3 which caused the dissociation of $NH_4{}^+$ to NH_3 and H^+. This results in the addition of H^+ ions to the cell interior, seen as a decrease in pH_i.

The pH_i was also lowered by the application of CO_2 (Thomas 1976; see

Figs. 3 and 4). The transition to CO_2-HCO_3^--buffered saline produced an intracellular acidification that was due to the permeation of CO_2 and its hydration to form carbonic acid, which dissociates into H^+ and HCO_3^- ions.

Na^+-free or amiloride-containing solutions and pH_i recovery. The acidification by exposure and subsequent removal of NH_4Cl was followed by pH_i recovery, which represents the movement of H^+, or its equivalent, against an electrochemical gradient. The pH_i recovery is taken as a measure of pH_i regulation in both neurons and glial cells. The mechanism of pH_i regulation in these cells in normal, Hepes-buffered saline may involve the exchange of external Na^+ for internal H^+. This Na^+/H^+ exchanger should be blocked by the removal of external Na^+ or by the application of amiloride.

The role of external Na^+ in the recovery of pH_i following acidification was studied in Retzius neurons (Fig. 2a). After the first exposure to NH_4Cl for 1.5 min, external Na^+ was removed completely for about 12 min and then replaced. There was no pH_i recovery in Na^+-free, Hepes-buffered saline. After the rapid acidification on removal of NH_4Cl, the pH_i continued to decrease more slowly for the entire period of Na^+ absence. As soon as external Na^+ was restored, the pH_i recovered. The control exposure to NH_4Cl was carried out at the end of the experiment.

The effect of amiloride supported the suggestion that Na^+/H^+ exchange occurs across the neuronal membrane. Fig. 2b shows an experiment in which a Retzius neuron was exposed twice to 20 mM NH_4Cl. The pH_i recovery in normal saline is shown after the first exposure to NH_4Cl. After the second exposure to NH_4Cl, the application of amiloride for 15 min blocked pH_i recovery.

The role of external Na^+ or amiloride in the recovery of pH_i in Hepes-buffered saline following acidification was also studied in neuropile glial cells. The experiment illustrated in Fig. 2c shows the effect of Na^+-free solution on the pH_i recovery. After the second exposure to NH_4Cl, there was no pH_i recovery in Na^+-free, Hepes-buffered saline; the pH_i continued to decrease. As soon as external Na^+ was restored, the pH_i recovered.

Fig. 2d shows an experiment in which a leech neuropile glial cell was exposed to amiloride. The control exposure to NH_4Cl was carried out at the beginning of the experiment; pH_i was allowed to recover in normal saline. The application of amiloride after the second exposure to NH_4Cl reduced the rate of pH_i recovery considerably.

The results suggest that in Hepes-buffered, HCO_3^--free solutions, pH_i recovery from an acid load in leech neurons and glial cells probably depends on an amiloride-sensitive Na^+/H^+ exchange.

HCO_3^- ions and pH_i recovery. In all invertebrate preparations so far investigated, HCO_3^- ions play a role in pH_i regulation. Fig. 3a shows the pH_i

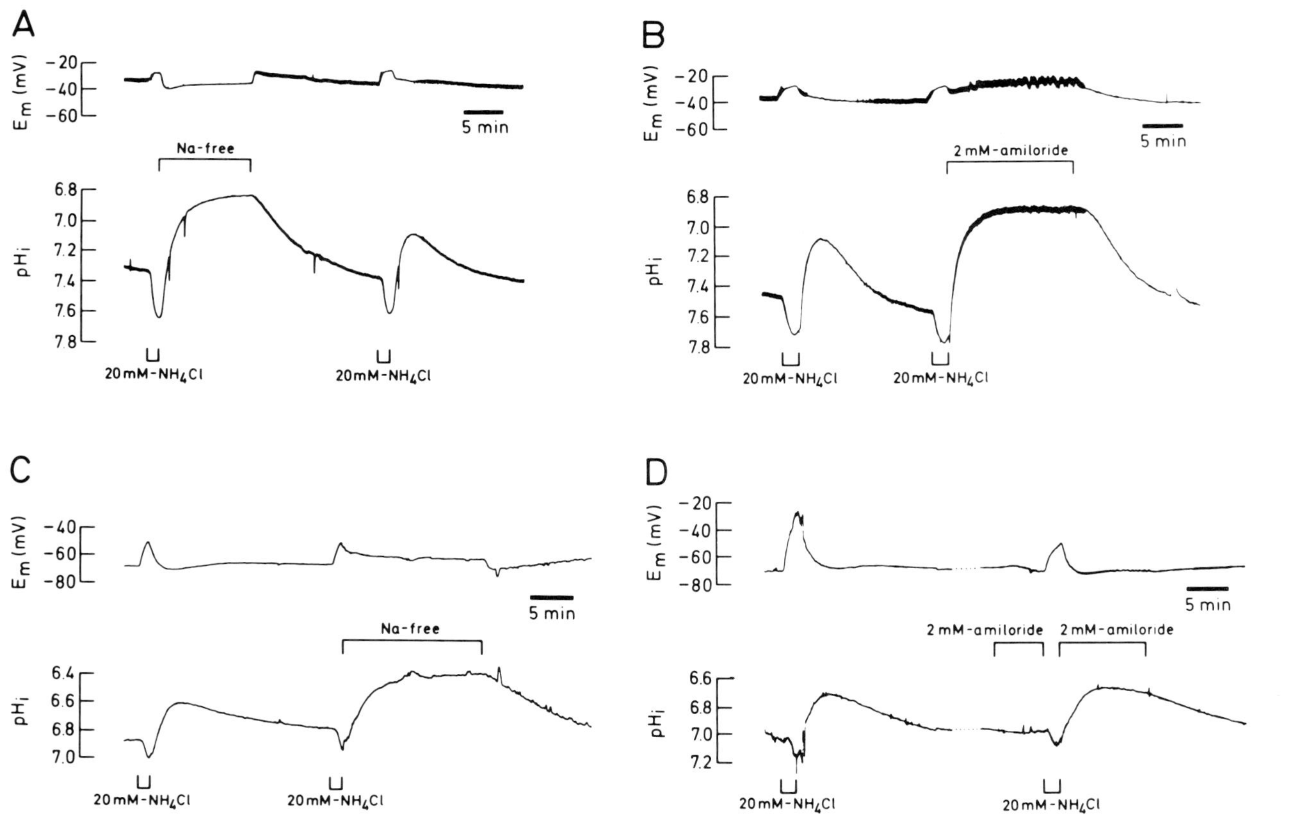

FIG. 2. Inhibition of pH$_i$ recovery from NH_4Cl-induced acidification in leech Retzius neurons (a,b) and neuropile glial cells (c,d) by removal of external Na^+ or by application of 2 mM amiloride. All solutions were HCO_3^- free.

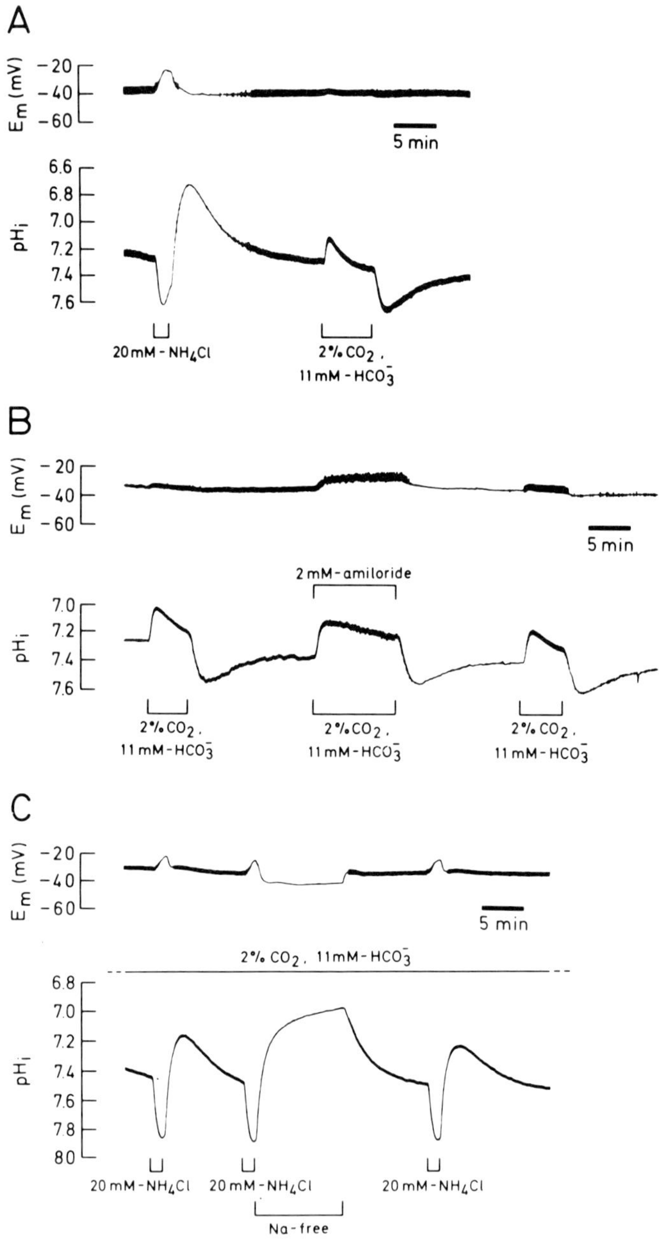

FIG. 3. Leech Retzius neuron pH_i recovery from two different types of acidification. (a) pH_i recovery from NH_4Cl-induced or CO_2-HCO_3^--induced acidification. (b) Effect of amiloride on pH_i recovery from acidification induced by CO_2-HCO_3^- buffer. Except for the periods indicated, the superfusate was HCO_3^- free. (c) Effect of removing external Na^+ on pH_i recovery from NH_4Cl-induced acidification. During this experiment the solutions were buffered with CO_2-HCO_3^-, except during NH_4Cl application.

recovery in 2% CO_2–11 mM HCO_3^- in a Retzius cell after an exposure to NH_4Cl. The rate of pH_i recovery is faster in this CO_2-HCO_3^- buffer than in Hepes-buffered saline. Since the cell membrane is freely permeable to CO_2, intracellular buffering is also much higher in the presence of CO_2 and HCO_3^- (Thomas 1976). Allowing for the increased buffering power in this case, acid extrusion was increased by a factor of 3.9 (see Schlue & Thomas 1985 for details).

Carbonic anhydrase inhibitors. Application of carbonic anhydrase inhibitors, the sulphonamides ethoxzolamide (10^{-8} M – 10^{-5} M) and acetazolamide (2×10^{-4} M), to the bathing medium reduced or even abolished the rapid transient acidification on addition of CO_2–HCO_3^- buffers and the transient alkalinization on their removal.

The effect of amiloride in HCO_3^-. To see whether amiloride caused any slowing of pH_i recovery in the presence of HCO_3^-, we have compared pH_i recovery rates in HCO_3^- with and without amiloride (Fig. 3b). It can be seen from Fig. 3b that amiloride slowed but did not block pH_i recovery.

Na^+-free HCO_3^- solutions and pH_i recovery. We have tested the effect of removing Na^+ on pH_i recovery of leech neurons in CO_2–HCO_3^--buffered solutions. In the experiment illustrated in Fig. 3c, a Retzius cell was acidified by addition and subsequent removal of 20 mM NH_4Cl. The recoveries of pH_i in normal saline are shown after the first and third exposure to NH_4Cl. After the second exposure to NH_4Cl, there was no pH_i recovery in Na^+-free, CO_2–HCO_3^--buffered saline; the pH_i continued to decrease. As soon as external Na^+ was restored, the pH_i recovered.

In glial cells, there was a transient increase of pH_i caused by the removal of external Na^+ immediately after the exposure to NH_4Cl, but subsequently the pH_i recovery was inhibited in Na^+-free saline.

SITS or intracellular Cl^- depletion and pH_i recovery. The results so far suggest that in CO_2–HCO_3^--buffered saline there is an additional mechanism responsible for pH_i recovery. The two most likely possibilities are simple HCO_3^-/Cl^- exchange or Na^+-dependent HCO_3^-/Cl^- exchange. Both mechanisms should be inhibited by the stilbene, SITS, but only one should be inhibited by Na^+ removal.

In the neurons and glial cells, SITS had only a moderate effect on the steady-state pH_i in either Hepes- or CO_2–HCO_3^--buffered saline. However, when applied during an intracellular acidification by application and removal of CO_2-HCO_3^--buffered saline, SITS slowed the rate of pH_i recovery considerably, as shown in Fig. 4a for a Retzius cell.

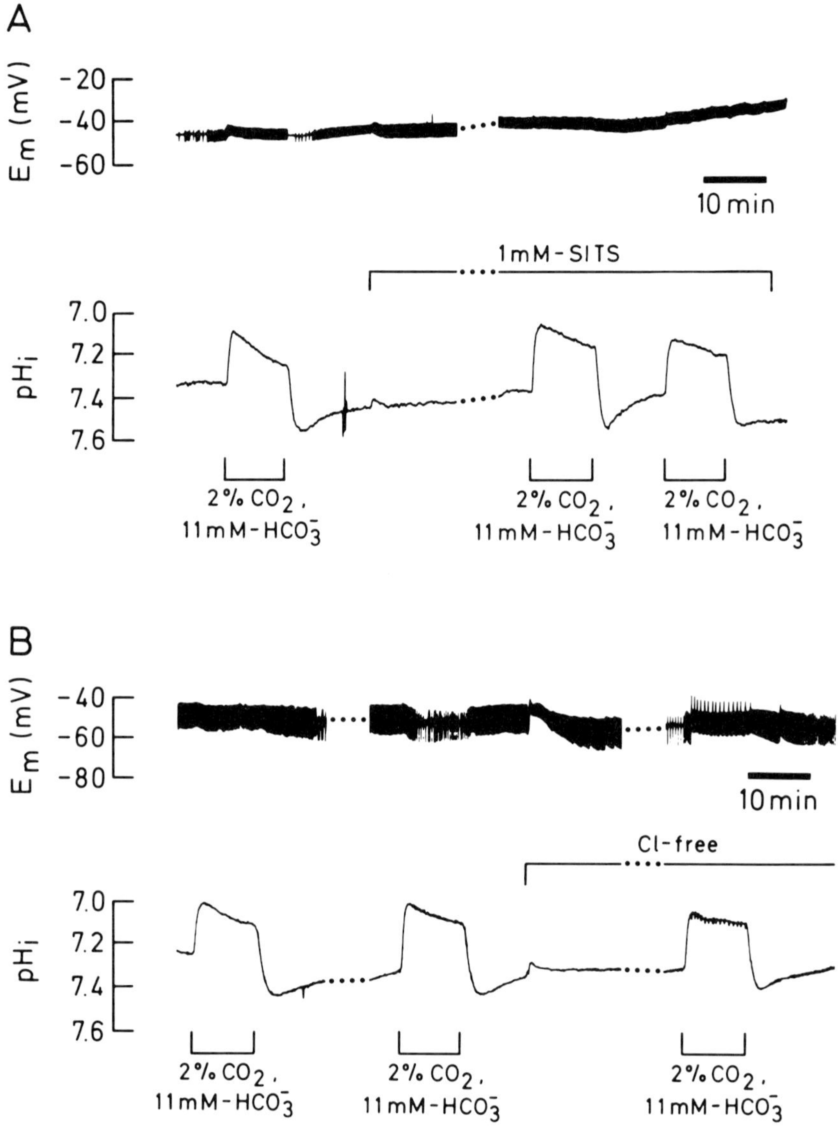

FIG. 4. The effect of application of 1 mM SITS (a) and removal of external Cl^- (b) on the pH_i recovery from acidification induced by CO_2-HCO_3^- in two different leech Retzius cells. Except where indicated solutions were HCO_3^- free.

By analogy with snail neurons (Thomas 1977), it should be possible to slow pH_i recovery in leech neurons by depleting internal Cl^-. We attempted to inhibit pH_i recovery from acidification by superfusing the cell with Cl^--free solutions. The rate of pH_i recovery was slowed considerably by the removal of external Cl^-, as shown in Fig. 4b for a Retzius cell.

Evidence for Na^+-HCO_3^- co-transport in glial cells

It had been shown that an alkalinization occurred in neuropile glial cells when a Hepes-buffered saline was exchanged for a saline buffered with 2% CO_2–11 mM HCO_3^- (Table 1). The alkalinization was maintained in the presence of CO_2-HCO_3^- buffer, and was reversed upon its removal (see Fig. 5). The carbonic anhydrase inhibitors ethoxzolamide (1 μM) and acetazolamide (0.2 mM) did not affect the CO_2-HCO_3^--induced alkalinization. It should be emphasized that the alkalinization was typical for the leech glial cells and was not observed in neurons, where pH_i decreased and then increased following the change from a Hepes- to a CO_2-HCO_3^--buffered saline (see Figs. 3 and 4).

The alkalinization could be due to the activity of an inwardly directed Na^+-HCO_3^- co-transporter (see Deitmer & Schlue 1987). A coupled transport of Na^+ and HCO_3^- seems to play an important role in transepithelial transport of bicarbonate in a variety of cells (Boron & Boulpaep 1983, Jentsch et al 1984, 1985a, b, Yoshitomi et al 1985, Alpern 1985, Biagi & Sohtell 1986).

Na^+-free solutions. Since the regulation of pH_i in neuropile glial cells following an acid load was dependent on external Na^+, the role of Na^+ during the CO_2-HCO_3^--induced alkalinization was studied (Fig. 5a). After replacing the Hepes buffer by 2% CO_2-11 mM HCO_3^-, the pH_i increased and then decreased again on return to Hepes-buffered saline. The following re-addition of CO_2-HCO_3^- was performed in the absence of external Na^+. In this Na^+-free saline the pH_i remained constant for the first four minutes and then decreased slowly, but when Na^+ was re-added the alkalinization occurred.

It is particularly clear in this experiment that the membrane hyperpolarized in CO_2-HCO_3^--buffered saline and that this hyperpolarization was reversed in the absence of external Na^+. The membrane hyperpolarization in CO_2-HCO_3^- buffer was also observed after a period in Na^+-free saline upon re-addition of external Na^+.

Stilbene derivatives. It was reported previously that SITS (0.5 mM) did not affect the CO_2-HCO_3^--induced alkalinization in neuropile glial cells (Deitmer & Schlue 1987).This result was unexpected, because it has been shown that the Na^+-HCO_3^- co-transporter is inhibited by SITS (Boron & Boulpaep 1983, Yoshitomi et al 1985). We therefore investigated the effect of the related stilbene derivative, DIDS, which binds covalently to the anion transport site (Cabantchik & Rothstein 1972). DIDS had been shown to inhibit the Na^+-HCO_3^- co-transporter in monkey kidney epithelial cells BSC-1, a continuous cell line derived from the African green monkey (Jentsch et al 1985b, 1986). Indeed, 0.5 mM DIDS caused an inhibition of the CO_2-HCO_3^--induced alkalinization in neuropile glial cells (Fig. 5b). The membrane hyper-

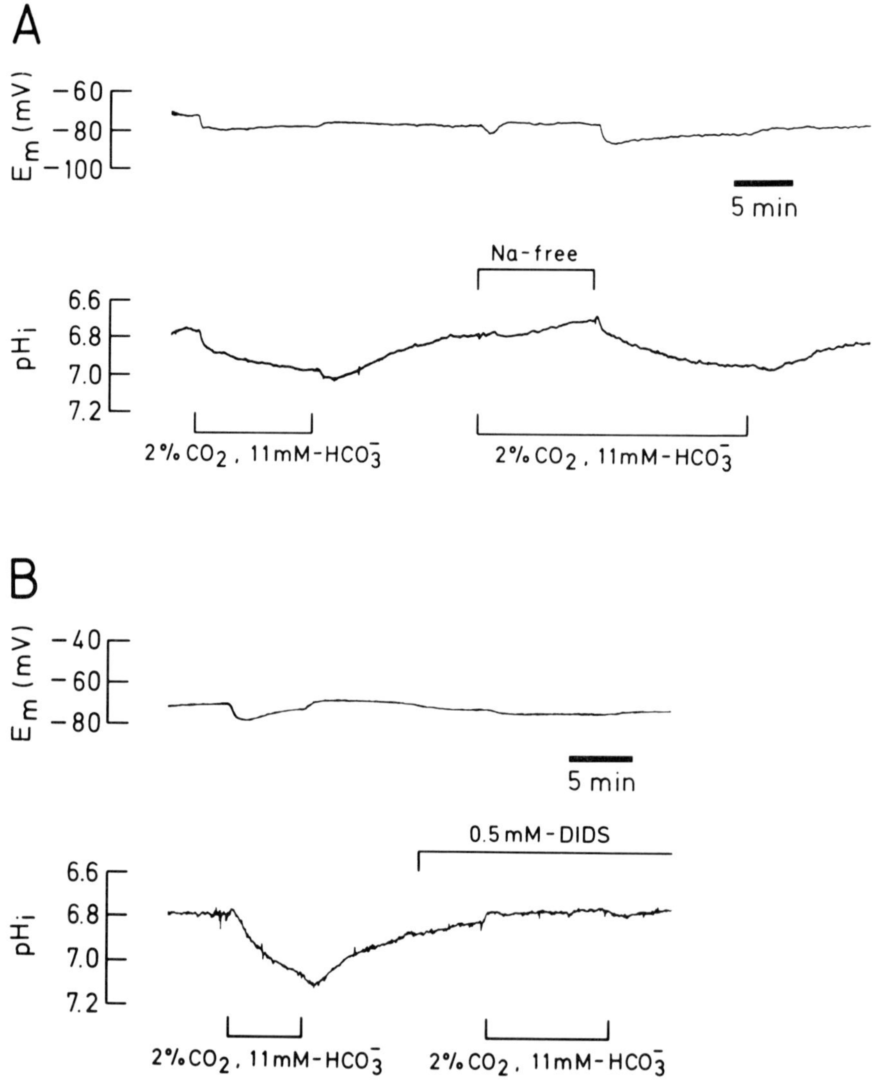

FIG. 5. The effect of Na^+-free saline (a) and application of 0.5 mM DIDS (b) on the CO_2-HCO_3^--induced alkalinization in two different neuropile glial cells.

polarization in CO_2-HCO_3^--buffered saline was also abolished by DIDS.

CO_2-HCO_3^- buffer and intracellular Na^+ activity. The fact that the alkalinization in CO_2-HCO_3^--buffered saline is inhibited by incubating the cells in Na^+-free solution could indicate that Na^+ influx is part of the underlying mechanism. An increase in intracellular Na^+ activity (aNa_i) should be detected, if Na^+ influx occurs during the development of the alkalinization. To

test this prediction, experiments were carried out in which aNa_i was recorded using Na^+-selective microelectrodes. When a nominally HCO_3^--free, Hepes-buffered saline was exchanged for one buffered with 2% CO_2-11 mM HCO_3^-, aNa_i increased and decreased again upon return to Hepes buffer. The aNa_i increase was also present in solutions containing 5 mM Ba^{2+} (K. Ballanyi, unpublished observation), but was blocked by 0.5 mM DIDS.

Amiloride. The membranes of leech neuropile glial cells possess a Na^+/H^+ exchanger which is inhibited by amiloride. However, the CO_2-HCO_3^--induced alkalinization was unaltered even in the presence of 2 mM amiloride. Therefore, the alkalinization cannot be explained by the activation of the Na^+/H^+ exchanger in the glial cell membrane.

Internal Cl^- depletion. An important question is the possible involvement of the Cl^- system in the generation of the glial cell alkalinization. When Cl^--free salines were used to deplete all extracellular and intracellular Cl^-, CO_2-HCO_3^--buffered saline still produced a reversible alkalinization.

Furosemide. The effect of the loop diuretic furosemide, which inhibits Cl^- fluxes across the cell membrane in a variety of cells (Brazy & Gunn 1976, Boron et al 1978, Russell 1979), including cultured primary rat astrocytes (Kimelberg & Frangakis 1985), on the alkalinization was investigated. Furosemide (0.5–1.0 mM) did not affect the alkalinization induced by CO_2-HCO_3^- in neuropile glial cells.

Discussion

In leech glial cells and neurons, as in other cells, the pH_i is maintained at values too high to be explained by passive ion movements. The H^+ distribution across the glial and neuronal membrane is not in equilibrium, and an alkaline pH_i has to be maintained against an electrochemical gradient.

Active pH_i regulation has been investigated by inducing an intracellular acidification by addition and subsequent removal of NH_4Cl or CO_2. For recovery from this acidification to normal, there appear to be two separate mechanisms in glial cells and neurons.

In HCO_3^--free saline, the recovery of pH_i from acidification is inhibited completely by removing external Na^+ or by applying amiloride; both treatments are rapidly reversible. We can conclude, therefore, that pH_i recovery occurs via a Na^+/H^+ exchanger in the cell membrane, as seen in many vertebrate preparations, e.g. mammalian muscle (Aickin & Thomas 1977), salamander kidney (Boron & Boulpaep 1983), mammalian fibroblasts (Frelin et al 1983, Moolenaar et al 1984), rat lymphocytes (Grinstein et al 1984), and also in invertebrate preparations e.g. crayfish central neurons (Moody 1981), crayfish muscle fibres (Galler & Moser 1986). The existence of a Na^+/H^+ exchanger in the membrane of cultured primary astrocytes has been inferred from the observation that addition of Na^+ ions to cells suspended in Na^+-free media specifically increased the rate of acidification of the culture medium (Kimelberg et al 1979).

In many invertebrate preparations, HCO_3^- ions may be involved in pH_i regulation: in snail neurons, squid axons and barnacle muscle they are essential (see Roos & Boron 1981, Boron 1983, Thomas 1984). It was shown that HCO_3^- greatly stimulates the rate of acid efflux in leech Retzius neurons. The stilbene SITS slowed pH_i recovery in the presence of HCO_3^-, not only in Retzius neurons but also in neuropile glial cells, and removal of external Na^+ blocked pH_i recovery. It was also shown that in solutions buffered with 2% CO_2 and 11 mM HCO_3^-, amiloride slowed but did not block pH_i recovery. These results suggest that pH_i regulation in the presence of CO_2-HCO_3^- occurs via an additional mechanism in these cells. This mechanism is Na^+ dependent, as had already been shown for snail neurons (Thomas 1977), squid axons (Boron & Russell 1983), and also for stretch receptor (Moser 1985), muscle fibres (Galler & Moser 1986), and neurons of crayfish (Moody 1981). Indirect evidence for the existence of simple HCO_3^-/Cl^- exchange has been obtained for cultured primary astrocytes (Kimelberg et al 1979) and LRM55 glioma cells originally isolated from a rat spinal tumour (Wolpaw & Martin 1984). In both preparations the uptake and efflux of Cl^- was inhibited by SITS.

Unlike neuronal pH_i, glial pH_i increases when Hepes is replaced by CO_2-HCO_3^- as the extracellular buffer, and decreases again on return to Hepes buffer. This unusual CO_2-HCO_3^--dependent increase in steady-state pH_i, termed here the alkalinization, has so far only been reported for mammalian smooth muscle cells (Aickin 1984). There, a pH_i change of similar magnitude was recorded when introducing a CO_2-HCO_3^--buffered saline.

The alkalinization in glial cells and smooth muscle occurs in the opposite direction to that which would be expected from the movement of CO_2 across the cell membrane and its hydration to carbonic acid, which dissociates into H^+ and HCO_3^- ions. The expected acidification, however, is observed in neurons investigated here, and is reduced by acetazolamide and ethoxzolamide, inhibitors of carbonic anhydrase which catalyses the formation of carbonic acid. On the other hand, these drugs are shown not to affect the CO_2-HCO_3^--induced alkalinization in neuropile glial cells. The observations suggest that Na^+-HCO_3^- co-transport across the glial cell membrane, mediating the influx of HCO_3^- ions into the cell interior, could be responsible for the unusual alkalinization. A Na^+-HCO_3^- co-transport has recently been suggested also for cultured oligodendrocytes from mouse spinal cord (H. Kettenmann & W.-R. Schlue, unpublished work).

Evidence for the activation of Na^+-HCO_3^- co-transport in leech glial cells is based on the following findings: it is 1) inhibited in Na^+-free saline or in the presence of DIDS; 2) accompanied by a DIDS-sensitive increase in intracellular Na^+ activity; 3) a membrane hyperpolarization which is abolished in either Na^+-free or DIDS-containing solution; 4) unaltered in the presence of amiloride; and 5) unaffected by removal of external Cl^- (thereby depleting internal Cl^-) or by the presence of furosemide.

By applying ion-selective microelectrodes to the isolated, perfused renal tubules of the tiger salamander, Boron & Boulpaep (1983) were the first to demonstrate that HCO_3^- does not move independently across the basolateral membrane but in coupling with Na^+. Their results were compatible with the presence of an electrogenic Na^+-HCO_3^- co-transporter that carries Na^+ and HCO_3^- (and net negative charge) in the same direction. Subsequently, a similar process was identified in the bovine corneal epithelium, a mammalian tissue, which is also involved in transepithelial transport of HCO_3^- (Jentsch et al 1984, 1985a). Recently, however, it has been suggested that a Na^+-HCO_3^- co-transporter is also present in the mammalian kidney: in the epithelial cell line BSC-1 derived from the kidney of the African green monkey (Jentsch et al 1985b) and in proximal tubules of the rat (Yoshitomi et al 1985, Alpern 1985) and rabbit (Biagi & Sohtell 1986).

As shown in the present investigation, the co-transport in the membrane of leech glial cells seems to mediate an influx of Na^+ and HCO_3^-, and thus might be different from the proximal tubular transporting mechanism. In the proximal tubule of the kidney, the physiological role of the co-transporter is assumed to be an efflux of Na^+ and HCO_3^- across the basolateral membrane, giving rise to transepithelial HCO_3^- reabsorption (Boron & Boulpaep 1983, Yoshitomi et al 1985, Alpern 1985). For the salamander kidney, Boron & Boulpaep (1983), knowing intracellular and extracellular ion concentrations, demonstrated that an apparent coupling ratio of HCO_3^- to Na^+ of 2:1 is compatible with this transport direction. Similar conclusions have been reached by Alpern (1985) for the rat, whereas Yoshitomi et al (1985) postulate a stoichiometry of 3:1 to account for an efflux of HCO_3^- and Na^+ in the same species.

Conclusion

Three separate mechanisms seem to be responsible for pH_i regulation in the leech nervous system. They are Na^+-H^+-HCO_3^-/Cl^- exchange, Na^+/H^+ exchange and Na^+-HCO_3^- co-transport. All three mechanisms have a profound effect on the maintenance of pH_i homeostasis in glial cells, whereas in leech nerve cells, as in other neuronal cells studied previously, the predominant mechanisms are Na^+/H^+ and Na^+-H^+-HCO_3^-/Cl^- exchange.

Acknowledgements

The studies on the leech nervous system were supported by equipment grants to W.R.S. (Schl 169/6–7) and a Heisenberg Fellowship to J.W.D. (De 231/4–1). We thank Dr M Szatkowski for comments on the manuscript, and E. Friedrich and M. Lessens for excellent assistance during the experiments.

REFERENCES

Aickin CC 1984 Direct measurement of intracellular pH and buffering power in smooth muscle cells of guinea-pig vas deferens. J Physiol (Lond) 349:571–585

Aickin CC, Thomas RC 1977 An investigation of the ionic mechanism of intracellular pH regulation in mouse soleus muscle fibres. J Physiol (Lond) 273:295–316

Alpern RJ 1985 Mechanism of basolateral membrane $H^+/OH^-/HCO_3^-$ transport in the rat proximal convoluted tubule. J Gen Physiol 86:613–636

Ammann D, Lanter F, Steiner RA, Schulthess P, Shijo Y, Simon W 1981 Neutral carrier based hydrogen ion selective microelectrode for extra- and intracellular studies. Anal Chem 53:2267–2269

Biagi BA, Sohtell M 1986 Electrophysiology of basolateral bicarbonate transport in the rabbit proximal tubule. Am J Physiol 250:267–272

Boron WF 1983 Transport of H^+ and of ionic weak acids and bases. J Membr Biol 72:1–16

Boron WF, Boulpaep EL 1983 Intracellular pH regulation in the renal proximal tubule of the salamander. Basolateral HCO_3^- transport. J Gen Physiol 81:53–94

Boron WF, Russell JM 1983 Stoichiometry and ion dependencies of the intracellular-pH-regulating mechanism in squid giant axons. J Gen Physiol 81:373–399

Boron WF, De Weer P 1976 Intracellular pH transients in squid giant axons caused by CO_2, NH_3, and metabolic inhibitors. J Gen Physiol 67:91–112

Boron WF, Russell JM, Brodwick MS, Keifer DW, Roos A 1978 Influence of cyclic AMP on intracellular pH regulation and chloride fluxes in barnacle muscle fibres. Nature (Lond) 276:511–513

Brazy PC, Gunn RE 1976 Furosemide inhibition of chloride transport in human red blood cells. J Gen Physiol 68:583–599

Cabantchik ZI, Rothstein A 1972 The nature of the membrane sites controlling anion permeability of human red blood cells as determined by studies with disulfonic stilbene derivatives. J Membr Biol 10:311–330

Chesler M 1986 Regulation of intracellular pH in reticulospinal neurones of the lamprey, *Petromyzon marinus*. J Physiol (Lond) 381:241–261

Deitmer JW, Schlue WR 1981 Measurements of the intracellular potassium activity of Retzius cells in the leech central nervous system. J Exp Biol 91:87–101

Deitmer JW, Schlue WR 1983 Intracellular Na^+ and Ca^{2+} in leech Retzius neurones during inhibition of the Na^+-K^+ pump. Pfluegers Arch Eur J Physiol 397:195–201

Deitmer JW, Schlue WR 1987 The regulation of intracellular pH by identified glial cells and neurones in the central nervous system of the leech. J Physiol (Lond) 388:261–283

Frelin C, Vigne P, Lazdunski M 1983 The amiloride-sensitive Na^+/H^+ antiport in 3T3 fibroblasts. J Biol Chem 258:6272–6276

Galler S, Moser H 1986 The ionic mechanism of intracellular pH regulation in crayfish muscle fibres. J Physiol (Lond) 374:137–151

Grinstein S, Cohen S, Rothstein A 1984 Cytoplasmic pH regulation in thymic lymphocytes by an amiloride-sensitive Na^+/H^+ antiport. J Gen Physiol 83:341–369

Kimelberg HK, Frangakis MV 1985 Furosemide- and bumetanide-sensitive ion transport and volume control in primary astrocyte cultures from rat brain. Brain Res 361:125–134

Kimelberg HK, Biddlecome S, Bourke RS 1979 SITS-inhibitable Cl^- transport and Na^+-dependent H^+ production in primary astroglial cultures. Brain Res 173:111–124

Jentsch TJ, Keller SK, Koch M, Wiederholt M 1984 Evidence for coupled transport of bicarbonate and sodium in cultured bovine corneal endothelial cells. J Membr Biol 81:189–204

Jentsch TJ, Stahlknecht TR, Hollwede H, Fischer DG, Keller SK, Wiederholt M 1985a A bicarbonate-dependent process inhibitable by disulfonic stilbenes and a Na^+/H^+ exchanger mediate $^{22}Na^+$ uptake into cultured bovine corneal endothelium. J Biol Chem 260:795–801

Jentsch TJ, Schill BS, Schwartz P, Matthes H, Keller SK, Wiederholt M 1985b Kidney epithelial cells of monkey origin (BSC-1) express a sodium bicarbonate cotransport. J Biol Chem 260:15554–15560

Jentsch TJ, Janicke I, Sorgenfrei D, Keller SK, Wiederholt M 1986 The regulation of intracellular pH in monkey kidney epithelial cells (BSC-1). J Biol Chem 261:12120–12127

Johnson JD, Epel D, Paul M 1976 Intracellular pH and activation of sea urchin eggs after fertilisation. Nature (Lond) 262:661–664

Moody W 1981 The ionic mechanism of intracellular pH regulation in crayfish neurones. J Physiol (Lond) 316:293–308

Moody W 1984 Effects of intracellular H^+ on the electrical properties of excitable cells. Annu Rev Neurosci 7:257–278

Moolenaar WH, Tertoolen LGJ, de Laat SW 1984 The regulation of cytoplasmic pH in human fibroblasts. J Biol Chem 259:7563–7569

Moser H 1985 Intracellular pH regulation in the sensory neurones of the stretch receptor of the crayfish (*Astacus fluviatilis*). J Physiol (Lond) 362:23–38

Roos A, Boron WF 1981 Intracellular pH. Physiol Rev 61:296–434

Russell JM 1979 Chloride and sodium influx: a coupled uptake mechanism in the squid giant axon. J Gen Physiol 73:801–818

Schlue WR, Wuttke W 1983 Potassium activity in leech neuropile glial cells changes with external potassium concentration. Brain Res 270:368–372

Schlue WR, Deitmer JW 1980 Extracellular potassium in neuropile and nerve cell body region of the leech central nervous system. J Exp Biol 87:23–43

Schlue WR, Deitmer JW 1987 Direct measurement of intracellular pH in identified glial cells and neurones of the leech central nervous system. Can J Physiol Pharmacol 65:978–985

Schlue WR, Thomas RC 1985 A dual mechanism for intracellular pH regulation by leech neurones. J Physiol (Lond) 364:327–338

Schlue WR, Walz W 1984 Electrophysiology of neuropile glial cells in the central nervous system of the leech: a model system for potassium homeostasis in the brain. In: Fedoroff S (ed) Advances in Cellular Neurobiology 5. Academic Press, New York, p 143–175

Steiner RA, Oehme M, Ammann D, Simon W 1979 Neutral carrier sodium ion-selective microelectrode for intracellular studies. Anal Chem 51:351–353

Thomas RC 1976 The effect of carbon dioxide on the intracellular pH and buffering power of snail neurones. J Physiol (Lond) 255:715–735

Thomas RC 1977 The role of bicarbonate, chloride and sodium ions in the regulation of intracellular pH in snail neurones. J Physiol (Lond) 273:317–338
Thomas RC 1978 Comparison of the mechanisms controlling intracellular pH and sodium in snail neurones. Respir Physiol 33:63–73
Thomas RC 1984 Experimental displacement of intracellular pH and the mechanism of its subsequent recovery. J Physiol (Lond) 354:3P–22P
Wolpaw E, Martin DL 1984 Cl^- transport in glioma cell line: evidence for two transport mechanisms. Brain Res 297:317–327
Yoshitomi K, Burckhardt BC, Frömter E 1985 Rheogenic sodium-bicarbonate co-transport in the peritubular cell membrane of rat renal proximal tubule. Pfluegers Arch Eur J Physiol 405:360–366
Zeuthen T 1980 How to make and use double-barrelled ion-selective microelectrodes. Curr Top Membr Transp 13:31–47

DISCUSSION

Frömter: What is the evidence for the supposed difference between your Na^+-HCO_3^- co-transporter and the one that we and others have found in epithelia (Boron & Boulpaep 1983, Yoshitomi et al 1985)? Have you been able to calculate a stoichiometry that differs from Na^+:3HCO_3^-; or do you know the direction of net flux and the electrochemical driving forces and find that these data require a different stoichiometry?

Schlue: We cannot say anything about the stoichiometry of the glial Na^+-HCO_3^- co-transporter at the moment. Because of the hyperpolarizing changes in glial membrane potential we expect that more HCO_3^- than Na^+ is transported into the cell interior. The difference is not the mechanism itself but the direction of the transport. In our preparation, the Na^+-HCO_3^- co-transporter seems to be inwardly directed.

Frömter: As I understand this, we are dealing with a passive mechanism whose direction of transport depends on the stoichiometry and on the sum of the electrochemical driving forces for each ion.

Schlue: But normally in epithelia, as a contribution to HCO_3^- reabsorption, the Na^+-HCO_3^- co-transporter is outwardly directed, the transport is from the inside of the cell into the blood across the basolateral membrane.

Frömter: This is true for the kidney but we have evidence for the same or a similar system in frog stomach (Curci et al 1987). We don't know the stoichiometry yet, but unpublished experiments with stimulation by histamine (S. Curci, Symposium on exokrine secretion, Hong Kong, February 1988) suggest that the direction of net transport may be opposite, i.e. it might preferentially move HCO_3^- into the cell.

Gillespie: The electrogenic hyperpolarization doesn't seem to be maintained during the alkalosis, yet the pH continues to alkalinize. Are there other conductances involved or is it more complicated than you described?

Schlue: The membrane hyperpolarization was rather variable in our preparation, and was not often transitory but more usually sustained.

Boron: The hyperpolarization would be an index of the flux through the electrogenic Na^+-HCO_3^- co-transporter. That flux is expected to be proportional to dpH/dt. So, assuming that membrane resistance is constant, at the instant that Dr Schlue adds Na^+ back, the hyperpolarization is expected to be maximal because that's the point at which dpH/dt is maximal. As the pH recovery continues and dpH/dt falls, the flux slows down and the amount of current flowing to the membrane decreases. Thus, the hyperpolarization should decrease with time.

Vaughan-Jones: Is the current–voltage relationship linear for these cells?

Thomas: You need to voltage-clamp to be sure what's happening to the current.

Schlue: We measured the current-voltage relationship for neuropile glial cells over a narrow range of membrane potential (Schlue & Walz 1984). The relationships were linear in both the hyperpolarizing and depolarizing regions over a range of plus or minus 20 mV.

Thomas: The rise in Na^+ will instantly stimulate the sodium pump, which is itself electrogenic. One needs to know a lot more about membrane resistance, whether membrane channels are open or closed, Ca^{2+} effects on Na^+ channels, pH effects on K^+ channels and so on.

Boron: The hyperpolarization that you see upon the addition of Na^+, is it HCO_3^- dependent? Have you acid loaded the cell in the absence of HCO_3^-? We did this experiment in salamander proximal tubules and we showed that this hyperpolarization was produced either by adding HCO_3^- or by adding Na^+, and the addition of either required the other in order to get a voltage change.

Schlue: Yes, the membrane hyperpolarization is HCO_3^- dependent and not pH dependent, since it does not occur during NH_4^+-induced alkalosis or acidosis on removal of NH_4^+. It is also Na^+ dependent. This is strong evidence that it is related to the Na^+-HCO_3^- co-transporter.

Thomas: In your experiment where you stimulated pH regulation then added barium, which increases membrane resistance, could the fall in membrane potential be due to an electrogenic process of some sort? It's much bigger with barium.

Schlue: Yes.

Vaughan-Jones: Presumably you can't rule out the fact that because Na^+ is entering the cell it is indeed stimulating the electrogenic Na^+ pump, which will contribute to the outward current.

Thomas: This decrease in Na^+ proves that it is being pumped out.

Boron: You should be able to see it in the presence of ouabain, have you tried that?

Schlue: Ouabain (0.5 mM) reduced the membrane hyperpolarization but did not block it completely.

Machen: Have you done an experiment in Cl^--free solutions containing HCO_3^- where you add DIDS? You have measured the cellular concentration of Na^+, the pH and the membrane potential, these data might give an indication of the stoichiometry of the Na^+/HCO_3^- mechanism. If DIDS caused an acidification under those conditions then it is likely that Na^+ and HCO_3^- were entering the cell. This directionality coupled with the data already mentioned could give some information about the coupling ratio of the Na^+/HCO_3^- exchanger.

Schlue: No, it hasn't been done.

Kaila: I wonder whether the CO_2-dependent intracellular alkalosis could be explained just by assuming that the activation curve of the HCO_3^--dependent acid extruder is on a more alkaline level than that of the HCO_3^- independent H^+ transporter. When you put on CO_2 there is a very slow hydration of CO_2 so you don't see the acid overshoot, but the steady-state pH shifts to the steady-state level of the HCO_3^-dependent mechanism and this will then produce an alkalosis.

Schlue: This is exactly what we think. The Na^+/HCO_3^- co-transporter shifts the pH_i to a more alkaline value than the Na^+/H^+ exchanger is able to do in these glial cells.

Boron: I counted only two transporters in the glial cells: a Na^+/H^+ exchanger and a proposed Na^+-HCO_3^- co-transporter, what's the third?

Schlue: The Na^+-dependent Cl^-/HCO_3^- exchanger. When we acidify the neuropile glial cell in Na^+-free solution using the NH_4Cl pre-pulse technique and re-add external Na^+, we observe a rapid recovery of intracellular pH (Deitmer & Schlue 1987). This pH_i recovery is slowed in solutions containing SITS, so we think that this is an additional HCO_3^--dependent mechanism.

Boron: How do you distinguish between the Na^+-dependent Cl^-/HCO_3^- exchanger and Na^+-HCO_3^- co-transporter, both of which are Na^+ and HCO_3^- dependent?

Schlue: In our preparation, the Na^+-HCO_3^- co-transporter is unaffected by SITS (Deitmer & Schlue 1987), we can only block it using DIDS. We don't know the reason but DIDS binds covalently to the anion transport site, whereas SITS does not (Cabantchik & Rothstein 1972).

Thomas: Isn't there a discrepancy, Claire (Aickin) cannot inhibit the inward Na^+-HCO_3^- co-transport in her smooth muscle preparation with DIDS. But Dr Schlue, you say that DIDS does inhibit that transporter in your preparation of leech glial cells, which is going in the same direction. In epithelial cells DIDS and SITS work perfectly on the outward Na^+-HCO_3^- co-transporter. Claire, this electrogenic mechanism is not inhibited by anything you have tried?

Aickin: Correct, but it does seem that there are varying sources of DIDS. Dr Schlue, what colour is your DIDS as a solid?

Schlue: Yellow, we buy it from Sigma or Serva and dissolve it immediately before use.

Aickin: We buy ours from Calbiochem-Behring Corp.

Al-Awqati: DIDS and SITS are unstable, the age and the pH of the solutions are important—the optimum pH for the reaction is 11!

Stone: I compared DIDS from several commercial sources by scanning their absorption spectra and they are radically different. The DIDS that I use from Pearce is bright yellow; it is the best commercial source in terms of colour and biological effect.

Steinmetz: There is a precedent for a difference: in turtle bladder the α-cell type is clearly inhibited over 90% by DIDS and SITS; in the same tissue and the same preparation the β-cells are not inhibited by SITS and only slightly (30%) by DIDS (Stetson et al 1985).

Thomas: So it could be that DIDS doesn't work on smooth muscles. Claire, have you tried SITS?

Aickin: No, just DIDS. Dr Schlue, I would like to return to your co-transport question. It seems that you have evidence for the Na^+-dependent Cl^-/HCO_3^- exchange from the NH_4^+ rebound in the presence of CO_2, and evidence for the co-transport system simply from adding CO_2. Do you have evidence for the co-transport system in the recovery from an NH_4^+ rebound in the presence of CO_2? Surely it ought to be working in both cases? If you do the NH_4 rebound in the absence of Cl^- and the presence of CO_2, you then should be left with the Na^+-HCO_3^- co-transporter.

Schlue: We have not tried that yet.

Al-Awqati: I would like to ask about carbonic anhydrase in these cells. Dr Schlue, have you done immunocytochemistry to see whether one cell type has the enzyme while the other does not?

Schlue: We used several histochemical methods to localize carbonic anhydrase in the leech central nervous system. The reaction of carbonic anhydrase with DNSA (dimethylaminonaphthalene-5-sulphonamide) forms a highly fluorescent complex (Chen & Kernohan 1967, Pochammer et al 1979, Dermietzel et al 1985). This method gave a specific fluorescence in neuropile glial cells. We have also used the Hansson method (Hansson 1967), in which cobalt carbonate is precipitated in a reaction catalysed by carbonic anhydrase, then converted to the black sulphide. We obtained a positive result with the neuropile glial cells, although the cobalt sulphide precipitate was not very dark.

We have just started to use an immunocytochemical method (Sternberger 1979). With antibodies against isoenzyme II coupled with fluorescein isothiocyanate-conjugated immunoglobulin we observed a specific, but weak fluorescence in the neuropile glial cells. Neuronal cell bodies did not stain at all. The gland cells of the body wall, as a control, showed a bright fluorescence.

Thomas: Why does ethoxzolamide block your recovery from CO_2?

Schlue: I don't think that ethoxzolamide blocks pH_i recovery from CO_2. There is only a slow shift of pHi to the acid, due to the slow formation of H^+ which is only partially buffered. It is the fast and large acidification due to the carbonic anhydrase-dependent H^+ formation which is blocked by ethoxzolamide.

Machen: Why do furosemide and amiloride both cause acidification in the neurons? If furosemide is not blocking anything, how is it still causing acidification? Is there another mechanism that furosemide might be affecting?

Schlue: Furosemide interacts with the carbonic anhydrase but it is not certain how the block of carbonic anhydrase would acidify the intracellular pH.

Machen: Is it possible that there is a Na^+-K^+-$2Cl^-$ co-transporter and also a Cl^-/HCO_3^- exchanger? In that case, if you inhibited the NaKCl transporter with furosemide, intracellular Cl^- would decrease then, because of the active anion exchanger, efflux of HCO_3^- would be increased, thereby causing an acidification. Therefore furosemide is affecting intracellular pH by interacting directly with Na^+ coupled Cl^- transport and indirectly with an anion exchanger.

Oberleithner: Dr Schlue, you said that glial cell pH_i acidifies when you add furosemide. If there is a furosemide-sensitive Na^+-K^+-$2Cl^-$ co-transport system in the glial cell membrane, then this explains the acidification induced by furosemide: Cl^- can no longer enter the cell, so there is no Cl^- available for the Cl^-/HCO_3^- exchange. Shouldn't you add to the proposed transport mechanisms a Na^+-K^+-$2Cl^-$ co-transport system?

Schlue: We have no direct evidence for such a transporting mechanism in the glial cell membrane at present.

Sabolic: Is furosemide a weak acid?

Schlue: Furosemide is an acid with a pK of 3.6 (Imbs et al 1987) and, although the concentration is 1 mM, it is possible that the undissociated form crosses the membrane, dissociates and acidifies the cell interior.

Steinmetz: It inhibits carbonic anhydrase.

Aickin: It inhibits Cl^-/HCO_3^- exchange very nicely (Brazy & Gunn 1976).

Russell: It also inhibits the Na^+-dependent Cl^-/HCO_3^- exchanger in squid axon and barnacle muscle.

Harvey: It's strange that the inhibitors of carbonic anhydrase, such as ethoxzolamide, cause acidification when alkalinization is the normal response observed in other cells.

Syková: Are the ion fluxes across the glial cell membrane as fast as those in the neuron membrane? Do the alkalinization and acidification have the same time course in each cell type?

Schlue: No, the time course is slower in glial cells than in neurons.

Thomas: I fear that there are difficult surface pH problems there too. I noticed that when your electrode was between the Retzius cell and the glial cell, the pH in the space outside the glial cell was 7, whereas the bulk pH was 7.4. Is that a common problem, that the space around the glial cell is inaccessible to your solutions?

Schlue: From measurements of other ions, for example external K^+, there is no diffusion barrier. The pH measured in the extracellular space around neurons and glial cells averaged 7.2.

Thomas: Then why is the pH outside between the two cells more acid than the intracellular pH? Doesn't that suggest that the surface pH of the glial cell may be a complicating extra factor?

Schlue: It may be, we don't differentiate between measurements in extracellular spaces or at cell surfaces.

References

Boron WF, Boulpaep EL 1983 Intracellular pH regulation in the renal proximal tubule of the salmander: basolateral HCO^-_3 transport. J Gen Physiol 81:53–94

Brazy PC, Gunn RB 1976 Furosemide inhibition of chloride transport in human red blood cells. J Gen Physiol 68:583–599

Cabantchik ZI, Rothstein A 1972 The nature of the membrane sites controlling anion permeability of human red blood cells as determined by studies with disulphonic stilbene derivatives. J Membr Biol 10:311-330

Chen RF, Kernohan JC 1967 Combination of bovine carbonic anhydrase with a flourescent sulfonamide. J Biol Chem 242:5813–5823

Curci S, Debellis L, Frömter E 1987 Evidence for rheogenic sodium bicarbonate cotransport on the basolateral membrane of oxyntic cells of frog gastric fundus. Pfluegers Arch Eur J Physiol 408:497–504

Deitmer JW, Schlue WR 1987 The regulation of intracellular pH by identified glial cells and neurones in the central nervous system of the leech. J Physiol (Lond) 388:261–283

Dermietzel R, Leibstein A, Siffert W, Zamboglou N, Gros G 1985 A fast screening method for histochemical localization of carbonic anhydrase. J Histochem Cytochem 33:93–98

Hansson HPJ 1967 Histochemical demonstration of carbonic anhydrase activity. Histochemistry 11:112–128

Imbs JL, Schmidt M, Giesen-Crouse E 1987 Pharmacology of loop diuretics: state of the art. Adv Nephrol 16:137–158

Pochhammer C, Dietsch P, Siegmund PR 1979 Histochemical detection of carbonic anhydrase with dimethylaminonaphthalene-5-sulfonamide. J Histochem Cytochem 27:1103–1107

Schlue WR, Walz W 1984 Electrophysiology of neuropile glial cells in the central nervous system of the leech: a model system for potassium homostasis in the brain. In: Fedoroff S (ed) Advances in cellular neurobiology. Academic Press, New York, vol 5:143–175

Sternberger LA 1979 Immunocytochemistry. John Wiley & Sons, New York, p 24–58

Stetson DL, Beauwens R, Palmisano J, Mitchell PP, Steinmetz PR 1985 A double membrane model for urinary bicarbonate secretion. Am J Physiol 18:F546–F552

Yoshitomi K, Burckhardt BC, Frömter E 1985 Rheogenic sodium-bicarbonate cotransport in the peritubular cell membrane of the rat renal proximal tubule. Pfluegers Arch Eur J Physiol 405:360–366

Differential role of cation and anion exchange in lymphocyte pH regulation

Sergio Grinstein*†, Jesùs Garcia-Soto* and Michael J. Mason*

*Division of Cell Biology, The Hospital for Sick Children, 555 University Avenue, Toronto, Canada, M5G 1X8 and †Department of Biochemistry, University of Toronto

Abstract. In lymphocytes, the Na^+/H^+ antiport is well suited to function in cytoplasmic pH (pH_i) regulation. It is activated by departures from the physiological pH_i and is thermodynamically poised to compensate for the tendency of the cells to become acidic. The driving force for H^+ (equivalent) efflux is indirectly provided by the Na^+ pump. Lymphocytes also possess a cation-independent anion (Cl^-/HCO_3^-) exchange system, which, under the appropriate conditions, tends to restore pH_i after an alkali load. Unlike the cation antiport, the source of energy driving the anion exchanger, i.e. the factors that determine the transmembrane Cl^- distribution, is not well understood. The contribution of conductive pathways appears to be minimal, resulting in a marked difference between the membrane potential and E_{Cl^-}. Instead, E_{Cl^-} is very similar to E_{H^+}. Moreover, changes in the distribution of Cl^- lead to alterations in the transmembrane ΔpH and vice versa, suggesting a relationship between these parameters. Evidence is presented which suggests that the transmembrane distribution of HCO_3^-, dictated by ΔpH, is a major determinant of the intracellular Cl^- concentration, a process mediated by the anion exchanger. Thus, if Cl^- is driven by the gradient of HCO_3^-, the cation-independent anion exchanger cannot play an active role in determining pH_i. Instead, Cl^-/HCO_3^- exchange may simply stabilize pH_i by increasing the dynamic buffering power of the cells. Cation-independent Cl^-/HCO_3^- exchange could be involved in pH_i regulation only if coupled to a separate mechanism of intracellular Cl^- accumulation, such as Na^+-K^+-$2Cl^-$ co-transport or an inward Cl^- pump, which have not been detected in lymphoid cells. Finally, a Na^+-dependent yet amiloride-insensitive pH_i recovery mechanism is present in lymphocytes suspended in HCO_3^--containing media. This system is greatly inhibited by disulphonic stilbenes and is likely to involve $NaCO_3^-/Cl^-$ (or equivalent) exchange.

1988 Proton passage across cell membranes. Wiley, Chichester (Ciba Foundation Symposium 139) p 70–86

An effective pH_i regulatory system must fulfil a number of criteria. It should be activated by departures from the physiological pH_i and, conversely, inactivated when the normal pH_i is restored. Activation by acidic pH is particularly important, in as much as mammalian cells have a tendency towards spontaneous acidification due to their internally negative membrane potential and the continuous production of metabolic acid equivalents. In contrast, the cells are rarely exposed to conditions promoting cytoplasmic alkalosis. It is important that pH_i, rather than the transmembrane ΔpH, be the parameter regulated, if the cells are to defend their cytoplasmic milieu against changes in the extracellular pH. To be effective, a pH_i regulatory system must be thermodynamically poised to compensate for the spontaneous acidification, implying that a continuous source of energy must be available for ejection of H^+ equivalents. Clearly, the energy source must be different from, and preferably independent of, the parameter being controlled. The purpose of this chapter is to analyse whether in lymphocytes these criteria apply to the cation and anion transporters that have been suggested to regulate pH_i in other cells.

Na^+/H^+ exchange

There is ample evidence that the plasma membranes of lymphocytes possess a Na^+/H^+ counter-transport system (Rosoff et al 1984, Grinstein et al 1984a). In the forward (physiological) mode, its activity can be detected as an uptake of Na^+ and/or as the complementary efflux of H^+. As in other cells, both fluxes are sensitive to amiloride and its analogues and have a one-to-one stoichiometry. As a result of the latter, the flux is electroneutral and insensitive to changes in membrane potential (summarized in Grinstein & Dixon 1988). Several properties of the Na^+/H^+ exchanger are indicative that it plays a central role in pH_i regulation. First, antiport activity is very high (in excess of 15 mmole/litre/min) in severely acid-loaded cells, yet the system becomes nearly quiescent when pH_i returns to physiological levels (Fig. 1 and Grinstein et al 1984a). Secondly, though reversible, the system is asymmetric, responding with stimulation of transport only when the cytoplasmic, but not the extracellular, medium is acidified (Fig. 1b and Grinstein et al 1984b). This implies that the effect of intracellular acidification on the rate of exchange is not merely thermodynamic and that the parameter that is regulated is pH_i and not ΔpH.

Under physiological conditions, the inward Na^+ gradient generated by the Na^+/K^+ pump provides the energy for H^+ extrusion. The intracellular Na^+ concentration, estimated from photometric determinations, is approximately 15 mM, nearly one order of magnitude lower than the extracellular concentration. Thus, an electroneutral exchanger could drive pH_i one unit above the extracellular pH (pH_o). Such an unphysiologically alkaline level is never attained because the activity of the antiport is kinetically curtailed at high pH_i by an allosteric site that determines the rate of operation of the transport moiety. In combination, the Na^+ gradient and the allosteric modifier site provide a means for the controlled delivery of a large energy store towards the regulation of pH_i. The properties of the lymphocyte Na^+/H^+ antiport have been reviewed in more detail elsewhere (Grinstein & Dixon 1988).

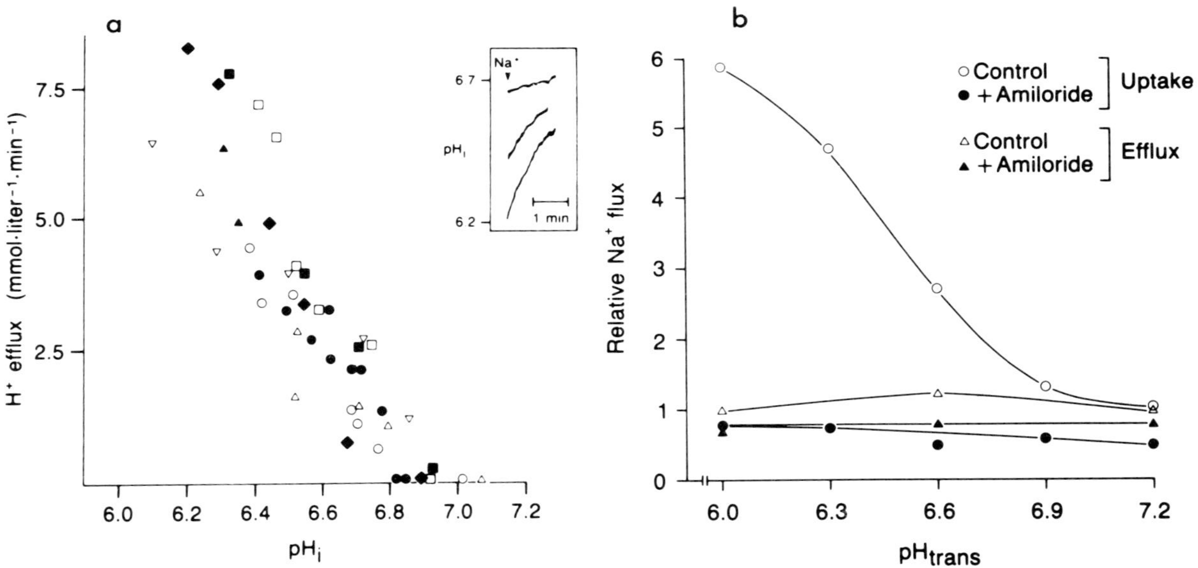

FIG. 1. a) Internal pH (pH_i) dependence of the rate of Na^+-induced acid extrusion from rat thymic lymphocytes. Acid extrusion was calculated from the rate of ΔpH_i measured fluorimetrically. The pH_i was brought to the desired value using nigericin, as described in Grinstein et al (1984a). b) Dependence of the rate of unidirectional Na^+ flux on the pH of the *trans* side of the membrane. The fluxes were normalized to facilitate comparison; unity is the control flux at pH_{trans} = 7.2. Reproduced from J Gen Physiol (1984) vol 83:341–369 and vol 84:585–600, by copyright permission of the Rockefeller University Press.

Cl^-/HCO_3^- exchange

Though direct proof was not available, the existence of an electroneutral anion antiport in lymphocytes had been suspected from the discrepancy between the total and conductive fluxes of Cl^- (Grinstein & Dixon 1988). Recently, we have obtained more direct evidence for the operation of a Cl^-/HCO_3^- exchanger in rat thymic lymphocytes. Both the uptake and efflux of isotopic Cl^- are greatly inhibited by disulphonic stilbene derivatives, such as DIDS. Moreover, these fluxes are also inhibited when the HCO_3^- concentration is increased, presumably by a competitive mechanism. In addition, the occurrence of stilbene disulphonate-sensitive HCO_3^- fluxes was inferred from measurements of intracellular pH, obtained using the fluorescent indicator 2′,7′-*bis*-(carboxyethyl)-5(6)-carboxyfluorescein (BCECF). As shown in Fig. 2, cells suspended in Cl^--free medium undergo a marked cytoplasmic alkalinization. This pH_i increase was eliminated by disulphonic stilbenes and was not observed in the nominal absence of HCO_3^-. The observed alkalinization is therefore consistent with uptake of HCO_3^- through the anion exchanger, driven by the downward Cl^- gradient. In keeping with this interpretation, re-introduction of Cl^- to the medium resulted in a return of pH_i towards the original level.

To determine if Cl^-/HCO_3^- exchange is involved in the regulation of pH_i in lymphocytes, we assessed whether this system conforms to the functional criteria listed above. The anion exchanger could only play a meaningful role

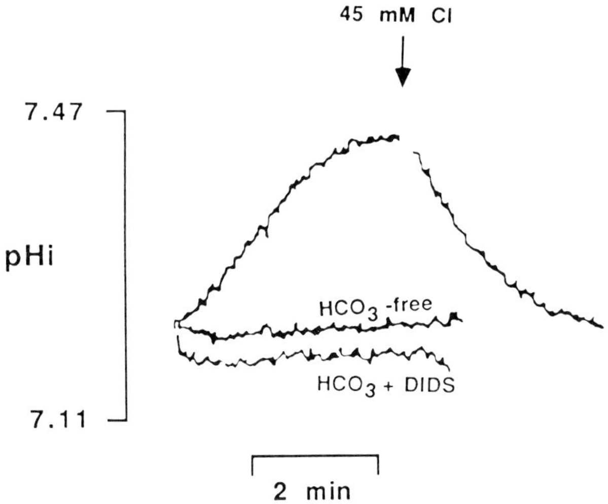

FIG. 2. pH_i changes of thymic lymphocytes suspended in low $[Cl^-]$ medium: dependence on extracellular HCO_3^- and effect of DIDS (200 μM). pH_i was monitored fluorimetrically using BCECF as in Grinstein et al (1984b). The traces start upon resuspension of otherwise untreated cells in medium containing 7 mM Cl^- and osmotically balanced with 4-aminohippurate. Where indicated by the arrow, 45 mM NaCl was added to the medium.

in the steady-state regulation of pH_i, if it mediated the continuous net uptake of HCO_3^- in order to compensate for the metabolic and electrically driven acid accumulation. Because at equilibrium $[Cl^-]_o/[Cl^-]_i=[HCO_3^-]_o/[HCO_3^-]_i$, this would require a transmembrane Cl^- gradient such that

$$\log\ [Cl^-]_o/[Cl^-]_i \leq pH_o - pH_i,$$

based on the assumption that CO_2 permeation and hydration are not rate limiting. Since the external Cl^- concentration ($[Cl^-]_o$) and the transmembrane ΔpH are known, the validity of this prediction can be established by measuring the intracellular Cl^- concentration ($[Cl^-]_i$). We have performed coulometric determinations of the Cl^- content of rat thymic lymphocytes and have calculated an intracellular concentration of 84.5±3.4 mM ($n=16$), based on a cellular volume of 114 μm^3. Using this concentration we calculate that for thymocytes $\log [Cl^-]_o/[Cl^-]_i$ is 0.24, approximately equal to $pH_o - pH_i$ when the cells are suspended in physiological medium. Thus, in principle, the transmembrane distribution of Cl^- could maintain pH_i in the physiological range by means of Cl^-/HCO_3^- exchange. Clearly, these calculations rest on the assumption that Cl^- is homogeneously distributed within a single intracellular (cytoplasmic) compartment.

Determinants of the intracellular Cl^- concentration

Though similar to that of other leucocytes (Baron & Ahmed 1969, Castranova et al 1979, Dissing et al 1984, Simchowitz & DeWeer 1986), the concentration of Cl^- determined for thymocytes is much higher than those reported for excitable and epithelial cells. In the latter cells, cation-independent Cl^-/HCO_3^- exchange is therefore unlikely to play a significant role in pH_i regulation.

In view of this difference, it is important to establish the mechanism(s) whereby Cl^- is accumulated in leucocytes, potentially enabling anion exchange to participate in the control of pH_i. Several anion transport pathways have been described in leucocytes including, in addition to the anion exchanger, Cl^- channels and Na^+-K^+-$2Cl^-$ co-transport. In thymocytes, Cl^- channels appear to play only a minor role in translocating Cl^- and in determining its transmembrane distribution. This conclusion is based on the following observations. First, the membrane potential is only marginally sensitive to alterations in the extracellular Cl^- concentration. Instead, the electrical behaviour of the membrane is similar to that of a K^+ electrode. A similar behaviour has been described for thymic and blood lymphocytes (Grinstein et al 1982). In the latter cells, the isotopic K^+ ($^{86}Rb^+$) fluxes are, surprisingly, much slower than those of Cl^-, indicating that the bulk of the $^{36}Cl^-$ flux occurs via electroneutral pathways. Further indication that Cl^- conductance is very low was obtained from measurements of net salt transport, detected as cellular volume changes. In the presence of a significant anion conductance, increasing cation conductive permeability would be expected to result in a sizeable translocation of salt and osmotically obliged water. This prediction was tested using valinomycin, a K^+-specific conductive ionophore. Under conditions where valinomycin increased K^+ conductance, as shown by an accompanying hyperpolarization, no significant volume loss was detected, despite the large outward electrochemical KCl gradient. Conversely, only marginal swelling was noted when gramicidin was used to increase monovalent cation conductance. This treatment is expected to increase NaCl uptake, provided an endogenous anion conductance is present. That the cells can indeed swell when both cation and anion conductive pathways are available can be shown using nystatin. This channel former increases the permeability to small ions such as Na^+, K^+ and Cl^-, regardless of their charge. Treatment with nystatin induced a marked, rapid swelling of thymocytes, doubling their size within 20 min. Taken together, these results indicate that, if present, Cl^- conductance is relatively low, constituting only a small fraction of the total Cl^- permeability.

Perhaps more importantly, a conductive pathway could not account for the comparatively high Cl^- content of thymic lymphocytes, given their membrane potential. Independent determinations of lymphocyte membrane potential obtained by a variety of methods yield values in the 50–60 mV range (summarized in Grinstein & Dixon 1988). If the distribution of Cl^- were dictated electrophoretically by the membrane potential, $[Cl^-]_i$ would equal $[Cl^-]_o \exp(zF.E_m/RT)$. Therefore, intracellular $[Cl^-]$ would be ≤ 20 mM, which is over fourfold lower than the experimentally determined values.

In contrast, Na^+-K^+-$2Cl^-$ co-transport could readily account for the observed $[Cl^-]_i$. If this were the sole (or the predominant) system transporting Cl^-, $[Cl^-]_i$ would be determined by the combined Na^+ and K^+ gradient, such that

$$[Cl^-]_i^2 = [Na^+]_o.[K^+]_o.[Cl^-]_o^2/[Na^+]_i.[K^+]_i.$$

Under physiological conditions ($[Na^+]_o = 140$ mM; $[K^+]_o = 5$ mM; $[Cl^-]_o = 147$ mM; $[Na^+]_i \simeq 15$ mM; $[K^+]_i \simeq 140$ mM) the co-transporter would drive $[Cl^-]_i$ to 85 mM, a figure that is virtually identical to the experimental value reported above. However, we have thus far failed to obtain evidence of significant co-transport activity on thymocytes. The Cl^- content of the cells was unaffected by a 60 min incubation in media devoid of Na^+ or K^+ or containing 1 mM furosemide or 50 μm bumetanide. Moreover, in two experiments, unidirectional ^{36}Cl uptake was not inhibited by either furosemide or bumetanide. It is therefore unlikely that co-transport is the main determinant of the intracellular Cl^- concentration.

Other transport mechanisms, such as an inward Cl^- pump, could also maintain $[Cl^-]_i$ above the level predicted by the electrochemical equilibrium. However, convincing evidence for an inward Cl^- pump in mammalian plasma membranes is not available. Instead, we considered the possibility that the anion exchanger itself is the main determinant of the intracellular Cl^- concentration. In this case, Cl^- would be driven across the membrane by the gradient of HCO_3^-, which is dictated by the transmembrane ΔpH. This hypothesis implies that pH_i (or ΔpH) is determined independently, by systems such as the Na^+/H^+ antiport, and is consistent with the finding that E_{Cl^-} and E_{H^+} are similar (see above).

Is the intracellular Cl^- concentration determined mainly by the gradient of HCO_3^-? This can be tested experimentally by varying the intracellular pH while keeping p_{CO2} constant. Under these conditions, Cl^- redistribution can be measured indirectly, as changes in cell volume, or directly, using the chloridometer. We have alkalinized the cytoplasm of cells suspended in HCO_3^--containing media by exchanging extracellular alkali cations for internal H^+, using exogenous ionophores such as monensin or nigericin. The alkalinization, which was confirmed fluorimetrically, was accompanied by marked cellular swelling and by a net Cl^- gain. When the cells were alkalinized by 0.36 pH units, volume increased by 55% in 20 min, while internal Cl^- rose by 83%. These changes were markedly inhibited in the nominal absence of HCO_3^- and were completely eliminated by disulphonic stilbenes and by omission of Cl^-, suggesting that they were indeed mediated by Cl^-/HCO_3^- exchange. Together, these results indicate that Cl^- can be driven by the gradient of HCO_3^- and suggest that the transmembrane ΔpH may be the major determinant of the intracellular Cl^- concentration.

Concluding remarks

It is clear from the above considerations that cation-independent Cl^-/HCO_3^- exchange could be involved in pH_i regulation only if coupled to a separate mechanism of intracellular Cl^- accumulation. We have thus far failed to detect such a mechanism in lymphocytes. Instead, we find that anion exchange itself is the predominant Cl^- transport system and possibly the major determinant of $[Cl^-]_i$. Hence, the anion exchanger, driven by the distribution of HCO_3^-, seems to play a central role in $[Cl^-]_i$ homeostasis. If HCO_3^- is indeed the ion defining the direction and magnitude of anion exchange, this mechanism cannot be an active determinant of pH_i. However, this does not rule out a passive role of the exchanger in the stabilization of pH_i. The establishment of a Cl^- gradient in accordance with ΔpH provides a reservoir of dynamic buffering power, which will oppose departures from the prevailing pH_i. Decreases in pH_i, which are accompanied by a reduction in $[HCO_3^-]_i$, will be compensated by intracellular Cl^- exit in exchange for external HCO_3^-; this tends to minimize the pH_i change. Conversely, some of the excess HCO_3^- formed upon alkalinization will exit the cell in exchange for outside Cl^-. This function is distinct from that of an active pH regulator in that: a) an independent source of energy for H^+ (equivalent) translocation does not exist, and b) the system is not responsive to the intracellular pH but to the transmembrane ΔpH. Such a mechanism could not sustain pH_i in the physiological range on a continued basis, nor could it protect the cell against alterations in extracellular pH.

It is notable that the cation-independent anion exchanger seems to play different roles in different cell types. In mammalian red cells, the intracellular Cl^- concentration is dictated by the electrical potential, which in these cells is close to that predicted by the Donnan equilibrium. This is due to the comparatively large anion conductance, which is orders of magnitude higher than the combined cation conductance. In addition, red cells have little Na^+/H^+ exchange and there is no evidence for the existence of alternative pH_i regulatory mechanisms. The abundant anion exchangers ensure that $[Cl^-]_o/[Cl^-]_i=[HCO_3^-]_o[HCO_3^-]_i$ at all times. Thus, the membrane potential, which dictates $[Cl^-]_i$, is the primary determinant of $[HCO_3^-]_i$ and of pH_i, a process mediated by anion exchange. In some excitable and epithelial cells, the Cl^- distribution is also largely determined by the transmembrane potential. However, the potentials are considerably more negative in these cells than in erythrocytes. Consequently, intracellular $[Cl^-]$ is much lower and $[Cl^-]_o/[Cl^-]_i > pH_o - pH_i$. Under such conditions, anion exchange would drive HCO_3^- out of cells, thereby acidifying the cytoplasm. This is detrimental to pH_i homeostasis, accentuating the natural tendency of the cells to become acidic. For these reasons, it is likely that the anion antiport functions only minimally in cells with high Cl^- conductance and large negative potentials, such as skeletal muscle. A mechanism that could account for this behaviour has been recently reported for *VERO* cells (Olsnes et al 1987). Anion exchange in these cultured cells displays a steep pH dependence, with inhibition at subphysiological levels. Therefore, Cl^-/HCO_3^- exchange may function to overcome a cytoplasmic alkalosis, without causing excessive acidification in the steady state. It is tempting to speculate that the differential role of Cl^-/HCO_3^- exchange in different cell types is attributable to their specific pH dependence. The pH_i sensitivity of the exchanger in leucocytes, where it seems to play a dynamic buffering role, has not been investigated.

Finally, it is important to point out that though cation-independent Cl^-/HCO_3^- exchange does not actively regulate pH_i in lymphocytes, a Na^+-dependent anion transport system is also present in these cells. As shown in Fig. 3, acid-loaded thymocytes readily recover near-normal pH_i even in the presence of powerful inhibitors of Na^+/H^+ exchange, provided Na^+ and HCO_3^- are present in the medium. Notice that some recovery is observed

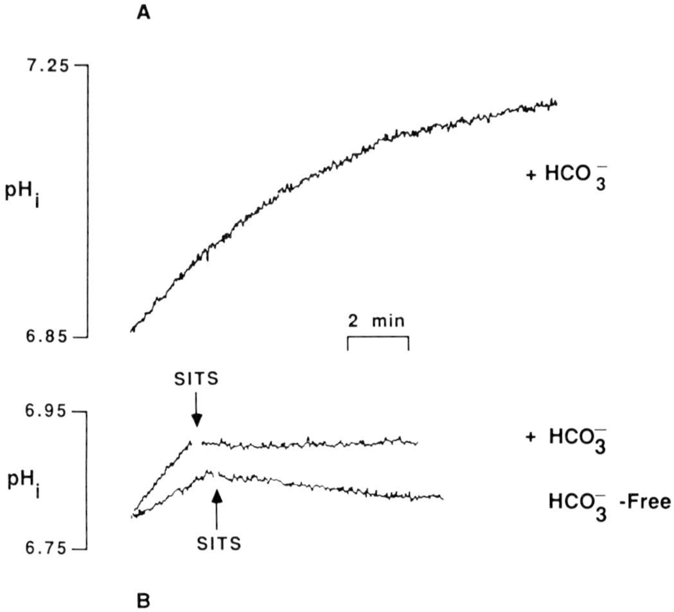

FIG. 3. pH$_i$ recovery of thymic lymphocytes acid-loaded by an ammonium pre-pulse. a) The extracellular medium contained 125 mM Na^+ and 21 mM HCO_3^- and was equilibriated with 5% CO_2 and 95% air. b) Effect of 1 mM SITS on the pH$_i$ recovery in HCO_3^--containing and in nominally HCO_3^--free medium. In both (a) and (b), Na^+/H^+ exchange was precluded by the addition of 17 μM *N*-ethyl-*N*-(1-methylethyl)amino amiloride. pH$_i$ was monitored fluorimetrically as in Fig. 2.

even in the nominal absence of HCO_3^- (approximately 200 μM). However, when the difference in intracellular buffering power is considered, the rate of H^+ equivalent extrusion is ≃2.5-fold slower than that in HCO_3^--containing solution. This bicarbonate-sensitive component is susceptible to inhibition by stilbene disulphonates and, by analogy with systems described in other cells (Roos & Boron 1981, Guggino et al 1983), probably represents the counter-transport of $NaCO_3^-$ for Cl^- or equivalent, e.g. Na^+ and $2HCO_3^-/Cl^-$ or $NaHCO_3/HCl$. The properties and relative role of this system in pH$_i$ regulation in lymphocytes is being investigated.

Acknowledgements

The work reviewed in this chapter was supported by the Cystic Fibrosis Foundation, the National Cancer Institute and the Medical Research Council (Canada). S. G. is the recipient of a Medical Research Council Scientist Award.

REFERENCES

Baron DN, Ahmed SA 1969 Intracellular concentrations of water and of the principal

electrolytes determined by the analysis of isolated human leukocytes. Clin Sci (Lond) 37:207–219
Castranova VL, Bowman L, Miles PR 1979 Transmembrane potential and ionic content of rat alveolar macrophages. J Cell Physiol 101:471–480
Dissing S, Hoffman R, Murnane MJ, Hoffman JF 1984 Chloride transport properties of human leukemic cell lines K562 and HL60. Am J Physiol 247:C53–C60
Grinstein S, Dixon SJ 1988 Membrane potential, ion transport and cytoplasmic pH in lymphocytes: changes during activation. Physiol Rev, in press
Grinstein S, Clarke CA, Dupre A, Rothstein A 1982 Volume-induced increase of anion permeability in human lymphocytes. J Gen Physiol 80:801–823
Grinstein S, Cohen S, Rothstein A 1984a Cytoplasmic pH regulation in thymic lymphocytes by an amiloride-sensitive Na/H antiport. J Gen Physiol 83:341–369
Grinstein S, Goetz JD, Rothstein A 1984b Na^+ fluxes in thymic lymphocytes. II Amiloride-sensitive Na/H exchange pathway; reversibility of transport and asymmetry of the modifier site. J Gen Physiol 84:585–600
Guggino WF, London R, Boulpaep EL, Giebisch G 1983 Chloride transport across the basolateral cell membrane of the Necturus proximal tubule: dependence on bicarbonate and sodium. J Membr Biol 71:227–240
Olsnes S, Ludt J, Tonnessen TI, Sandvig K 1987 Bicarbonate/chloride antiport in Vero cells: Mechanism for bicarbonate-dependent regulation of intracellular pH. J Cell Physiol 132:192–202
Roos A, Boron WF 1981 Intracellular pH. Physiol Rev 61:296–434
Rosoff PM, Stein L, Cantley L 1984 Phorbol esters induce differentiation in a pre-B lymphocyte cell line by enhancing Na/H exchange. J Biol Chem 259:7056–7061
Simchowitz L, de Weer P 1986 Chloride movements in human neutrophils. Diffusion, exchange and active transport. J Gen Physiol 88:167–194

DISCUSSION

Thomas: How nice to have fairly simple number of transporters. It's a pity you still don't know whether it is a Na^+-HCO_3^-- Cl^- system there or just a Na^+-HCO_3^-, possibly electrogenic, system. Why do you think these cells want a high Cl^- level? Is it just to provide this buffering?

Grinstein: No one has addressed this question. Certainly high buffering capacity would be advantageous.

Vaughan-Jones: The system that Sergio (Grinstein) describes is almost identical to that which I proposed for the sheep heart Purkinje fibre (see Vaughan-Jones 1982). In this tissue, the high intracellular concentration of Cl^- is achieved, in part, by Cl^- uptake in exchange for HCO_3^- efflux, which thus acidifies the cell interior. This anion exchanger can also act as an intracellular 'alkaline regulator', i.e. it helps to reduce pH_i in response to an internal metabolic alkalosis by transporting HCO_3^- out of the cell. The system also functions as a two-way acid-equivalent transporter across the membrane. For example, it brings acid-equivalents into the cell (HCO_3^- efflux) in response to an external metabolic acidosis. Since 60% of whole-body buffering of external acid disturbances occurs inside cells, there must be a way of getting that acid into the cells to be buffered. This is one method—certainly Na^+/H^+ exchange in heart can't do it.

Thomas: Does anyone know of any other role, apart from this one in pH regulation, for this high internal Cl^-? Is there an osmotic requirement for high Cl^-?

Malnic: Guggino et al (1983) have described a Na^+-dependent Cl^-/HCO_3^- exchanger in the basolateral membrane of *Necturus* proximal tubule which transports Cl^- from cell to interstitium.

Aickin: In smooth muscle the high Cl^- is used as an ionic battery for depolarization; there are Cl^- channels activated by noradrenaline and histamine.

Grinstein: There are no resting channels in this system but there may be hormone-regulated channels, in which case, the cells could depolarize in response to the hormone.

Harvey: You seem to have a one-way traffic for Cl^-, which is inward, what about osmotic equilibrium?

Grinstein: Cl^- can leave in exchange for HCO_3^-.

Machen: I disagree. The observed pH gradients mean that the HCO_3^- has to be going out and the Cl^- coming in. Where is the Cl^- going? Does it stay inside the cell until the intracellular concentration equals 85 mM and the anion exchanger stops working because it is at equilibrium?

Grinstein: At constant pH there will be no net movement of Cl^-, if the Cl^- ratio is similar to that of bicarbonate and the conductance is very low.

Harvey: The amount of Cl^- entering the cell must be high; if it's in equilibrium, the amount of Cl^- leaving through the channel must be the same.

Thomas: Don't forget the Na^+/H^+ exchanger is still working perfectly normally. It is only in some circumstances that the Cl^-/HCO_3^- exchanger becomes important.

Al-Awqati: The Cl^-/HCO_3^- exchanger is reversible, so if the intracellular concentration of Cl^- is too high, some can leave via the exchanger. If the HCO_3^- change is based on the Na^+/H^+ exchanger, the HCO_3^- concentration inside the cell may fall and that of Cl^- increase.

Grinstein: If the cell becomes acidic because of metabolic acid production, then the level of HCO_3^- will fall and the amount of intracellular Cl^- will fall.

Harvey: There is a problem with osmotic regulation, for example, in alkalosis when HCO_3^- is leaving the cell, the cell will accumulate Cl^- and swell because HCO_3^- is osmotically inactive.

Thomas: That is not a problem because there is no reason why these cells can't swell, it's not like a nerve cell in the brain which is not allowed to swell.

Harvey: Have you seen regulatory volume decrease?

Grinstein: Yes, these cells do alter their volume in a regulatory manner. That may be another function of the high Cl^- concentration in T cells, they have very large K^+ and Cl^- independent conductances. Whether these cells are ever exposed to hypotonic conditions where they need to regulate volume is a different issue but if they are artificially suspended in hypotonic media, Cl^- channels open and the cells lose a lot of Cl^-.

de Hemptinne: You said that when the cell goes alkaline it will swell. The red blood cell swells when it goes acid (taking up CO_2 in the metabolizing tissues) and shrinks when it goes alkaline (in the lung). I didn't understand why your cells swell when they become alkaline.

Grinstein: The red cell is a very peculiar case: the distribution of Cl^- is determined by the Donnan potential created by haemoglobin. As you increase the pH, you increase the negative charge on haemoglobin which will keep Cl^- out of the cell, causing the cell to shrink. In lymphocytes, when a cell becomes alkaline, the concentration of HCO_3^- increases inside, which drives the influx of Cl^-. As the cell gains Cl^-, it will swell.

Thomas: If there is a 1:1 exchange of Cl^-/HCO_3^-, why should a cell change volume?

Boron: Because the HCO_3^- is buffered by association with H^+ to form CO_2 which can diffuse across the cell membrane.

Grinstein: Cl^- is osmotically active, whereas HCO_3^- is partially inactive.

Harvey: We have seen that there is a role for uptake of Cl^- by the Cl^-/HCO_3^- exchanger and it seems to be separate from the pH regulatory role. Is there evidence that the Na^+/H^+ exchanger can also cause Na^+ uptake independent of pH control?

Grinstein: We have argued in the past that Na^+/H^+ exchange can be involved in Na^+ homeostasis. In principle there is no reason why it couldn't, especially if Na^+ may be regulating the antiport itself.

Aickin: Why should the cell want to use Na^+/H^+ exchange to regulate internal Na^+? The cell would then be gaining Na^+.

Grinstein: There are occasions when the cell wants to gain Na^+, for example, after a cell has shrunk and it wants to swell.

Aickin: So why not via a conductance?

Grinstein: Most osmotically shrunken cells respond with activation of Na^+/H^+ exchange, not with changes in conductance.

Vairo: Increased Na^+ influx also causes the Na^+/K^+-ATPase to be activated.

Boron: Concerning the lymphocyte high Cl^- concentrations, you said that $E_{Cl} = E_H$; what are the implications of that?

Grinstein: It means that the HCO_3^- and the Cl^- are at or near equilibrium. In the red cell they certainly are at equilibrium.

Boron: What is the membrane potential of these lymphocytes?

Grinstein: About -60 mV.

Boron: So if $E_{Cl} = E_H$, they are not equal to Em.

Thomas: If $E_H = E_{Cl}$, it means that the Cl^-/HCO_3^- exchanger is at equilibrium, no net transport will occur.

Grinstein: In a lymphocyte the membrane potential is very different from E_{Cl}, and E_{Cl} is very similar to E_H. In the nerve cell the membrane potential is approximately equal to E_{Cl} but very different from E_H.

Boron: So you are saying that the Cl^-/HCO_3^- exchanger is in equilibrium?

Grinstein: In the red cell it is certainly very near equilibrium, in the lymphocyte if it is not at equilibrium, it's probably because of chloride channels. The lower the Cl^- conductance and the higher the exchange, the closer Cl^- and HCO_3^- will be to equilibrium.

Stone: Does V_{max} for Na^+/H^+ exchange change after the endocytotic event? If endocytosis incorporates up to one-third of the plasma membrane into vesicles, and that membrane contains significant numbers of Na^+/H^+ antiporters, is there then a change in the rate at which the cell can correct an imposed pH gradient?

Grinstein: We haven't done those measurements properly; the main reason is that the volume of the cell changes. The membrane internalized by endocytosis is replaced by more membrane coming from inside the cell and that may or may not contain antiporters. We would have to make careful measurements of cell surface area and see if the total number of antiporters increases or decreases.

Stone: Endocytosis through the clathrin-coated pits is very selective: there is not uniform incorporation of the whole cell membrane.

Grinstein: Comparisons of either protein composition or lipid composition of the vesicles with that of the plasma membrane show that there is some degree of selectivity but not a great deal. For a protein like this, which could be 0.1% of total protein, we wouldn't have seen it anyway.

Morad: Concerning the elegant experiment you did, putting the cell in N-methylglucamine and then loading the phagosomes with dye, did you do the opposite experiment where you would first put the cell into Na^+, allow the phagosome to take up Na^+ and then put N-methylglucamine in the external medium?

Grinstein: We cannot get N-methylglucamine into the cells. We tried to run the antiport in the forward mode by making the plasma membrane permeable with nystatin then acidifying the cytoplasm. In the presence of nystatin, you expect an amiloride-sensitive acidification of the phagosome, which we don't see. Conversely, when we impose a Na^+ gradient, we have Na^+-loaded cells with Na^+-loaded phagosomes, we wash out the Na^+ using nystatin, replace it with K^+ and you would expect the Na^+/H^+ to create an amiloride-sensitive acidification; we don't see that. This supports the idea that the antiports are not active in the phagosome.

Thomas: So there is no evidence for Na^+/H^+ exchange in the phagosome. Do the phagosome membranes digest bacteria by exposing them to acid or is it some other mechanism?

Grinstein: It's a combination—there is the acid, there are lysosomal enzymes and the main mechanism is oxidation using superoxide and hydrogen peroxide produced in the phagosome.

Sabolic: The absence of Na^+/H^+ exchange from the phagosomes agrees completely with our findings in endocytotic vesicles in proximal tubules in the rat kidney (Sabolic et al 1985) but disagrees with the findings of Gurich & Warnock (1986) in endosomes from rabbit renal cortex. We tried many times to demonstrate active Na^+/H^+ exchange in endosomes from rat renal cortex but it worked only in some exceptional experiments. Endosomes in nearly all cells have a very powerful proton pump, which can be measured by the quenching method using the pH-sensitive fluorescent dye, acridine orange: the endosomal proton pump pumps protons into the vesicles. In the presence of ATP and Cl^-, as protons accumulate in the intravesicular space, acridine orange also moves into the vesicles, since it is a weak base. This interacts with protons causing a decrease in the fluorescence of acridine orange. After ΔpH had fully developed, we added Na^+ to the outside buffer. If a Na^+/H^+ exchanger were present in these vesicles, Na^+ would go in and protons would come out, so one would expect an increase in fluorescence. When we observed very little quenching, i.e. the pH gradient was small, there was no significant change after the addition of Na^+, which suggested the absence of a Na^+/H^+ exchanger in endosomes. In a few experiments, when the transmembrane pH gradient was very big, there was a small increase in fluorescence. In these cases, Na^+ had a transient effect. Thus it could be that the number of Na^+/H^+ exchangers in these vesicles is very small and the capacity of this mechanism is therefore limited. Furthermore, after the small increase in fluorescence caused by Na^+, there was always a further decrease. This suggests that the proton pump is pH sensitive and when there is a decrease in the proton gradient, pumping is activated. So there could be a Na^+/H^+ exchanger in the phagosomes, which is difficult to demonstrate because of its low activity or it may be different from the one in the plasma membrane.

Machen: So there is no Na^+/H^+ exchange in the endosome and your data indicate that there is no proton pump from the endosome on the plasma membrane because when you acidify the cells they don't recover unless there is Na^+ outside.

Grinstein: The lymphocyte recovers very slowly in the absence of Na^+; neutrophils and macrophages behave slightly differently. In macrophages there is substantial recovery in Na^+ and HCO_3^- containing media; this recovery is sensitive to N-ethyl maleimide and to DCCD. So macrophages may have a pump in the membrane but in the lymphocyte there is no good evidence for a proton pump in the plasma membrane.

Sabolic: Are there any other kinds of endocytotic vesicles, besides phagosomes, in macrophages and neutrophils?

Grinstein: There is normal endocytosis—turnover of membrane receptors and so on.

Sabolic: So there could still be Na^+/H^+ exchange in some other endocytotic vesicle.

Grinstein: We haven't measured these compartments, so I couldn't tell you.

Machen: The membrane from the endosome starts in the plasma membrane and then it pinches off to form an endosome, which has a proton pump that, at least in some cells, was not present in the surface membrane. Where does the proton pump come from?

Grinstein: That's a very good question; I don't think anybody knows. There is no doubt that a proton pump is present in endosomes and phagosomes. Whether that pump is present and active in the plasma membrane of non-epithelial cells is not clear. The only, very indirect, evidence is that clathrin-coated vesicles, which supposedly originated directly from the plasma membrane, have a proton pump. However, there is now evidence that there are two kinds of clathrin-coated vesicles, some which originate from the Golgi and some from the plasma membrane, and it seems that only those from the Golgi have the proton pump. So if we have a pump active in the phagosome, it may mean that it's the result of fusion of phagosomes with lysosomes but it is very hard to determine when that fusion occurs. Alternatively, the pump could have been in the plasma membrane, active or inactive. If it is active, it is very hard to measure; if it's inactive, it could easily be activated as the phagosome is formed, in the same way as the other transporters are inactivated.

Sabolic: There is a possibility that clathrin-coated vesicles which come from the Golgi are just vectors, which deliver proton pump units to the vesicles that have been endocytosed.

Grinstein: George Sachs claims that he can see an electrogenic proton pump in the plasma membrane of Ehrlich ascites cells but the evidence is very indirect (Heinz et al 1981).

Vaughan-Jones: The anion exchanger in the lymphocyte is interesting because it is similar to the one in heart in that it has a steep activation curve as a function of pH_i. The one in heart seems to switch off when pH_i becomes acid but it switches on when pH_i becomes alkaline, which makes it effectively an 'alkaline regulator'. The anion exchanger in the lymphocyte has a steep activation curve; where does normal pH_i sit on that curve, is it half-way up on the pK? In other words, can you activate the system in an acid and an alkaline direction? If you can, then it is acting like a buffer. If you are not on the pK but right at the bottom of the curve, it is more like a regulator.

Grinstein: I can't give you firm data. We have done the experiments only twice, I think it is going to be about one-third of the way up that activation curve.

Vaughan-Jones: So one could argue that it's a pH_i regulator—it copes with alkaline loads more easily than with acid loads.

Grinstein: But it's not a regulator in the sense that it doesn't have an independent source of energy dictating the Cl^- distribution. In other words, once the Cl^- for redistribution is exhausted, if you imposed a further alkaline load, the anion exchanger couldn't cope with it—unless there is a Cl^- pump, which has not been ruled out.

Oberleithner: If there is 100 mM Cl^- in the cytoplasm of the lymphocyte, what is the accompanying cation?

Grinstein: It is largely potassium.

Gillespie: You mentioned that when you take away Cl^-, the alkalinization response has a sigmoid time course. Could you elaborate on that.

Grinstein: I think that's a result of the very steep pH dependence of the anion exchanger. When we remove the Cl^- from the medium we are at or near to physiological pH where the anion exchanger is only moderately active. As the cell becomes progressively more alkaline, the exchanger is activated.

Gillespie: Do you see the same effect when you add Cl^- back?

Grinstein: When we add Cl^- back the system is activated because the pH is high. It goes very fast and then it slows down. It slows down partly because of kinetics, partly because of thermodynamics: when it turns on the thermodynamics is such that it should be faster initially.

Vaughan-Jones: Do you see that? Is there a slow tail on recovery?

Grinstein: There is a tail but that could be because the cells are equilibrating with Cl^-.

Gillespie: The alkalinization on removal of Cl^- in pancreatic acinar cells also has a sigmoid time course.

Grinstein: In experiments to displace the Cl^- content of the cells by changing pH, this works very well when the cell is made alkaline, the cell gains Cl^-. We spent a month trying to make the cell acidic and lose Cl^- but we could not. I think the reason is that the system is inactivated.

Thomas: So the Cl^-/HCO_3^- exchanger only works really well at alkaline pH where there is plenty of HCO_3^-, which physiologically there is.

Machen: I think that is true only for some anion exchangers. The red blood cell anion exchanger seems to be relatively insensitive to pH, and there may be others.

Grinstein: It's not the presence or absence of HCO_3^- that is the major determinant of the pH dependence. The curve is much steeper than you would predict from the change in HCO_3^- concentration.

Thomas: All anion exchangers are inhibited by SITS, but they vary in their pH sensitivity. I grew up with the Na^+ pump which is always inhibited by removing K^+, always inhibited by ouabain, it's always electrogenic. The Na^+/H^+ exchanger is always inhibited by amiloride—is that still true?

Vaughan-Jones: No, Sergio Grinstein has described an amiloride-insensitive Na^+/H^+ exchanger (Grinstein et al 1984).

Grinstein: That may be a different system altogether or it may be some other electroneutral Na^+ transporting mechanism. We may have been looking at a Na^+-HCO_3^- co-transporter and thinking that it was a Na^+/H^+ exchanger.

Aickin: There is also the problem in smooth muscle that amiloride inhibits in the absence of CO_2 but not in the presence, when the Na^+/H^+ exchanger ought to be still active.

References

Grinstein S, Goetz JD, Rothstein A 1984 $^{22}Na^+$ fluxes in thymic lymphocytes. I. Na^+/Na^+ and Na^+/H^+ exchange through an amiloride-insensitive pathway. J Gen Physiol 84:565–584

Guggino WB, London R, Boulpaep EL, Giebisch G 1983 Chloride transport across the basolateral cell membrane of the Necturus proximal tubule: dependence on bicarbonate and sodium. J Membr Biol 71:227–240

Gurich RW, Warnock DG 1986 Electrically neutral Na^+-H^+ exchange in endosomes obtained from rabbit renal cortex. Am J Physiol 251:F702–F709

Heinz A, Sachs G, Schager JA 1981 Evidence for activation of an active electrogenic proton pump in Ehrlich ascites cells during glycolysis. J Membr Biol 61:143–153

Sabolic I, Haase W, Burckhardt G 1985 ATP-dependent H^+ pump in membrane vesicles from rat kidney cortex. Am J Physiol 248:F835–F844

Vaughan-Jones RD 1982 Chloride-bicarbonate exchange in the sheep cardiac Purkinje fibre. In: Nuccitelli R, Deamer D (eds) Intracellular pH: its measurement, regulation and utilization in cellular functions. Kroc Foundation Series vol 15, p239–252

General discussion I

Effects of growth factors on intracellular pH

Moolenaar: We did some experiments on fibroblasts in culture in monolayer. We found that most, if not all, fibroblasts, including human fibroblasts, mouse embryo fibroblasts and 3T3 cells, have a steady-state pH in medium containing physiological concentrations (24 mM) of HCO_3^- under 5% CO_2 that is substantially higher than their pH in the absence of HCO_3^-. This steady-state pH is mediated by the Na^+-dependent Cl^-/HCO_3^- exchanger: it's sensitive to DIDS, it requires HCO_3^- and it requires Na^+. If you deplete the cells of internal Cl^- and shift them to HCO_3^-/CO_2-containing media, there is an initial acidification but the cell does not establish a new steady-state internal pH. So HCO_3^- in fibroblastic cells raises steady-state pH to a level that is normally seen when growth factors are applied in the absence of HCO_3^-. If we then add growth factors (usually epidermal growth factor, EGF, at 50 ng/ml), nothing further happens to pH, unless we artifically lower intracellular pH. Then we see apparent activation of Na^+/H^+ exchange and the pH goes up. That suggests very strongly that the Na^+-dependent Cl^-/HCO_3^- exchanger raises pH to a level that is beyond the operating range of the Na^+/H^+ exchanger—at least in fibroblastic cells.

Conversely, in the absence of HCO_3^-, using Hepes as a buffer, when we artificially raise the steady-state pH to a level normally observed in the presence of HCO_3^-, growth factors fail to increase the pH any further. That's consistent with this hypothesis. I don't want to imply that this is true for all cell types but for fibroblasts, which are widely used in growth factor studies, this is apparently what happens.

This emphasizes the importance of cytoplasmic pH: it should be maintained within a very narrow range and the cell does everything possible to keep it within that range.

Thomas: I am delighted to hear that the same Na^+-HCO_3^--Cl^- dependent pH_i regulating system that was first found in snail neurons and squid axon is also found in mammalian cells.

Moolenaar: Another implication is that, unlike Na^+/H^+ exchange, Na^+-dependent Cl^-/HCO_3^- exchange is not activated by growth factors.

Grinstein: What is the significance of the activation of the antiport physiologically when there is HCO_3^- present?

Moolenaar: I have no idea—it happens with all mitogenic stimuli. Perhaps it enhances the housekeeping capacity to regulate pH following metabolic activation of the cell.

Boron: We have made similar observations in renal mesangeal cells. These are smooth muscle-like cells which respond to arginine vasopressin (AVP), a hormone which is also a growth factor for these cells, by dividing. In the absence of HCO_3^-, AVP causes an increase in pH_i, exactly as Wouter (Moolenaar) has shown. In the presence of HCO_3^-, AVP causes only a small pH_i decrease. Nevertheless, the Na^+/H^+ exchanger is stimulated, even in the presence of HCO_3^-. If you add ethylisopropylamiloride, an inhibitor of Na^+/H^+ exchange, in the presence of AVP and HCO_3^-, the cell acidifies at a very high rate. So it seems that the Na^+/H^+ exchanger must be stimulated just to keep the pH within the normal range.

Moolenaar: The initial acidification is due to Ca^{2+} mobilization.

Boron: And that's gone, too, in the mesangeal cells. Presumably, the open-system CO_2/HCO_3^- buffering power is so high that the H^+ dumping caused by release of Ca^{2+} cannot cause a substantial decrease in pH_i. In addition, a Na^+-dependent Cl^-/HCO_3^- exchanger is active in the presence of HCO_3^- and this would also minimize the fall in pH_i.

Gillespie: Concerning what happens in HCO_3^- containing solutions, in chick embryonic somatic cells there is no Na^+/H^+ exchange in the early stages. But if you culture them in serum for 5–6 hours, the Na^+-HCO_3^- co-transport mechanism disappears and Na^+/H^+ exchange is activated. This mechanism may be more important developmentally than in cancer cells.

Grinstein: In some lymphocytes with mitogens we don't get an alkalinization—in fact we get a slight acidification, which is Ca^{2+} dependent. This acidification is greatly magnified when we take away Na^+ or add amiloride analogues. So perhaps the activation of the antiport is just to compensate for increased metabolic acid production.

Vairo: Even if cytoplasmic alkalinization is not obvious in the presence of HCO_3^-, growth factor-mediated activation of the antiport will still result in an influx of Na^+ into the cell. This, in turn, will stimulate the Na^+/K^+-ATPase, thereby increasing the concentration of intracellular K^+, which is thought to be important for modulating various metabolic processes. Therefore it is possible that in the presence of HCO_3^- the increased cellular K^+ may be the more important consequence of activation of the antiport by growth factors.

Harvey: I have a problem with the apparent inactivation of the Na^+/H^+ exchanger by pH. Do you think that the exchanger does not respond to growth hormones in the presence of CO_2/HCO_3^- because the pH is alkaline?

Moolenaar: The exchange rate increases with increasing H^+ concentration. At alkaline pH the exchanger is virtually inactive.

Grinstein: I think the activation of the exchanger is a shift in the set point of the antiporter. The antiporter is nearly quiescent at normal pH; when you give the mitogen, it goes to a higher pH and then it becomes nearly quiescent again.

If you artifically increase the pH, the antiport will be quiescent and the mitogen will not turn it on.

Thomas: You can't always see that it's activated, if the pH_i is already relatively alkaline.

Grinstein: That's why the external pH experiments don't show anything.

Oberleithner: I think this is a very interesting observation. We saw something similar recently in cultured dog kidney cells (MDCK cells). In these cells, aldosterone activates the Na^+/H^+ exchanger, as it does in the amphibian kidney. We made these measurements in Hepes-buffered solution. We have not yet measured intracellular pH. However, there is a striking observation: if we use CO_2/HCO_3^-, the pH gradient maintained by the epithelium in the absence of HCO_3^- disappears. The working hypothesis we have at the moment is that aldosterone activates the Na^+/H^+ exchanger. However, as pH_i goes alkaline, in the presence of HCO_3^-, there is an activation of the Cl^-/HCO_3^- exchanger, which limits the aldosterone-induced pH change.

Sabolic: I still do not fully understand the role of Na^+/H^+ exchange in alkaline conditions. I remember the experiments by Aronson et al (1982) which showed that there is a pH dependence of Na^+/H^+ exchange. Aronson proposed that the Na^+/H^+ exchange would switch off at pH 7.3. We measured it in rat renal brush border membranes at pH 7.4, 7.8 and it was as active as at pH 7 or 6.5. Recently, I measured Na^+/H^+ exchange by the same method in red blood cell membrane vesicles and it was abolished at pH 7.

Grinstein: In epithelial cells the purpose of the antiport is to transport salt across the tissue. In other cells it is intended to regulate pH. It is very likely that the set point is shifted upwards so that it is transporting Na^+ all the time rather than being shut off at physiological pH.

Boron: In intact proximal tubule cells in the kidney there is no evidence of Na^+/H^+ exchange being active above pH about 7.3 or 7.4. I think that there is a discrepancy between vesicle data and data obtained from whole cells. This may mean that when the vesicles are formed something happens to the exchanger that removes some of this inhibition. It certainly doesn't happen completely because Aronson and his colleagues demonstrate the effect that H^+ *per se* have on the transporter. But clearly those Na^+/H^+ exchangers are active at an alkaline pH, at which in an intact cell they would be inactive. We have done an experiment recently on mesangeal cells in which we increased the pH of the cell to 8. When we remove the Na^+ from a cell at pH 8 nothing happens to the pH_i; it is stuck because the Na^+/H^+ exchanger is shut off. If you do that experiment with brush border membrane vesicles those will possess Na^+/H^+ exchange activity. Perhaps when the vesicles are formed some critical group is dephosphorylated or something like that happens and it just changes the properties of the exchanger.

Thomas: In your cell stuck at pH 8, is there HCO_3^- present?

Boron: This is in Hepes. If you take Na^+ away at pH 7, the cell becomes very acid.

References

Aronson PS, Nee J, Suhm MA 1982 Modifier role of internal H^+ with the Na^+-H^+ exchanger in renal microvillus membrane vesicles. Nature (Lond) 299:161–163

Role of monocarboxylate transport in the regulation of intracellular pH of renal proximal tubule cells

Walter F. Boron, Arthur W. Siebens* and Nazih L. Nakhoul

Department of Cellular and Molecular Physiology, Yale University School of Medicine, New Haven, Connecticut 06510, USA and **Laboratory of Kidney and Electrolyte Metabolism, National Heart, Lung and Blood Institute, National Institutes of Health, Bethesda, Maryland 20205, USA*

Abstract. Traditional models of acid-base transport and intracellular pH (pH_i) regulation in the renal proximal tubule have been based on the existence of a Na^+/H^+ exchanger at the luminal membrane and a simple HCO_3^- conductance at the basolateral membrane. Our recent work, in which we used pH-sensitive microelectrodes or dyes to monitor pH_i in isolated renal tubules perfused in the nominal absence of HCO_3^-, has demonstrated the existence of a novel mechanism of acid extrusion in amphibian and mammalian proximal tubule cells. The salamander proximal tubule, for example, possesses an electroneutral Na^+ monocarboxylate (Na^+-X^-) co-transporter, but only at the luminal membrane. It also possesses an electroneutral H^+-X^- co-transporter, but only at the blood side or basolateral membrane. In the presence of lactate, the luminal Na^+-lactate co-transporter mediates a net influx of lactate, driven by the Na^+ gradient. The cell-to-blood lactate gradient, in turn, drives the coupled efflux of H^+ and lactate across the basolateral membrane. The net effect is the reabsorption of lactate, the luminal uptake of Na^+ and the basolateral extrusion of H^+. Acid extrusion mediated by this monocarboxylate system in the salamander is comparable in magnitude to that mediated by the Na^+/H^+ exchanger. In the S3 segment of the rabbit proximal tubule, a similar monocarboxylate system (studied with acetate instead of lactate) extrudes acetate at twice the rate of the Na^+/H^+ exchanger. Thus, monocarboxylate transport, at least in the nominal absence of HCO_3^-, can have a major impact on pH_i regulation.

1988 Proton passage across cell membranes. Wiley, Chichester (Ciba Foundation Symposium 139) p 91–105

Traditional models of acid-base transport in the renal proximal tubule have been based on the existence of a Na^+/H^+ exchanger at the luminal membrane and a simple HCO_3^- conductance at the basolateral, i.e. blood-side, membrane. Recent work, however, indicates that proximal tubule acid-base transport is far more complex. For example, the pathway for basolateral HCO_3^-

transport appears not to be a simple HCO_3^- channel, but rather an electrogenic Na^+-HCO_3^- co-transporter (Boron & Boulpaep 1983b, Yoshitomi et al 1985, Soleimani et al 1987). A second example is formate transport in the proximal tubule. Work with both membrane vesicles and intact tubules indicates that luminal formate/Cl^- exchange, in combination with the non-ionic diffusion of formic acid, results in the net uptake of HCl across the luminal membrane of the proximal tubule (Karniski & Aronson 1985, Schild et al 1987, Alpern 1987).

Recent work in our laboratory has provided evidence for yet another new class of proximal tubule acid-base transport, one involving two separate pathways for the movement of monocarboxylates. Together, these two monocarboxylate pathways can cause a profound increase in intracellular pH (pH_i), and presumably also mediate the reabsorption of the monocarboxylate across the epithelium. In this paper, we describe the initial work performed with microelectrodes on the proximal tubule of the tiger salamander, *Ambystoma tigrinum*, that led to the description of the effects of monocarboxylate transport on pH_i in renal tubules (Siebens & Boron 1987). We also briefly summarize experiments that have extended these observations to the S3 segment of the rabbit proximal tubule (Nakhoul & Boron 1988).

Lactate transport in the salamander proximal tubule

Lactate transport across the basolateral membrane

One might expect that for a monocarboxylate (X^-) in equilibrium with its conjugate weak acid (HX) the likeliest mechanism of transport across a cell membrane might be the diffusion of the uncharged HX. If this were true, exposing the basolateral, i.e. blood-side, membrane of a salamander proximal tubule to lactate would cause a rapid and reversible acidification. As illustrated in Fig. 1 (Siebens & Boron 1987), this is indeed the case. In this and the other salamander proximal tubule experiments illustrated here, one cell of an isolated perfused tubule was penetrated with a liquid-membrane pH-sensitive microelectrode (Ammann et al 1981), and another cell was penetrated with a voltage microelectrode filled with 3 M KCl. The solutions were nominally HCO_3^- free and at pH 7.5 throughout. As a preliminary step, lactate (L-lactate unless stated otherwise) was removed bilaterally, i.e. from both the basolateral or bath solution and from the lumen, causing a sustained fall in pH_i (*ab* in Fig. 1). The subsequent application of 3.6 mM lactate to the bath caused pH_i to fall rapidly (segment *bc*) by an average of 0.08 units. If the non-HCO_3^- buffering power is 36 mM (Boron & Boulpaep 1983a), then equilibration of lactic acid across the basolateral membrane should have caused pH_i to fall by 0.07 units. The initial rates of the pH_i decline (in pH units/min) are given by the numbers adjacent to the broken lines.

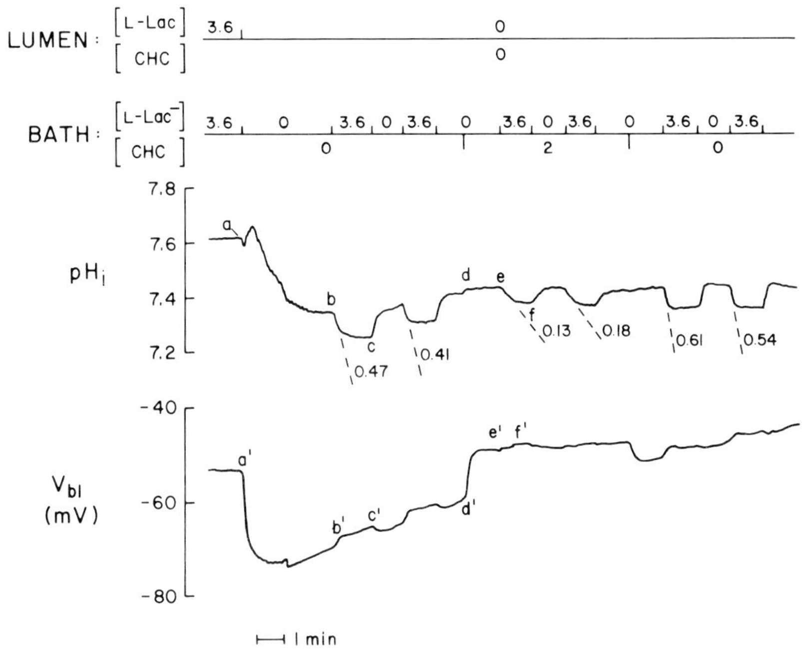

FIG. 1. Effects of basolateral lactate (Lac^-) on intracellular pH (pH_i) and basolateral membrane potential (V_{bl}), in both the presence and absence of basolateral α-cyano-4-hydroxycinnamate (CHC). Broken lines indicate the slopes of the pH_i versus time record on basolateral addition of lactate, and the associated numbers indicate the initial rates in pH units per min. From Siebens and Boron (1987); reproduced with permission of the Rockefeller University Press.

Although the above result is consistent with the simple non-ionic diffusion of lactic acid, two lines of evidence indicate that basolateral lactate transport in this preparation is carrier mediated. First, the initial rate of intracellular acidification exhibited saturation kinetics with respect to basolateral lactate concentration, the apparent K_m being 11.5 mM (Siebens & Boron 1987). Second, the initial rate of the lactate-induced acidification was reversibly inhibited by α-cyano-4-hydroxycinnamate (CHC), a competitive inhibitor of H^+-lactate co-transport (or lactate/base exchange) in other systems (Halestrap & Denton 1974, Spencer & Lehninger 1976). This effect also is illustrated by the experiment of Fig. 1. Application of 2 mM CHC to the bath (at *d*) caused a substantial reduction in the initial rate of acidification induced by the basolateral application of 3.6 mM lactate (compare *ef* to *bc*). In six similar experiments, CHC reduced the initial rate of pH_i recovery by 75%.

It is evident from the V_{bl} record of Fig. 1 that the application of basolateral lactate is accompanied by only rather small depolarizations (points *b′* and *e′*). This observation suggests that basolateral lactate transport is electroneutral.

In other experiments (not shown), we found that D-lactate and pyruvate are transported across the basolateral membrane about as well as L-lactate (Siebens & Boron 1987). Thus, lactate transport across the basolateral membrane of the salamander proximal tubule is mediated by a relatively nonspecific H^+-monocarboxylate co-transporter. Similar transporters have been identified in Ehrlich ascites tumour cells (Spencer & Lehninger 1976), erythrocytes (Deuticke 1982) and cardiac Purkinje fibres (de Hemptinne et al 1983).

Lactate transport across the luminal membrane

The effect of adding lactate to the luminal side of the proximal tubule cell contrasts sharply with that of adding lactate to the basolateral side. Whereas 3.6 mM lactate elicited a rapid acidification of nearly 0.1 pH units when added to the bath (see Fig. 1), it caused a rapid alkalinization of nearly 0.4 pH units when added to the lumen. This effect is illustrated by segment *ab* of Fig. 2. However, the simple influx of lactate cannot explain the pH_i increase, since the pK governing the equilibrium Hlac $\rightleftharpoons H^+ + lac^-$ is ~ 3.7, substantially below the initial pH_i of $\sim$7.4. Given an intrinsic intracellular buffering power of 36 mM (Boron & Boulpaep 1983a), and the fact that less than 0.2% of the entering lactate is expected to combine with H^+ to form lactic acid, approximately 72 *molar* lactate would have to enter the cell over a period of less than two minutes in order to account for the observed pH_i increase.

On the other hand, the intracellular alkalinization induced by luminal lactate could be easily accounted for by a biochemical or transport process that would consume the entering lactate along with H^+. An example of such a biochemical reaction is the formation of glucose from lactate and H^+, which would be expected to alkalinize the cell. However, 1 mM 3-mercaptopicolinate, an inhibitor of gluconeogenesis, had little effect on the pH_i increase induced by luminal lactate (Siebens & Boron 1987). Thus, gluconeogenesis does not appear to play a significant role in the alkalinization.

Another process that could consume H^+ and incoming lactate is the basolateral efflux of these solutes, mediated by the H^+-lactate co-transporter described above. If the basolateral efflux of lactic acid were indeed responsible for the pH_i increase elicited by the addition of luminal lactate, then one would expect the luminal lactate-induced alkalinization to be inhibited by basolateral CHC. As illustrated by segment *de* of Fig. 2, basolateral CHC greatly slows the initial rate of alkalinization elicited by luminal lactate. The mean inhibition of the initial rate was 69% in five experiments. Not shown is an example of the experiments in which we found that CHC has no effect when added to the lumen (Siebens & Boron 1987), suggesting that an H^+-lactate co-transporter is not present at this side of the cell. In summary, the

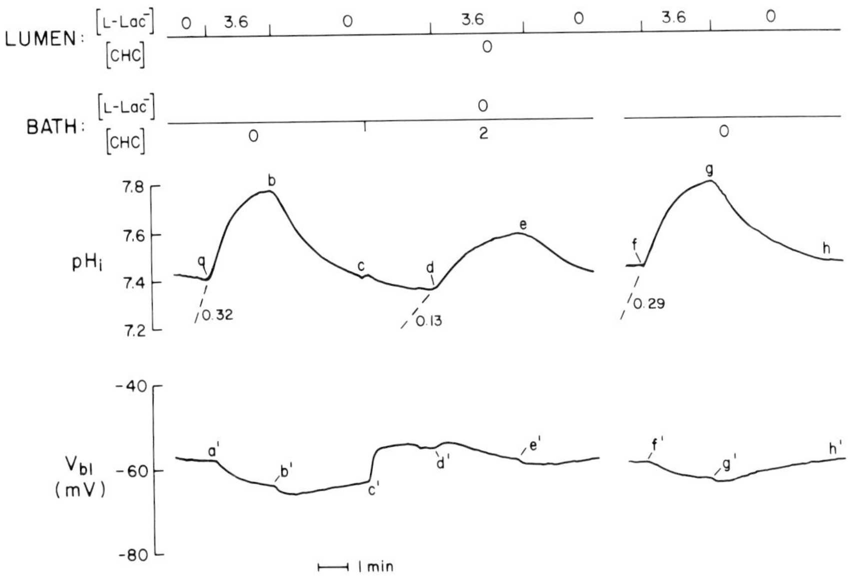

FIG. 2. Effects of luminal lactate on pH_i and V_{bl}, in both the presence and absence of basolateral CHC. Abbreviations as for Fig. 1. From Siebens and Boron (1987); reproduced with permission of the Rockefeller University Press.

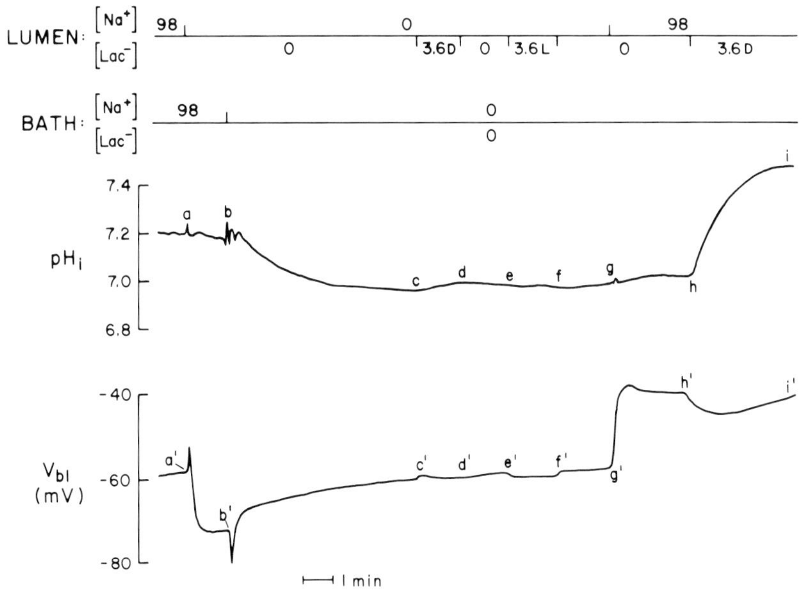

FIG. 3. Effect of Na^+ removal on the pH_i and V_{bl} changes induced by the luminal addition of D- or L-lactate. Abbreviations as for Fig. 1. From Siebens and Boron (1987); reproduced with permission of the Rockefeller University Press.

results suggest that the addition of luminal lactate produces a pH_i increase as lactate that enters across the luminal membrane exits with H^+ across the basolateral membrane.

If lactate crossed the luminal membrane as a monovalent anion, then one would expect the lactate-induced alkalinization to be accompanied by an immediate and substantial hyperpolarization. On the contrary, the addition of lactate to the lumen produced only a small and slow hyperpolarization (*a′b′* in Fig. 2), suggesting that luminal lactate transport is mediated by an electroneutral transport mechanism. One might imagine that such a transporter could be either an electroneutral lactate/anion exchanger or a lactate-cation co-transporter. The experiment shown in Fig. 3 was performed to test the latter possibility, specifically, that the luminal entry of lactate is tightly coupled to the entry of Na^+. Na^+ was removed first from the lumen (point *a*) and then from the bath (*b*). The latter caused a sustained fall in pH_i (*bc*). In the continued bilateral absence of Na^+, the addition of 3.6 mM D- or L-lactate had little if any effect on pH_i (*cd* and *ef*, respectively). Thus, the luminal lactate uptake mechanism is presumably an electroneutral Na^+-lactate co-transporter. The fact that the luminal application of lactate did not cause an acidification in the absence of luminal Na^+ confirms the hypothesis (see above) that there is no H^+-lactate co-transporter at the luminal membrane.

Model of acid-base transport in the salamander proximal tubule

Although Na^+-monocarboxylate co-transporters and H^+-monocarboxylate co-transporters had been described in other cells, the two transporters had previously not been identified in the same cell. Because the proximal tubule is an epithelium, there are, in principle, several combinations in which the Na^+-monocarboxylate and H^+-monocarboxylate co-transporters could have been incorporated into the proximal tubule cell, each of which would have its own unique combination of effects on pH_i and transepithelial transport. The combination of monocarboxylate co-transporters present in the salamander proximal tubule (see Fig. 4) is expected to have three major effects:

(i) *Intracellular alkalinization.* Because the bilateral removal of lactate causes a substantial acidification (see Fig. 1), we can conclude that the net effect of luminal and basolateral lactate transport is normally to raise pH_i. Such an alkalinization can be accounted for if the usual direction of Na^+-lactate co-transport across the luminal membrane is an influx, and if that of H^+-lactate co-transport across the basolateral membrane is normally an efflux. Note that although the application of luminal lactate initiates an alkalinization, it is the basolateral exit of H^+ and lactate that actually constitutes the acid extrusion step responsible for the increase in pH_i.

(ii) *Na^+ reabsorption.* The Na^+-lactate co-transporter is expected to con-

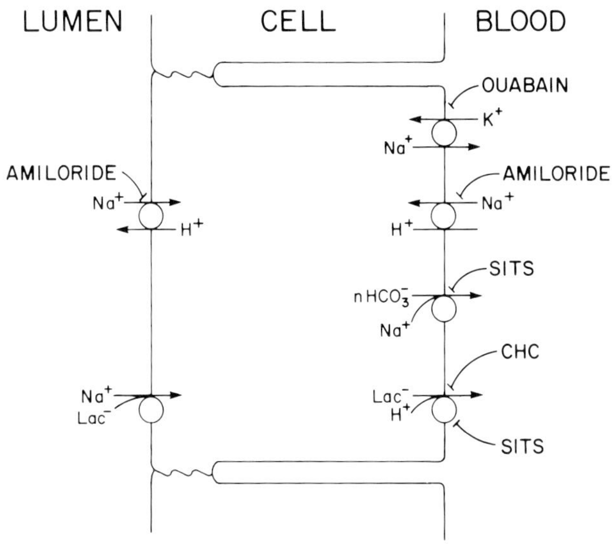

FIG. 4. Model of acid-base transport in the proximal tubule of the tiger salamander. From Siebens and Boron (1987); reproduced with permission of the Rockefeller University Press.

tribute to the luminal uptake of Na^+, and thus to the luminal step of transepithelial Na^+ reabsorption. (Of course, there are other mechanisms of luminal Na^+ uptake, such as Na^+-glucose co-transport, Na^+-amino acid co-transport, and uncoupled Na^+ transport.) The basolateral step of Na^+ reabsorption is mediated predominantly by the Na^+/K^+ pump, although the electrogenic Na^+-HCO_3^- co-transporter probably also makes a contribution. The contribution of monocarboxylate transport to transepithelial Na^+ reabsorption has been documented in mammalian proximal tubules (Schäfer & Andreoli 1976).

(iii) *Monocarboxylate reabsorption.* The pH_i data presented above and the model of Fig. 4 predict that normally the luminal Na^+-monocarboxylate co-transporter should be operating in the direction of monocarboxylate influx, whereas the basolateral H^+-monocarboxylate co-transporter should be operating in the direction of monocarboxylate efflux. The predicted net effect is the transepithelial reabsorption of monocarboxylate.

Acetate transport in the S3 segment of the rabbit proximal tubule

In more recent work, we studied the effect of acetate transport on pH_i in isolated perfused S3 segments of the rabbit proximal tubule (Nakhoul et al

1988, Nakhoul & Boron 1988). Tubule cells were loaded with a pH-sensitive dye, and pH_i calculated from the absorbance spectrum of intracellular dye. The latter was obtained by passing a beam of white light through the dyed tubule, focusing the transmitted light on a diffraction grating, and then focusing the resulting spectrum on a linear array of photodiodes (Chaillet & Boron 1985). All experiments were conducted in the nominal absence of HCO_3^- to minimize the possible contribution of HCO_3^- transport to the pH_i changes. The following is a brief summary of the results.

The first indication that monocarboxylate transport is important for pH_i regulation in the S3 segment came from experiments in which we examined the recovery of pH_i from acute intracellular acid loads. The latter were imposed either by a brief pulse of NH_4^+ (Boron & De Weer 1976) or the bilateral removal of Na^+. We found that, although the majority of the pH_i recovery depended upon the presence of Na^+ in the lumen, inhibiting Na^+/H^+ exchange with high doses of luminal amiloride reduced the rate of pH_i recovery by only about 30%. The solutions in these experiments, however, contained 10 mM acetate. The bilateral removal of acetate greatly diminished the rate of pH_i recovery from acid loads. Moreover, in the absence of acetate, amiloride blocked nearly 70% of the pH_i recovery. Thus, S3 cells apparently possess an amiloride-insensitive acid extrusion mechanism requiring luminal Na^+ and acetate (Nakhoul et al 1988).

Further experiments, in which pH_i was monitored with the aforementioned dye technique (Nakhoul & Boron 1988), demonstrated that S3 cells possess Na^+-independent mechanisms for the transport of acetic acid across both luminal and basolateral membranes. Thus, acetate transport in the rabbit S3 segment differs from lactate transport in the salamander in that this mammalian proximal tubule possesses a mechanism for the transport of the protonated form of the monocarboxylate at the luminal membrane in addition to the one at the basolateral membrane. Whether the movement of acetic acid in the rabbit S3 segment is carrier mediated or proceeds via non-ionic diffusion has not been established. In addition to the bilateral acetic acid transport mechanisms, the S3 cells also possess a Na^+-acetate co-transport system. As is the case for the salamander, this Na^+-monocarboxylate co-transporter is confined to the luminal membrane. Similar Na^+-monocarboxylate co-transport mechanisms have been previously identified in the luminal membrane of mammalian proximal tubules, on the basis of membrane vesicle experiments (see Wright 1985). In contrast to the salamander proximal tubule, the mammalian proximal tubule is believed to have an electrogenic Na^+-monocarboxylate co-transporter. Our data, obtained only with dyes sensitive to pH, do not address the issue of electrogenicity.

The consequences of acetate transport in the S3 segment of the rabbit proximal tubule are expected to be similar to those of the salamander proximal tubule. Cells in both possess a Na^+-monocarboxylate co-transporter at the luminal membrane and an acetic acid system at the basolateral membrane. Thus, acetate transport in the rabbit proximal tubule should produce an increase in pH_i, mediate the uptake of Na^+ across the luminal membrane, and cause the transepithelial reabsorption of acetate. Indeed, we found that the bilateral removal of acetate causes steady-state pH_i to fall by nearly 0.4 units. As noted above, others have found that acetate removal decreases Na^+ and volume reabsorption, and that acetate itself is reabsorbed (Schäfer & Andreoli 1976).

S3 cells differ from salamander proximal tubule cells in that the former possess an acetic acid transport mechanism at the luminal as well as the basolateral membrane. The presence of a luminal Na^+-acetate co-transporter and an acetic acid transport pathway is expected to give rise to acetate recycling, i.e. Na^+-acetate uptake plus acetic acid efflux, across the luminal membrane which would contribute to luminal acidification (Kahn & Aronson 1983). Indeed, experiments in which the pH of collected luminal fluid was measured in the absence of HCO_3^- (Geibel et al 1987) demonstrated that the bilateral addition of acetate does indeed increase luminal acidity. Finally, it should be noted that although our original observations with acetate were made using concentrations of 10 mM, the rate of pH_i increase induced by the luminal addition of acetate appears to be half maximal at acetate concentrations of about 0.2 mM. Similarly, the luminal acidification induced by 1 mM acetate is approximately equal to that caused by 10 mM acetate (Geibel et al 1987). Thus, the results indicate that, at least in the nominal absence of HCO_3^-, acetate transport has important effects both on pH_i and acid secretion.

Acknowledgements

This work was supported by grants from the National Institutes of Health (DK 30344 and DK 17433).

REFERENCES

Alpern RJ 1987 Apical membrane Cl/base exchange in the rat proximal convoluted tubule (PCT). Kidney Int 31:428

Ammann D, Lanter F, Steiner RA, Schulthess P, Shijo Y, Simon W 1981 Neutral carrier based hydrogen ion selective microelectrode for extra- and intracellular studies. Anal Chem 53:2267–2269

Boron WF, Boulpaep EL 1983a Intracellular pH regulation in the renal proximal tubule of the salamander. Na-H exchange. J Gen Physiol 81:29–52

Boron WF, Boulpaep El 1983b Intracellular pH regulation in the renal proximal tubule of the salamander: basolateral HCO_3^- transport. J Gen Physiol 81:53–94

Boron WF, De Weer P 1976 Intracellular pH transients in squid giant axons caused by CO_2, NH_3 and metabolic inhibitors. J Gen Physiol 67:91–112

Chaillet JR, Boron WF 1985 Intracellular calibration of a pH-sensitive dye in isolated

perfused salamander proximal tubules. J Gen Physiol 86:765–794
de Hemptinne AR, Marrannes R, Vanheel B 1983 Influence of organic acids on intracellular pH. Am J Physiol 245:C178–C183
Deuticke B 1982 Monocarboxylate transport in erythocytes. J Membr Biol 70:89–103
Geibel J, Boron WF, Giebisch G 1987 Effects of acetate on luminal acidification in the S-3 segment of the rabbit proximal tubule. Proc Xth Int Congr Nephrol p 539
Halestrap AP, Denton RM 1974 Specific inhibition of pyruvate transport in rat liver mitochondria and human erythrocytes by *α*-cyano-4-hydroxycinnamate. Biochem J 138:313–316
Kahn A, Aronson PS 1983 Urate transport via anion exchange in dog renal microvillus membrane vesicles. Am J Physiol 244:F56–F63
Karniski LP, Aronson PS 1985 Chloride/formate exchange with formic acid recycling: a mechanism of active chloride transport across epithelial membranes. Proc Natl Acad Sci USA 82:6362–6375
Nakhoul NL, Boron WF 1988 Acetate transport in the S3 segment of the rabbit proximal tubule and its effect on intracellular pH. J Gen Physiol, in press
Nakhoul NL, Lopes AG, Chaillet JR, Boron WF 1988 Intracellular pH regulation in the S3 segment of the rabbit proximal tubule in HCO_3^--free solutions. J Gen Physiol, in press
Schäfer JA, Andreoli TE 1976 Anion transport processes in the mammalian superficial proximal straight tubule. J Clin Invest 58:500–513
Schild L, Giebisch G, Karniski P, Aronson PS 1987 Effect of formate on volume reabsorption in the rabbit proximal tubule. J Clin Invest 79:32–38
Siebens AW, Boron WF 1987 Effect of electroneutral luminal and basolateral lactate transport on intracellular pH in salamander proximal tubules. J Gen Physiol 90:799–831
Soleimani M, Grassl SM, Aronson PS 1987 Stoichiometry of Na^+-HCO_3^- cotransport in basolateral membrane vesicles isolated from rabbit renal cortex. J Clin Invest 79:1276–1280
Spencer TL, Lehninger AL 1976 L-Lactate transport in Ehrlich ascites-tumour cells. Biochem J 154:405–414
Wright EM 1985 Transport of carboxylic acids by renal membrane vesicles. Annu Rev Physiol 47:127–141
Yoshitomi K, Burckhardt B-Ch, Fromter E 1985 Rheogenic sodium-bicarbonate cotransport in the peritubular cell membrane of rat renal proximal tubule. Pfluegers Arch 405:360–366

DISCUSSION

Thomas: How much acetate is normally present in the rabbit ultrafiltrate?

Boron: Certainly not 10 mM. We did a series of experiments in which we removed acetate from both the lumen and the bath, causing pH to fall to around 7.2. We then restored acetate just to the lumen, at concentrations ranging from 0.2 mM to 10 mM. We found that the rate of alkalinization is nearly maximal at about 0.5 mM acetate. A study by Siebens and myself (1987) on the salamander proximal tubule, as well as work by others, suggests that the affinity of the Na^+-monocarboxylate co-transporter for acetate is lower than that for lactate; the concentration of lactate normally present is 1–2 mM. I propose that because this transport system works with low concentrations of acetate, it is likely to transport lactate under physiological conditions.

Thomas: So would you say that in physiological concentrations of acetate and lactate, this system would be transporting acid equivalents of the same order as the Na^+/H^+ exchanger?

Boron: In the nominal absence of HCO_3^-, acid extrusion would be about twice as fast as that mediated by the Na^+/H^+ exchanger.

Oberleithner: But under physiological conditions, there will be HCO_3^- present in kidney perfusate. What percentage of H^+ would be extruded by this mechanism in those conditions?

Boron: I can't answer that question precisely. We conducted these experiments in the nominal absence of HCO_3^- in order to make the interpretation more straightforward. Our aim was to identify and characterize the transporters; this has been accomplished. The next step is to understand how they function under physiological conditions, i.e. in the presence of HCO_3^-. Our final goal is to derive a mathematical model which can take all our reductionist data and synthesize them into a model which can explain life. We have only done a few experiments in the presence of HCO_3^-, finding that removal of acetate causes a slower and smaller pH_i decrease than in the absence of HCO_3^-. The fact that the changes are slower probably reflects the substantial increase in buffering power provided by CO_2/HCO_3^-. The pH_i changes are probably smaller because the increase in pH_i caused by acetate functionally competes with a HCO_3^--dependent acid extrusion process present in these proximal tubule cells. Thus, from the steady-state pH changes, it appears that acetate (and presumably all monocarboxylates) are somewhat less important in the presence of HCO_3^- than in its absence.

Incidently, from experiments on transepithelial HCO_3^- reabsorption (in collaboration with John Geibel and Gerhard Giebisch), we have found that acetate inhibits HCO_3^- reabsorption. So there may be a trade-off between the cells' reabsorption of monocarboxylate and HCO_3^-.

Steels: Are all these monocarboxylate transport systems electroneutral?

Boron: In the salamander proximal tubule, the luminal Na^+ lactate system and the basolateral H^+ system appear to be electroneutral. The Na^+-monocarboxylate co-transporter in mammalian cells has been studied by others in brush border membrane vesicles from either the gut or the kidney. Those experiments suggest that the Na^+-monocarboxylate co-transporter is electrogenic. We have not measured membrane voltage in these experiments, and therefore cannot address the issue of electrogenicity.

Al-Awqati: Is it the same carrier for all the monocarboxylates? Do they compete with each other or are there multiple carriers?

Boron: Evidence from experiments in membrane vesicles conducted by other people (Wright 1985) indicate that there is a competition amongst a wide range of monocarboxylates: D- and L-lactate, pyruvate, butyrate, acetate and so forth. In the salamander proximal tubule, we have seen that the transporters take D- and L-lactate, pyruvate and acetate, but we haven't studied competition *per se*. The exceptions are urate and formate, which have their own transport system, which is a Cl^-/urate or formate exchanger (Kahn & Aronson 1983). This Na^+-monocarboxylate transport system seems to be rather non-specific.

Al-Awqati: How much do you think that metabolism of the acetate or lactate coming in from the apical membrane contributes to alkalinization?

Boron: In the salamander proximal tubule, blocking gluconeogenesis with 3-mercaptopicolinate had a minimal effect on the alkalinization induced by lactate addition. Furthermore, the alkalinization was reduced by about two thirds by inhibiting basolateral H^+-lactate co-transport with CHC. Thus, a metabolic component is likely to be minor in the salamander. As far as the rabbit proximal tubule is concerned, our observation that adding acetate to only the lumen causes a large Na^+-dependent increase in pH_i, whereas adding acetate to only the bath causes a large Na^+-dependent pH_i decrease, argues that the majority of the pH_i changes are due to transport. However, we cannot rule out a small metabolic component.

Malnic: Can this transporter carry formate or carbonic acid?

Boron: The evidence obtained from others on these vesicles is that formate cannot be carried by the Na^+-monocarboxylate co-transporter. As far as HCO_3^- is concerned, we have added HCO_3^- to only the lumen of rabbit S3 proximal tubule and seen a progressive decrease in pH_i, rather than the increase expected if HCO_3^- was carried on the Na^+-carboxylate transporter.

de Hemptinne: Do you know of a blocking agent which is able to block that Na^+-monocarboxylate transporter?

Boron: No, I don't. In the salamander proximal tubule, the basolateral H^+-lactate co-transporter is inhibited by CHC. Indeed, basolateral CHC inhibits the alkalinization produced by adding lactate to the lumen. Adding CHC to the luminal membrane had no effect.

de Hemptinne: You showed a reversal of that transport, with Na^+ leaving the cell. Is that thermodynamically possible with the large concentration of Na^+ outside the cell? Could it not be that lactate or acetate is leaving with K^+? On one side it enters with Na^+ and on the other side it leaves with K^+.

Boron: We have no evidence that would rule that out. We haven't tried to deplete the cells of K^+, for example. Nevertheless, I think it highly unlikely that the co-transporter is asymmetrical with respect to the cation. We have observed that the progressive decline in pH_i elicited by addition of acetate to the bath is blocked by bilateral Na^+ removal. This implies that Na^+ and acetate must exit together across the luminal membrane. Although the cell-to-lumen Na^+ gradient is unfavourable for Na^+-acetate exit, keep in mind that there is no acetate in the lumen and so theoretically at least there is an infinite gradient for acetate exit.

Harvey: You mentioned that proton extrusion could possibly occur also via the electrogenic proton pump. If there are membrane potential changes associated with removing Na^+, or adding amiloride or even acetate, this would affect the proton extrusion capacity of the pump because it's electrogenic. Another point is that lactate is an energy source for this proton pump in epithelia.

Boron: Lactate is probably an energy source for the proximal tubule cell in general. I don't know whether changing extracellular monocarboxylate concentrations produces changes in the internal ATP concentration large enough to affect a proton pump. Although the proton pump is electrogenic, I know of no direct evidence that modest changes in membrane potential produce major effects on the proton pump rate. I should emphasize that we don't even know that it is a proton pump. Unfortunately, there is no magic bullet analogous to ouabain that will knock out the proton pump. DCCD inhibits the proton pump in vesicle preparations, but when we applied DCCD to the proximal tubule, it killed the cells before we could see an inhibition of the proposed pump. It is inhibited about 50% by the combination of cyanide and iodoacetate, suggesting that there may be an ATP dependence.

Qais Al-Awqati and George Schwartz have suggested that in response to intracellular acidosis proton pumps are inserted into the luminal membrane of the cortical collecting tubule. Such an insertion could be happening here. I would like to emphasize that the Na^+-independent mechanism may not be operating in the normal steady-state; it is possible that this mechanism is only turned on when the cell is artificially acidified and held at a low pH for a long time.

Grinstein: Ira Kurtz has good evidence for proton pump activity in the S3 segment of the proximal tubule.

Boron: He has repeated our experiments with similar results, except that he was able to achieve a higher degree of inhibition with cyanide and iodoacetate than we were.

Grinstein: Does Kurtz have data with DCCD or anything like that?

Boron: He used DCCD at 10^{-3} M and has shown inhibition. But when we use such concentrations of DCCD for even 2–3 minutes, the tubule cells look so granular and deteriorated that we do not regard them as reliable.

Machen: Most of your pH values in the presence of lactate are very high. Do you think that under these conditions normal absorption of Na^+ across the apical membrane is going to be mostly by the Na^+-acetate or Na^+-lactate co-transporter and not by Na^+/H^+ exchange?

Boron: Schaefer and Andreoli (1976) have shown that the presence of acetate substantially increases Na^+ reabsorption in rabbit proximal tubules, even in the presence of HCO_3^-.

The reason why pH_i is so alkaline in our experiments is that these cells are incubated in nominally HCO_3^- free solution, so the acetate transport mechanism was able to drive pH to a very alkaline level. If we add CO_2, the pH_i falls by about 0.15–0.2 pH units to the normal steady-state value of about 7.3.

Finally, we have done experiments in collaboration with John Geibel and Gerhard Giebisch, in which we measured acid secretion into the lumen of the S3 proximal tubule in the nominal absence of HCO_3^-. We found that adding 1 mM acetate causes a decrease in luminal pH, consistent with our pH_i data, which predict that acetate recycling across the luminal membrane contributes to luminal acidification. I would assume that it also makes a major contribution to Na^+ reabsorption.

Machen: If you add amiloride to the luminal side in those HCO_3^--free conditions, does anything happen to the pH of the cell?

Boron: I don't think we have ever added amiloride to proximal tubule cells that were not acid loaded.

Grinstein: Does it make sense for a cell to reabsorb a major fraction of the Na^+ when there is 1 mM lactate and 140 mM Na^+?

Boron: You are probably worried that the tubule cells are not going to have enough lactate to reabsorb more than 1 mM Na^+. This would be a problem if the only luminal pathway for acetate were the Na^+-monocarboxylate co-transporter. However, our data indicate that acetic acid leaves the cell as Na^+-acetate enters. Thus, there is a recycling of acetate across the luminal membrane, such that a little bit of monocarboxylate can buy you a lot in terms of Na^+ reabsorption and luminal acid secretion.

Sabolic: Would you expect that this system would be useful in situations like diabetes, where the concentration of acetate is increased?

Boron: That's a possibility, not only for acetate but also for other monocarboxylates. It would be interesting to know whether the number of transporters increases when plasma levels of monocarboxylates rise.

Sabolic: This means that the Na^+-carboxylate co-transporter could compete effectively for substrates with the Na^+/H^+ exchanger. However, it is known that in conditions where there is increased secretion, such as diabetes or animals treated with glucocorticoids, the activity of the renal Na^+/H^+ exchanger is stimulated. This suggests that the monocarboxylate transporter has a limited capacity for H^+ secretion.

Boron: This monocarboxylate transporter looks like an amiloride-insensitive Na^+/H^+ exchanger, except the Na^+ goes in across the luminal membrane and the H^+ can come out across either side. Functionally they do the same thing—they both make the cell alkaline and import Na^+ across the luminal membrane.

Vaughan-Jones: Does DIDS or SITS affect the Na^+-acetate transporter?

Boron: We have not examined this in the rabbit proximal tubule. In the amphibian proximal tubule, we have looked at the effect of SITS on the basolateral H^+-lactate co-transporter, which is very similar to the one Alex de Hemptinne has worked on. This amphibian co-transporter is about 50% inhibited by SITS, though more than 10 minutes is required for maximal inhibition. This is in agreement with the findings of other people (Deuticke 1982). I don't think we have examined the effect of SITS on the luminal Na^+-monocarboxylate co-transporter.

Vaughan-Jones: What concentration is that?

Boron: We have used 0.5 mM SITS, which is very high. On the other hand, the CHC works very quickly, although about 2 mM CHC inhibits the basolateral H^+-lactate co-transporter by only about 75% in the presence of 3.6 mM lactate.

Steels: Is this type of intracellular pH regulation of renal proximal tubule cells restricted to the S3 segment?

Boron: So far we have only done this experiment in the S3 segment. John Geibel has begun to repeat these experiments on cells from S1 and S2 segments.

References

Deuticke B 1982 Monocarboxylate transport in erythrocytes. J Membr Biol 70:89–103

Kahn A, Aronson PS 1983 Urate transport via anion exchange in dog renal microvillus membrane vesicles. Am J Physiol 244:F56–F63

Schaefer JA, Andreoli TE 1976 Anion transport processes in the mammalian superficial proximal straight tubule. J Clin Invest 58:500–513

Siebens AW, Boron WF 1987 Effect of electroneutral luminal and basolateral lactate transport on intracellular pH in salamander proximal tubules. J Gen Physiol 90:799–831

Wright EM 1985 Transport of carboxylic acids by renal membrane vesicles. Ann Rev Physiol 47:-141

Mechanisms of basolateral base transport in the renal proximal tubule

E. Frömter, B.C. Burckhardt, Y. Kondo

Zentrum der Physiologie, J.W. Goethe Universität, Theodor Stern Kai 7, 6000 Frankfurt 70, FRG

Abstract. Renal proximal tubules absorb HCO_3^- by secretion of H^+ into the tubular lumen. This paper focuses on the mechanisms of HCO_3^- exit across the basolateral cell membrane. The major exit pathway is rheogenic sodium bicarbonate co-transport. This system transports Na^+ and HCO_3^-, but not Cl^-, in obligatory coupling at a fixed overall stoichiometry of three HCO_3^- to one Na^+. The fact that HCO_3^- flux is reduced after inhibition of cytoplasmic and/or membrane-bound peritubular carbonic anhydrase indicates that HCO_3^- is not transported as such but is split during permeation into its buffer subspecies from which it is regenerated on the other side of the membrane. Since flow of OH^- or of H^+ (in opposite directions) can be excluded, it appears most likely that one HCO_3^- and one CO_3^{2-} move together with one Na^+. Besides carbonic anhydrase inhibitors, disulphonic stilbenes and harmaline are known to block the co-transporter. In addition to rheogenic $Na^+ (HCO_3^-)_3$ co-transport, Na^+-dependent and Na^+-independent electroneutral Cl^-/HCO_3^- exchange have been identified. The latter mechanisms are particularly important in S3 segments of proximal tubule where $Na^+ (HCO_3^-)_3$ co-transport is missing. Further mechanisms which operate in parallel, but at lower rates, are electroneutral SO_4^{2-}/HCO_3^- exchange and, in some species, lactate/HCO_3^- exchange. Moreover, there may be some uncoupled OH^- flux and it is reasonable to assume that OH^- (or H^+) flux is involved in the transport of dicarboxylic acids across the basolateral cell membrane.

1988 Proton passage across cell membranes. Wiley, Chichester (Ciba Foundation Symposium 139) p 106–121

Under normal acid/base conditions, a great amount of bicarbonate appears in the glomerular filtrate (4.5 mol/day in humans) which is almost fully reabsorbed (≈90%) in the first nephron segment, the proximal tubule. Absorption is a transcellular active transport process which involves HCO_3^- uptake against an electrochemical gradient across the brush border membrane into the cell and passive exit along a favourable electrochemical gradient across the basolateral cell membrane towards the interstitium.

Bicarbonate uptake into the cell is achieved by H^+ secretion into the tubular lumen which leads to splitting of luminal HCO_3^- into CO_2 and H_2O, followed by diffusion of CO_2 into the cell with subsequent regeneration of

HCO_3^- in the cytoplasm catalysed by carbonic anhydrase (EC 4.2.1.1) (CA). Two mechanisms of H^+ secretion have been identified: 1) secondary active, Na^+ gradient-driven Na^+/H^+ antiport (Murer et al 1976), and 2) primary active ATP-driven H^+-pumping (Kinne-Saffran et al 1982). The relative contribution of each mechanism to net H^+ secretion, which may be expected to vary in different segments of the proximal tubule, is still a matter of debate.

Bicarbonate exit from cell to interstitium proceeds via different transport pathways. The main pathway, at least in S1 and S2 segments, is a rheogenic sodium bicarbonate co-transport mechanism which is missing from S3 segments (Kondo & Frömter 1987). In addition, Na^+-dependent and Na^+-independent Cl^-/HCO_3^- exchange have been described, as well as other coupled or non-coupled base transport mechanisms. In this paper we briefly review the known properties of the individual transporters.

Na^+-HCO_3^- co-transport

The existence of this transport mechanism was initially recognized by its rheogenicity (Frömter & Sato 1976): sudden changes of peritubular HCO_3^- concentration induced transient potential changes across the peritubular cell membrane indicative of a high anionic conductance for bicarbonate (Fig. 1). The coupling of Na^+ to net anionic HCO_3^- transfer was noted only later in experiments on proximal tubule of *Ambystoma* (Boron & Boulpaep 1983). It was eventually also confirmed in mammalian proximal tubule (Alpern 1985, Yoshitomi et al 1985, Biagi & Sohtell 1986, Grassl & Aronson 1986, Akiba et al 1986, Sasaki et al 1987) and in other renal cells (Jentsch et al 1986, Lopes et al 1987, Wang et al 1987). Thus far the following features have been identified:

Ion flux coupling. The system performs coupled transport of at least two ions, Na^+ and HCO_3^-. Flow coupling is obligatory and proceeds at a fixed stoichiometry (see below). In the absence of Na^+, the electrical signals and the expected cellular pH changes in response to peritubular HCO_3^- concentration steps are strongly suppressed or abolished (Yoshitomi et al 1985). Evidence for the possible involvement of Cl^- which was previously discussed seems to have originated from the parallel existence of Cl^-/HCO_3^- exchange (see below). In rat proximal tubule, where omission of Cl^- strongly reduces the electrical response to peritubular HCO_3^- concentration steps (Burckhardt et al 1984b), we hypothesize that the substituent anions inhibit CA (E. Grell, personal communication 1988).

Coupling ratio. The rheogenic nature of coupled Na^+ and HCO_3^- flow implies a coupling ratio (n) of HCO_3^- to Na^+ greater than unity. Assuming stoichiometric numbers to be integers, the lowest possible estimate would be

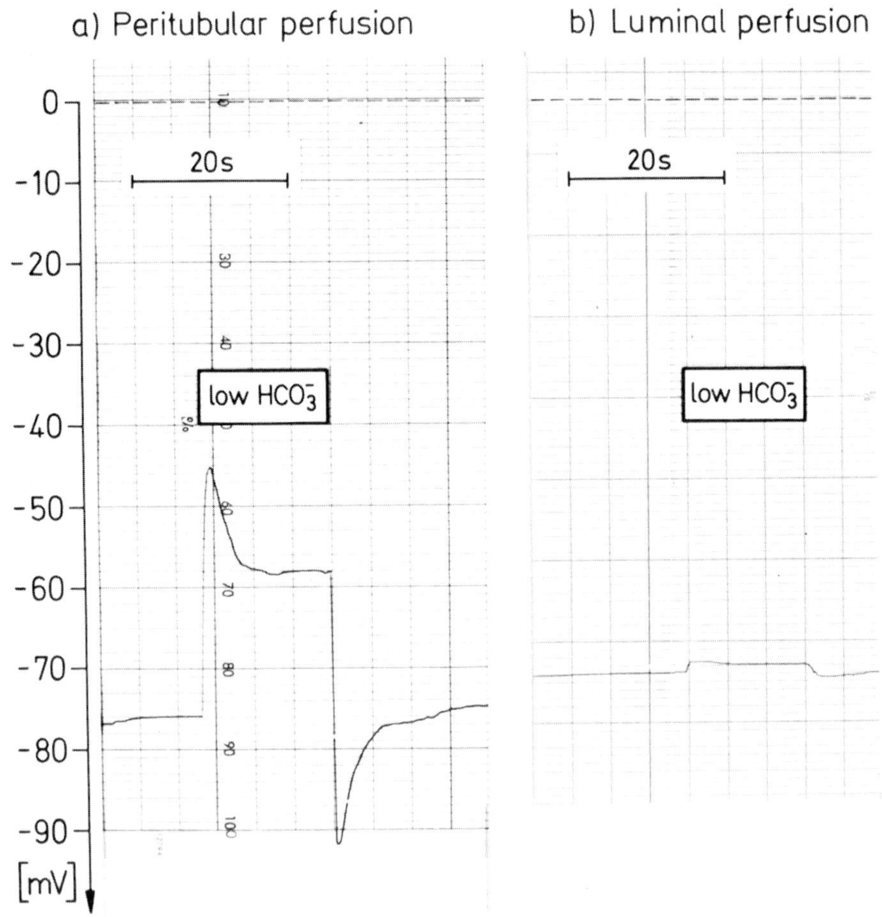

FIG. 1. Trace record of peritubular cell membrane potential (Pd) in response to peritubular capillary perfusion or luminal perfusion with low HCO_3^- solution. The experiment was performed on rat kidney proximal tubule *in vivo*. Where indicated the HCO_3^- concentration was lowered from 30 to 3 mM. Note electrical response to peritubular perfusion indicating conductive HCO_3^- transport, and absence of this transport mechanism in the brushborder membrane (from Burckhardt et al 1984a).

n = 2. This value has also been suggested from kinetic data (Hill plot analysis) of flux measurements on membrane vesicles (Akiba et al 1986) and cultured renal BSC-1 cell monolayers (Jentsch et al 1986). On the other hand, using ion-selective intracellular microelectrodes to measure the initial rate of change of cellular Na^+ and HCO_3^- (or H^+) concentration after sudden changes in peritubular HCO_3^- concentration, Yoshitomi et al (1985) arrived at a coupling ratio of three HCO_3^- to one Na^+. This value was confirmed by quantitatively comparing the peak depolarizations in response to sudden

TABLE 1 Electrochemical driving force of Na^+ and HCO_3^- across the basolateral cell membrane of rat proximal tubule under free flow

	Ion concentrations		E_i	*Pd*	E_i + *Pd*
	c_i' *(cell)* *mmol/l*	c_i'' *(interstitium)* *mmol/l*	*mV*	*mV*	*mV*
Na^+	17.5±2.8 (n = 20)	144	−56.3	−73.5	−129.6
HCO_3^-	13.4±2.7 (n = 19)	28	+19.5	−73.5	−53.8

E_i is the electrical equivalent of the chemical potential difference (Pd) of ion i calculated as:

$$E_i = \frac{RT}{z_i F} \ln \frac{c_i'}{c_i''}$$

where R, T, z_i, F are gas constant, absolute temperature, valency and Faraday constant, respectively. Note that the coupling ratio of n = 2 yields $| E_{Na} + Pd | > | E_{HCO_3^-} + Pd |$ and that n = 3 is required to achieve net HCO_3^- efflux (from Yoshitomi & Frömter 1985).

peritubular tenfold reduction of either Na^+ or HCO_3^- concentration. Moreover, Soleimani et al (1987) recently obtained the same coupling ratio in experiments on basolateral membrane vesicles in which the magnitude of HCO_3^- gradients required to counterbalance Na^+ gradients in abolishing net $^{22}Na^+$ flow was assessed. A stoichiometric ratio of three agrees also with estimates of the electrochemical driving forces for Na^+ and HCO_3^- across the peritubular cell membrane of rat renal proximal tubule *in vivo* (Table 1). From those data, one can deduce that the Na^+ gradient would pull HCO_3^- back into the cell, if the coupling ratio was only n = 2, and that a ratio of n = 3 or greater is required to allow net HCO_3^- efflux.

Transported buffer species. Because HCO_3^- is involved in the following chemical net reactions:

$$HCO_3^- + H^+ \rightleftharpoons H_2CO_3 \rightleftharpoons H_2O + CO_2$$
$$HCO_3^- \rightleftharpoons OH^- + CO_2$$
$$HCO_3^- \rightleftharpoons H^+ + CO_3^{2-} \text{ and } CO_3^{2-} + Na^+ \rightleftharpoons NaCO_3^-$$

any observed net flow of HCO_3^- does not necessarily mean that HCO_3^- ions as such move across the membrane. Instead they may be transformed on one side of the membrane into one or more of the other buffer components, which then move across the membrane and from which bicarbonate is regenerated on the other side. The first evidence in favour of such transformation was obtained in 1976, when Frömter and Sato observed that the classical CA inhibitor, acetazolamide, strongly reduced the electrical effects of bicarbonate substitution on the peritubular cell membrane potential. This indicated that CA played a role in peritubular bicarbonate transport and hence that bicarbonate was most likely transformed during permeation.

Generally speaking, the following buffer species other than HCO_3^- might be transported: 1) OH^- or (in opposite direction) H^+ with separate diffusion of CO_2, 2) $NaCO_3^-$, and/or 3) CO_3^{2-}.

Initially, we favoured the first possibility, OH^- or H^+ permeation, since it readily explained the ability of highly lipid-soluble weak acids to substitute for HCO_3^- in tubular buffer transport (Ullrich et al 1975) and in the electrical effects at the peritubular cell membrane (Burckhardt et al 1984a). This possibility, however, was excluded recently, because the electrical effects of pH changes in 'HCO_3^--free' solutions could not be inhibited by SITS*, indicating that any component of OH^- permeation which might be present did not utilize the sodium bicarbonate co-transporter (Burckhardt & Frömter 1987, Krapf et al 1987).

The second possibility, transport as the ion pair $NaCO_3^-$, was suggested from a series of kinetic measurements on BSC-1 renal epithelial cell cultures (Jentsch et al 1986, see below).

The third possibility is suggested by recent observations of $^{22}Na^+$ influx into basolateral membrane vesicles (Soleimani & Aronson 1988). At constant HCO_3^- concentration, $^{22}Na^+$ uptake was stimulated by CO_3^{2-} and by SO_3^{2-} in a way suggesting competition at a specific binding site. Furthermore, transport as $NaCO_3^-$ appeared unlikely because harmaline, a well known inhibitor of various Na^+ transport sites, also inhibited sodium bicarbonate co-transport.

Taking all observations together, the most probable model is currently one in which three species are transported together: HCO_3^-, CO_3^{2-} and Na^+ (see Fig 2). Note that this model agrees with the overall stoichiometry of three HCO_3^- plus one Na^+. It also explains the involvement of CA, which must be present on both membrane surfaces. Using a dextrane-bound CA inhibitor, we have recently demonstrated that CA activity is present not only in the cytoplasm but also on the outer surface of the peritubular cell membrane (Burckhardt et al 1985).

Transport kinetics. At present, little is known about how the transport rate depends on ambient substrate concentrations. Measuring initial $^{22}Na^+$ influx into basolateral membrane vesicles under maximal gradient conditions, Akiba et al (1986) and Grassl & Aronson (1986) observed typical saturation kinetics of the transport rate as a function of external Na^+ concentration, with an apparent K_m value of approximately 10 mM, but Akiba et al (1986) found that the dependence on external HCO_3^- concentration was more complex. This was taken as evidence in favour of more than one anion binding site on

* SITS: 4-acetamido-4′-isothiocyano-2,2′disulphonic stilbene. It inhibits rheogenic Na^+-HCO_3^- co-transport in the millimolar concentration range (Boron & Boulpaep 1983, Burckhardt et al 1984b)

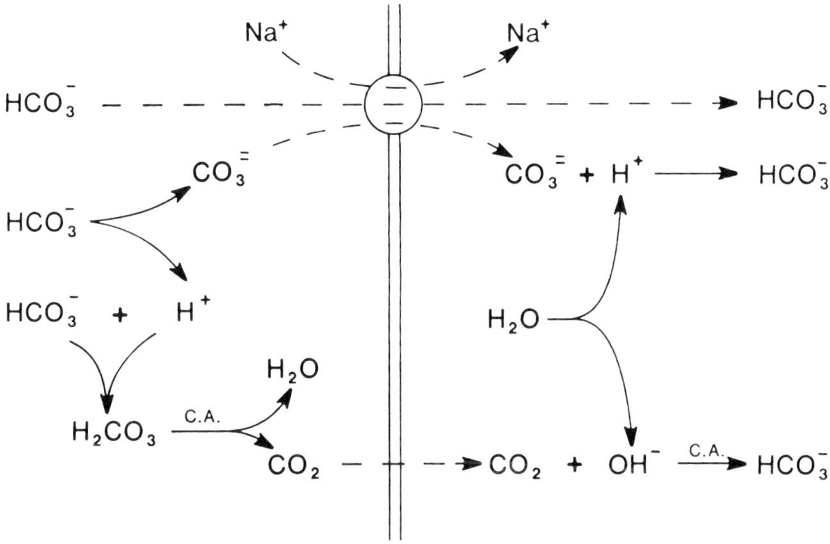

FIG. 2. Suggested mechanism for Na^+ $(HCO_3^-)_3$ co-transport. Note involvement of carbonic anhydrase (C.A.) on either surface of the membrane.

the co-transporter. In contrast to this, working with Na^+-loaded BSC-1 cells, Jentsch et al (1986) observed that the SITS-inhibitable $^{22}Na^+$ uptake obeyed simple saturation kinetics, as a function of both Na^+ and HCO_3^- concentration, with apparent K_m values for HCO_3^- between nine and 36 mM (increasing with decreasing ipsilateral Na^+ concentration), and apparent K_m values for Na^+ between 14 and 73 mM (increasing with decreasing ipsilateral HCO_3^- concentration). Plotting the flux data as a function of ambient $NaCO_3^-$ concentration, Jentsch et al noticed that all data could be represented by a single saturation curve, with a single K_m value for $NaCO_3^-$ of approximately 15 μM. From this they favoured a model in which co-transport proceeds as an ion pair. A possible explanation for the disparate conclusions is that the different preparations contained different co-transporters or that experiments with concentrations below 5 mM were not included in the latter study.

Cl^-/HCO_3^- exchange

The first evidence for Cl^-/HCO_3^- exchange in the basolateral cell membrane was obtained by Edelman et al (1981) in *Necturus* proximal tubule. Working with the same preparation, Guggino et al (1983) reported that this transport system was coupled to Na^+ in the sense of $NaHCO_3/Cl$ exchange. In mammalia, the first evidence for a small component of Cl^-/HCO_3^- exchange was obtained in studies on basolateral membrane vesicles (Grassl & Aronson

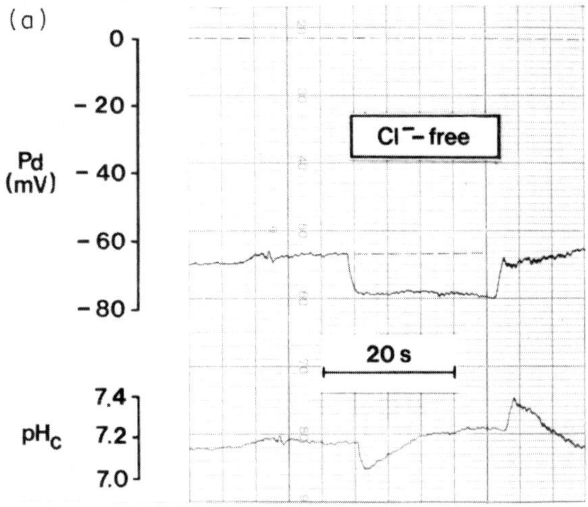

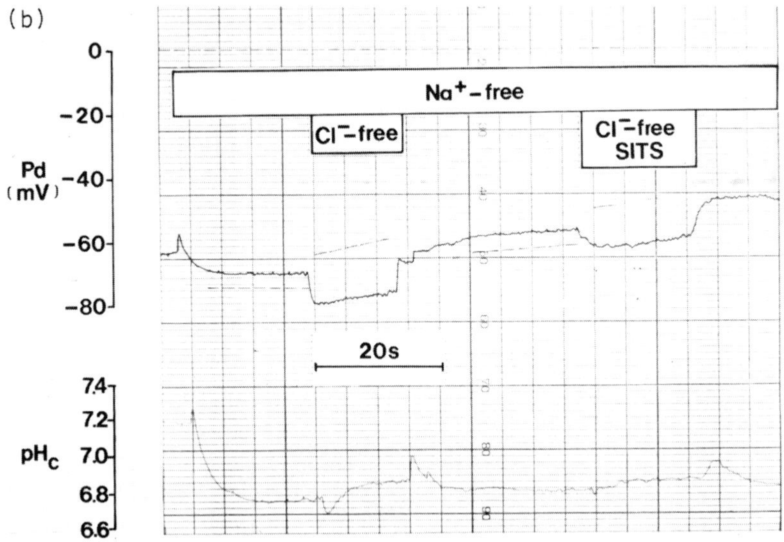

FIG. 3. Trace record of Pd and cellular pH (pH_c) of rat proximal tubular cell during peritubular capillary perfusion with Cl^--free Ringer solution. (a) Control conditions, (b) Na^+-free solutions. Dashed line indicates correction for liquid junction potential. Transient changes in pH_c trace are considered as capacitive coupling artifacts of the double-barrelled pH microelectrode. Note cell alkalinization, indicative of Cl^-/HCO_3^- exchange; partial persistence of exchange in Na^+-free solutions; and complete inhibition by SITS (1 mM).

TABLE 2 Effect of SITS on cell Pd and cell pH during peritubular replacement of chloride by gluconate in Na^+ Ringer and Na^+-free solution

Control perfusion	*Pd mV*	pH_c	*Test perfusion*	*Pd mV*	pH_c	*n*	*P*
Cl^-	−67.2 ±1.2	7.16 ±0.03	glc^-	−69.5 ±1.6	7.22 ±0.03	9	0.001
Na^+-free Cl^-	−57.7 ±5.1	6.85 ±0.03	Na^+-free glc^-	−55.9 ±5.1	6.90 ±0.03	4	0.05
Na^+-free Cl^-	−57.7 ±5.1	6.85 ±0.03	Na^+-free glc^-, SITS	−52.2 ±3.8	6.84 ±0.04	5	NS

Mean values ± SE. n, number of observations: P, probability of error for assumed difference of pH_c values, calculated from paired observations; NS, not significant; glc^-, gluconate. SITS 1 mmol/l.

1986, Grassl et al 1987). Our initial attempt to identify this transport system in rat proximal tubule *in vivo* failed (Burckhardt et al 1984b), because our Cl^--sensitive intracellular microelectrodes (based on Corning No. 477913 resin) were not selective enough to recognize an increase in cellular Cl^- concentration in response to a sudden reduction of peritubular HCO_3^-, the most probable reason being that the Cl^- concentration change was too small and was disguised by a simultaneous decrease in cellular HCO_3^- concentration. Recently, however, we have performed the opposite experiment, measuring cellular pH (or HCO_3^- concentration) while changing peritubular Cl^- concentration. As shown in Fig. 3a, this experiment gave a positive result. Fig 3b and Table 2 show further data demonstrating that Cl^-/HCO_3^- exchange is inhibited by SITS and that it may include a component which is coupled to Na^+. These observations agree with a recent report by Alpern & Chambers (1987).

At present, virtually nothing is known about the coupling ratio, transported buffer species and kinetics of this transport system. However, the similarity of the Na^+-dependent system with a transporter in non-epithelial cells (Roos & Boron 1981) suggests that the HCO_3^- transport possibly proceeds as the ion pair, $NaCO_3^-$ (Boron 1985). Besides SITS, it has been reported that pyridoxal phosphate and picrylsulphonic acid are able to block Na^+-dependent Cl^-/HCO_3^- exchange (Madshus & Olsnes 1987).

Other base transport mechanisms

Besides the transport mechanism reported above, the basolateral cell membrane of *Necturus* proximal tubule contains a lactate/OH^- exchanger (or H^+-lactate co-transporter) which functions electroneutrally, is independent of Na^+ and can be inhibited by α-cyano-4-hydroxycinnamate (Siebens & Boron 1987). In mammalia, however, the data are somewhat conflicting. Working with membrane vesicles, Barac-Nieto et al (1982) presented evidence for a stereospecific lactate transport system which exhibited tracerflux trans-stimulation and was sensitive to membrane potential but independent of Na^+. On the other hand, working with rat proximal tubule *in vivo*, Ullrich & Papavassiliou (1986) concluded that lactate and other small hydrophilic fatty acids traverse the basolateral cell membrane by simple diffusion across unspecific anion channels.

Another pathway which contributes to acid/base transport across the peritubular cell membrane is the electroneutral SO_4^{2-}/HCO_3^- exchanger or H_2^+-SO_4^{2-} co-transporter. It has been identified on membrane vesicles of rabbit and rat (Brazy & Dennis 1981, Pritchard & Renfro 1983, Löw et al 1984) and on rat tubules *in vivo* (Ullrich et al 1984). Recently, the kinetic parameters of this transporter have been defined (Shimada & Burckhardt 1986).

In addition to the various co-transporters and exchangers described above, there is some evidence for uncoupled permeation of OH^- (or H^+) across the basolateral cell membrane of rat proximal tubule *in vivo* (Burckhardt & Frömter 1987, Ullrich & Papavassiliou 1987). The significance of these observations, however, remains to be further elucidated.

Finally, we may expect that the transport of dicarboxylic and tricarboxylic acids, which depends on pH, is also associated with the transport of OH^- or HCO_3^- (Burckhardt 1984). However, this transport system is not yet fully characterized.

Acknowledgements

The technical assistance of Mrs N. Stallmach and the secretarial assistance of Mrs I. Harward and U. Merseburg are gratefully acknowledged.

References

Akiba T, Alpern RJ, Eveloff J, Calamina J, Warnock DG 1986 Electrogenic sodium/bicarbonate cotransport in rabbit renal cortical basolateral membrane vesicles. J Clin Invest 78:1472–1478

Alpern RJ 1985 Mechanism of basolateral membrane $H^+/OH^-/HCO_3^-$ transport in the rat convoluted tubule. A sodium coupled electrogenic process. J Gen Physiol 86:613–636

Alpern RJ, Chambers M 1987 Basolateral membrane Cl/HCO_3 exchange in the rat proximal convoluted tubule. J Gen Physiol 89:581–598

Barac-Nieto M, Murer H, Kinne R 1982 Asymmetry in the transport of lactate by basolateral and brush border membranes of rat kidney cortex. Pflügers Arch Eur J Physiol 392:366–371

Biagi BA, Sohtell M 1986 Electrophysiology of bicarbonate transport in the rabbit proximal tubule. Am J Physiol 250:F267–F272

Boron WF, Boulpaep EL 1983 Intracellular pH regulation in the renal proximal tubule of the salamander. Basolateral HCO_3^- transport. J Gen Physiol 81:53–94

Boron WF 1985 Intracellular pH-regulating mechanism of the squid axon. J Gen Physiol 85:325–345

Brazy PC, Dennis VW 1981 Sulfate transport in rabbit proximal convoluted tubules: presence of anion exchange. Am J Physiol 241:F300–F307

Burckhardt BC, Sato K, Frömter E 1984a Electrophysiological analysis of bicarbonate permeation across the peritubular cell membrane of rat kidney proximal tubule. I. Basic observations. Pflügers Arch Eur J Physiol 401:34–42

Burckhardt BC, Cassola AC, Frömter E 1984b Electrophysiological analysis of bicarbonate permeation across the peritubular cell membrane of rat kidney proximal tubule. II. Exclusion of HCO_3^--effects on other ion permeabilities and of coupled electroneutral HCO_3^--transport. Pflügers Arch Eur J Physiol 401:43–51

Burckhardt BC, Geers C, Frömter E 1985 Role of membrane-bound carbonic anhydrases in HCO_3^- transport across rat renal proximal tubular cell membranes. Pflügers Arch Eur J Physiol 405:R31

Burckhardt BC, Frömter E 1987 Evidence for OH^-/H^+ permeation across the peritubular cell membrane of rat renal proximal tubule in HCO_3^--free solutions.Pflügers Arch Eur J Physiol 409:132–137

Burckhardt G 1984 Sodium-dependent dicarboxylate transport in rat renal basolateral membrane vesicles. Pflügers Arch Eur J Physiol 401:254–261

Edelman A, Bouthier M, Anagnostopoulos T 1981 Chloride distribution in the proximal convoluted tubule of Necturus kidney. J Membr Biol 62:7–17

Frömter E, Sato K 1976 Electrical events in active H^+/HCO_3^- transport across rat kidney proximal tubular epithelium. In: Kasbekar DK, Sachs G, Rehm WS (eds) Gastric hydrogen ion secretion. Dekker, New York p 382–403

Grassl SM, Aronson PS 1986 Na^+/HCO_3^- co-transport in basolateral membrane vesicles isolated from rabbit renal cortex. J Biol Chem 261:8778–8783

Grassl SM, Holohan PD, Ross CR 1987 Cl^--HCO_3^- exchange in rat renal basolateral membrane vesicles. Biochim Biophys Acta 905:475–484

Guggino WB, London R, Boulpaep EL, Giebisch G 1983 Chloride transport across the basolateral cell membrane of the Necturus proximal tubule: Dependence on bicarbonate and sodium. J Membr Biol 71:227–240

Jentsch TJ, Schwartz P, Schill S et al 1986 Kinetic properties of the sodium bicarbonate (carbonate) symport in monkey kidney epithelial cells (BSC-1). J Biol Chem 261:10673–10679

Kinne-Saffran E, Beauwens R, Kinne R 1982 An ATP-driven proton pump in brushborder membranes from rat renal cortex. J Membr Biol 64:67–76

Kondo Y, Frömter E 1987 Axial heterogeneity of sodium-bicarbonate cotransport in proximal straight tubule of rabbit kidney. Pflügers Arch Eur J Physiol 410:481–486

Krapf R, Alpern RJ, Rector Jr FG, Berry CA 1987 Basolateral membrane Na/base cotransport is dependent on CO_2/HCO_3 in the proximal convoluted tubule. J Gen Physiol 90:833–853

Löw I, Friedrich T, Burckhardt G 1984 Properties of an anion exchanger in rat renal basolateral membrane vesicles. Am J Physiol 246:F334–F342

Lopes AG, Siebens AW, Giebisch G, Boron WF 1987 Electrogenic Na/HCO_3 cotransport across basolateral membrane of isolated perfused Necturus proximal tubule. Am J Physiol 253:F340–F350

Madshus IH, Olsnes S 1987 Selective inhibition of sodium-linked and sodium-independent bicarbonate/chloride antiport in vero cells. J Biol Chem 262:7486–7491

Murer H, Hopfer U, Kinne R 1976 Sodium/proton antiport in brushborder membrane vesicles isolated from rat small intestine and rat kidney. Biochem J 154:597–604

Pritchard JB, Renfro JL 1983 Renal sulfate transport at the basolateral membrane is mediated by anion exchange. Proc Natl Acad Sci USA 80:2603–2607

Roos A, Boron WF 1981 Intracellular pH. Physiol Rev 61:296–434
Sasaki S, Shigai T, Yoshiyama N, Takeuchi J 1987 Mechanism of bicarbonate exit across basolateral membrane of rabbit proximal straight tubule. Am J Physiol 252:F11–F18
Shimada H, Burckhardt G 1986 Kinetic studies of sulfate transport in basolateral membrane vesicles from rat renal cortex. Pflügers Arch Eur J Physiol 407:S160–S167
Siebens AW, Boron WF 1987 Effect of electroneutral luminal and basolateral lactate transport on intracellular pH in Salamander proximal tubules. J Gen Physiol 90:799–831
Soleimani M, Grassl SM, Aronson PS 1987 Stoichiometry of Na^+-HCO_3^- cotransport in basolateral membrane vesicles isolated from rabbit renal cortex. J Clin Invest 79:1276–1280
Soleimani M, Aronson PS 1988 Ionic mechanism of Na^+:HCO_3^- cotransport in renal basolateral membrane vesicles. Kidney Int 33:407
Ullrich KJ, Rumrich G, Baumann K 1975 Renal proximal tubular buffer (glycodiazine) transport. Pflügers Arch Eur J Physiol 357:149–163
Ullrich KJ, Rumrich G, Klöss S 1984 Contraluminal sulfate transport in the proximal tubule of the rat kidney. Pflügers Arch Eur J Physiol 402:264–271
Ullrich KJ, Papavassiliou F 1986 Contraluminal transport of small aliphatic carboxylates in the proximal tubule of the rat kidney in situ. Pflügers Arch Eur J Physiol 407:488–492
Ullrich KJ, Papavassiliou 1987 Contraluminal bicarbonate transport in the proximal tubule of the rat kidney. Pflügers Arch Eur J Physiol 410:501–504
Wang W, Dietl P, Oberleithner H 1987 Evidence for Na^+ dependent rheogenic HCO_3^- transport in fused cells of frog distal tubules. Pflügers Arch Eur J Physiol 408:291–299
Yoshitomi K, Burckhardt BC, Frömter E 1985 Rheogenic sodium-bicarbonate cotransport in the peritubular cell membrane of rat renal proximal tubule. Pflügers Arch Eur J Physiol 405:360–366
Yoshitomi K, Frömter E 1985 How big is the electrochemical potential difference of Na^+ across rat renal proximal tubular cell membranes in vivo? Pflügers Arch Eur J Physiol 405:S121–S126

DISCUSSION

Malnic: Ullrich et al (1984a,b, 1987) described anion-exchange mechanisms in the basolateral membrane that are not very specific. Do you think that the mechanism you have described is part of that system or is it a more specific transporter?

Frömter: Ullrich and his collaborators have described three anion transport systems in the basolateral cell membrane: one for dicarboxylic acids, e.g. methylsuccinate, one for sulphate and one for para-aminohippurate (Ullrich et al 1984a,b,1987). The first system is coupled to Na^+ but independent of HCO_3^-, which suggest that it differs from Na^+-$(HCO_3^-)_3$ symport, as is the second system which transports sulphate better in the absence of Na^+ than in its presence. The third system is independent of Na^+. On the other hand, at present I cannot exclude that the Cl^-/HCO_3^- exchanger that I have described in my paper might be identical with the sulphate transporter (SO_4^{2-}/HCO_3^- exchanger).

Thomas: Walter, do you have any data on the coupling ratio of the Na^+-HCO_3^- co-transporter in the salamander tubule?

Boron: The original experiments were consistent with a stoichiometry of 2:1: we saw no reason to postulate a stoichiometry of 3:1, because 2:1 worked. More recently, Drs Frömter and Aronson have found that in order to explain their data they need to invoke a ratio of 3:1. We have now done more experiments in the salamander and in *Necturus*, and I think that in both cases a stoichiometry of 3:1 best explains the data.

Thomas: I worked on an electrogenic pump over 20 years ago and it is nice to know that there are more of them. Though this one differs from the electrogenic Na^+/K^+ pump in the sense that it's going downhill—the HCO_3^- is going down an electrical gradient and presumably dragging the Na^+ with it. Is there any involvement of ATP ?

Frömter: No, Krapf et al (1987) did *in vitro* microperfusion experiments on rabbit proximal tubules, in which they poisoned cell metabolism with cyanide and found that the Na^+-$(HCO_3^-)_3$ co-transport was still operative, if appropriate ion concentration gradients were induced.

Sabolic: Did you say that sulphate competes with HCO_3^- on the Na^+/HCO_3^- co-transporter in the cells?

Frömter: I have no evidence for competition between sulphate and bicarbonate in our experiments. The potential changes that I described are elicited by replacing peritubular HCO_3^- suddenly with Cl^-. Instead of Cl^- we used several other substitutes, such as phosphate, gluconate, cyclamate or sulphate, and obtained the same response. We cannot use lipid-soluble buffers, such as butyrate or glycodiazine, for which the cell membrane seems to be highly permeable. We conclude that sulphate acts as a poorly permeating or non-permeating anion (Burckhardt et al 1984a). On the other hand, substitution of Cl^-, e.g. by sulphate, may inhibit carbonic anhydrase and thereby impede Na^+-$(HCO_3^-)_3$ co-transport.

Sabolic: In your experiments on the S2 segment with Cl^- you always have an alkalinization after removal of external Cl^-, but there is no corresponding acidification when you add the Cl^- at the beginning.

Frömter: I am not sure whether I fully understand your question. The time course of the cellular pH response to peritubular substitution of Cl^- is shown in Fig. 3. On removal of Cl^- there is a fast transient 'acidification' which is followed by a sustained alkalinization; on returning to Cl^--containing solutions, a mirror image-like time course can be seen. I do not attach much importance to these observations because I fear that the initial 'acidification' may simply reflect capacitive coupling from the sudden potential change in the reference barrel to the selective barrel.

Gillespie: We also see that initial acidification in pancreatic acinar cells with the dye techniques so it can't be a simple electrical phenomenon as you suspect. It looks as if it is to do with low Ca^{2+} in the Cl^- free solution because you can mimic it to a certain extent with zero Ca^{2+} in the perfusion solution.

Frömter: In our perfusion solutions, the ionized Ca^{2+} concentration is always kept constant.

Vaughan-Jones: Are you compensating for Ca^{2+} binding by the gluconate?

Frömter: Yes, we increase the Ca^{2+} concentration from 1.25 mM to 11.0 mM.

Gillespie: We tried that too but still we see this small acidification, to a greater or lesser extent, depending on the Ca^{2+} concentration.

Frömter: I cannot say for sure that this initial peak acidification is indeed a technical artifact, but this is at present the most likely explanation.

Aickin: Were you using the same type of electrode in the S2 and the S3 segments? In the S3 segment you have a very clear alkalinization with no acidification first. In the S2 segment there is a very predominant acidification before a very small alkalinization—in some of those records it's hard to believe that the cell did become alkaline. The time course of that acidification I would have thought was too long to be an artifact problem.

Frömter: We used the same type of electrode but otherwise the experimental procedure is completely different. For example, whereas in the *in vivo* preparation the kidney surface is covered with a layer of mineral oil, the isolated tubule is mounted inside a bath filled with Ringer solution. Also, the reference electrodes are placed differently and have different shapes, so that the resistance of the preparation with respect to the ground state may differ by up to one order of magnitude.

Gillespie: It seemed to me that SITS was blocking that transient acidosis.

Frömter: No, if you look very carefully, the SITS experiment did not have the rapid solution changes (Fig. 3b). While the first change to Cl^--free solution was fast, as judged from the rapid development of the liquid junction potential, the second change (to Cl^--free solution with SITS) was slower and this may be the reason why no artifact developed. In support of this, in the latter case, on returning to Cl^--containing solution, the potential change was again faster and a small transient, probably spurious, alkalinization showed up simultaneously on the pH trace.

Thomas: Were the solutions flowing at the same speed?

Frömter: Unfortunately, we cannot ensure constant perfusion speed. The perfusion is fed under gravity through triple-barrelled pipettes inserted into single blood capillaries. One of the major problems with this technique is to prevent backflow and fluid mixing in the tip of the barrels while they are not perfusing. If this occurs, it delays the exchange time and may produce traces such as that on the right hand side of Fig. 3b.

Oberleithner: Concerning the Cl^-/HCO_3^- exchanger in the S3 segment, you mentioned that Cl^- is taken up by the cell and has to recirculate via a Cl^- conductance. Did you ever use a Cl^- permeability blocker to inhibit this Cl^- circulation. This should render the intracellular pH alkaline.

Frömter: We have unpublished results (Y. Kondo & E. Frömter) in which we see changes of cell potential and of intracellular Cl^- concentration that point to the presence of Cl^- channels in the basolateral cell membrane of this segment. Unfortunately, we have not been able to block this conductance by Cl^- channel blockers. This may not be significant, because in my laboratory we have tested these blockers in many different preparations, e.g. sweat glands (Bijman et al 1987), *Necturus* gall bladder (H. Ewald & E. Frömter, unpublished observations), collecting duct principal cell epithelium (Gross et al 1988) and others which are known to have Cl^- channels, but we have never seen the correct inhibitory responses.

Thomas: Which ones are those?

Frömter: These are DPC (diphenylamine carboxylate) and its derivatives, that have been developed by Dr Greger in collaboration with the Hoechst company in Frankfurt (Wangemann et al 1986). In some preparations these blockers work beautifully but in others they do not; they are also not as specific as I would like. As a result I am reluctant to place much emphasis on such blocker experiments at the moment.

Malnic: You have shown this beautiful relationship between the activity of the Na^+-HCO_3^- co-transporter and its location along the tubule. Would there be a similar but inverse relationship of the Cl^-/HCO_3^- exchanger—highest in the S3 and lowest in the S1?

Frömter: I cannot fully answer your question. We found the Cl^-/HCO_3^- exchanger first in the S2 segment of rat kidney and later in the S2 and S3 segments of rabbit kidney. We have not looked at the S1 segment at all. Whether the number of transporters or their capacity increases when going from S1 to S3, I don't know.

Steels: How do you explain the small hyperpolarization that occurred in the S3 segment when Cl^- was taken away?

Frömter: These small changes are very difficult to interpret. In contrast to the experiments on the rat kidney *in vivo*, the experiments on the S3 segments of rabbit kidney *in vitro* were performed under conditions where no liquid junction potential artifacts should develop, i.e. we used saturated KCl electrodes which were connected to the bath fluid under flowing boundary conditions. Nevertheless, it remains questionable whether the small initial depolarization, of the order of only 4 mV, that we observed on replacing Cl^- with gluconate can be safely taken as evidence for a Cl^- conductance.

Interpretation of the small steady-state hyperpolarization of the cell is even more difficult, because under those conditions intracellular ion concentrations and intracellular pH may change which may have secondary effects on the membrane potential.

Oberleithner: The explanation might be that because of the increase in intracellular pH, the pH-sensitive K^+ conductance rises and shifts the membrane potential closer to the equilibrium potential of potassium.

Frömter: Absolutely.

Steels: So is this barium sensitive?

Frömter: I don't know.

Aickin: I agree it's difficult to pin down what underlies potential changes but the changes you see in the S3 segment on removal of external Cl^- look very like those in mammalian skeletal muscle. If there is a finite Cl^- permeability and an intracellular accumulation of Cl^-, removal of external Cl^- will cause a transient depolarization, but then because you have removed a depolarizing influence you end up with a steady-state hyperpolarization. Of course there are lots of other explanations.

Frömter: The steady state is always uninterpretable unless you make a great number of further measurements. Hence I am reluctant to speculate on the meaning of steady-state responses. The situation is different with initial potential changes which are more readily interpretable in terms of ion conductances, as long as intracellular ion concentrations do not change. But again, additional experiments are necessary to prove that the interpretation is correct (see Burckhardt et al 1984b).

References

Bijman J, Englert HC, Lang HJ, Greger R, Frömter E 1987 Characterization of human sweat duct chloride conductance by chloride channel blockers. Pfluegers Arch Eur J Physiol 408:511–514

Burckhardt BC, Sato K, Frömter E 1984a Electrophysiological analysis of bicarbonate permeation across the peritubular cell membrane of rat kidney proximal tubule. Pfluegers Arch Eur J Physiol 401:34–42

Burckhardt BC, Cassola AC, Frömter E 1984b Electrophysiological analysis of bicarbonate permeation across the peritubular cell membrane of rat kidney proximal tubule. II. Exclusion of HCO_3^- effects on other ion permeabilities and of coupled electroneutral HCO_3^- transport. Pfluegers Arch Eur J Physiol 401:43–51

Gross P, Minuth WW, Ketteler M, Frömter E 1988 Ionic conductances of cultured principal cell epithelium of renal collecting duct. Pfluegers Arch Eur J Physiol, in press

Krapf R, Alpern RJ, Rector FC, Berry CA 1987 Basolateral membrane Na/base cotransport is dependent on CO_2/HCO_3 in the proximal convoluted tubule. J Gen Physiol 90:833–853

Ullrich KJ, Fasold H, Rumrich G, Kloess S 1984a Secretion and contraluminal uptake of dicarboxylic acids in the proximal convolution of rat kidney. Pfluegers Arch Eur J Physiol 400:241–249

Ullrich KJ, Rumrich G, Kloess S 1984b Contraluminal sulfate transport in the proximal tubule of the rat kidney. I. Kinetics, effects of K^+, Na^+, Ca^{2+}, H^+ and anions. Pfluegers Arch Eur J Physiol 402:264–271

Ullrich KJ, Rumrich G, Fritzsch G, Kloess S 1987 Contraluminal para-aminohippurate (PAH) transport in the proximal tubule of the rat kidney. I. Kinetics, influence of cations, anions, and capillary preperfusion. Pfluegers Arch Eur J Physiol 409:229–235

Wangemann P, Wittner M, Di Stefano A et al 1986 Cl^--channel blockers in the thick ascending limb of the loop of Henle. Structure activity relationship. Pfluegers Arch Eur J Physiol 407:S128–S141

Electrogenic proton transport by intercalated cells of tight urinary epithelia

Philip R. Steinmetz

Department of Medicine, University of Connecticut School of Medicine, Farmington, Connecticut 06032, USA

Abstract. Structure-function studies of the turtle bladder indicate that electrogenic proton secretion into the urinary compartment is accomplished by α-type intercalated cells which are rich in carbonic anhydrase. In the absence of electrochemical potential gradients ($\Delta\tilde{\mu}_H = 0$), the rate of H^+ secretion (J_H) is a function of the number of H^+ pumps in position at the apical cell membrane, as judged from morphometric and freeze-fracture studies of apical membrane area characterized by a cytoplasmic coating with studs and by rod-shaped intramembrane particles (RSP). At a given pump population, J_H is a sigmoid function of $\Delta\tilde{\mu}_H$, with ΔpH and $\Delta\psi$ having equivalent effects on J_H. The J_H versus $\Delta\tilde{\mu}_H$ relation reflects the intrinsic properties of the H^+ pump and suggests a H^+ pump model consisting of two components, a channel through the apical membrane across which $\Delta\ \tilde{\mu}_H$ falls, and a catalytic unit located within the cytoplasm (outside of $\Delta\tilde{\mu}_H$). Each intramembrane RSP is associated with several cytoplasmic studs, but the precise relations between the two remain to be clarified.

1988 Proton passage across cell membranes. Wiley, Chichester (Ciba Foundation Symposium 139) p 122–138

In many epithelial and non-epithelial cells, acid-base transport is accomplished by antiporters or symporters driven by chemical or electrical forces which originate from ATP-dependent sodium transport. This paper deals with certain tight urinary epithelia which are capable of electrogenic proton secretion via a H^+-translocating ATPase located in the apical cell membrane. These tight urinary epithelia are derived from the interaction of the mesonephric duct and the mesenchyme. They contain carbonic anhydrase-rich cells that are described as 'intercalated' among the common principal or granular cells. Considerable evidence indicates that the intercalated cells are responsible for the active secretion of protons and that this process is electrogenic and not directly affected by abolishing sodium transport (Steinmetz et al 1967, Steinmetz 1974).

The characteristics of active H^+ transport have been defined in a number of urinary epithelia that can be studied *in vitro* under controlled and simplified conditions, such as the sheet preparations of the turtle and toad urinary bladders and isolated perfused segments of the rabbit collecting duct. Recent biochemical studies of membrane vesicles of bovine medulla, turtle bladder (Gluck & Al-Awqati 1984, Gluck & Caldwell 1987) and endosomal preparations of brain (Stone et al 1983) and other organs have characterized a group of very similar H^+-translocating ATPases. One form of these ATPases appears to be responsible for H^+ secretion by the intercalated cells of distal urinary structures. In this paper I describe new understanding of electrogenic proton secretion, focusing on some of the structure–function correlations that appear to be involved in the control of proton transport at the cellular and molecular levels.

Active H^+ transport in turtle bladder

The sheet preparation of the urinary bladder of the water turtle, *Pseudemys scripta*, has been studied extensively *in vitro* to explore the cellular mechanisms of urinary acidification. In this preparation, when Na^+ transport is abolished by ouabain and CO_2 and HCO_3^- are kept at low levels, the secretion of H^+ (J_H) into the urinary compartment can be measured either by pH stat titration or as the short-circuit current (Steinmetz 1974). Under such simplified conditions, the short-circuit current matches the titration rate. Several studies have shown that at a given cellular acid-base status J_H is directly influenced by the electrochemical potential gradient, $\Delta\tilde{\mu}_H$, across the apical cell membrane (Steinmetz & Lawson 1971, Al-Awqati et al 1977, Andersen et al 1985). In Fig. 1, J_H approaches zero when the luminal pH is lowered to 4.5. When luminal pH is increased at constant serosal pH, J_H increases almost linearly up to pH 7.0 (at which point the transepithelial pH difference is zero). At higher luminal pH saturation becomes apparent, and between pH 8 and 9 a maximal J_H is reached.

The J_H versus ΔpH relationship becomes considerably more interesting, if we make two additional simplifications, the validity of which is supported by a large body of evidence. First of all, studies by Steinmetz & Lawson (1971) and Beauwens & Al-Awqati (1976) have shown that the apical cell membrane of the turtle bladder has a low passive H^+ permeability, such that passive H^+ flows are negligible compared to the active flows. Hence, J_H directly reflects the active component of H^+ transport. The second simplification is based on the high acid-base permeability of the basolateral cell membrane (Cohen et al 1978, Andersen et al 1985) which means that cell pH is closely linked to the extracellular pH of the serosal loose tissue layer. If the pH of the serosal compartment is maintained in a physiological range near pH 7.0, the cell pH tends to remain relatively constant at a slightly higher level. Under these

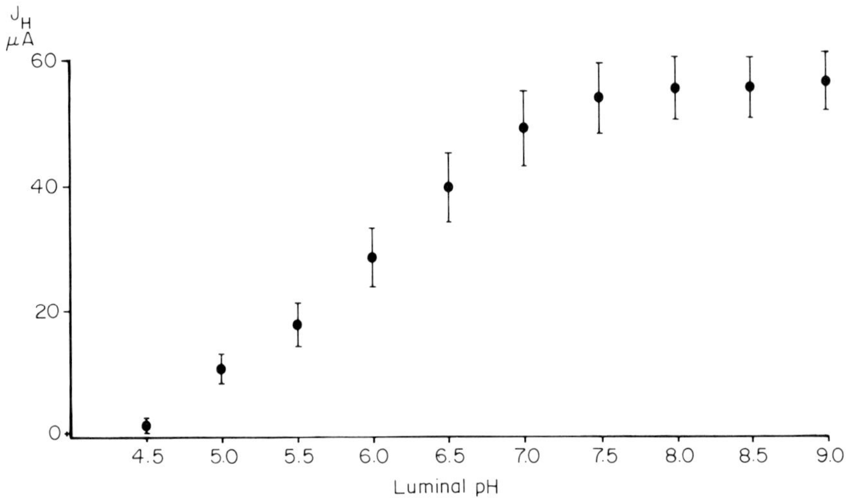

FIG. 1. Relation between J_H and luminal pH in a group of eight bladders at constant serosal pH. Reproduced from Andersen et al (1985) by copyright permission of The Rockefeller University Press.

conditions, changes in luminal pH impose pH differences primarily across the tight apical cell membrane. Hence, the behaviour of J_H in Fig. 1 represents the behaviour of the active H^+ transport rate in response to a changing ΔpH across the apical cell membrane.

The maximal value for J_H depends not only on a favourable ΔpH but also on the acid-base and hormonal status of the epithelium. For example, in intercalated cells the apical membrane area with microplicae is highly variable. It shrinks when H^+ secretion is inhibited by disulphonic stilbenes (Husted et al 1981), by acetazolamide or by lowering the CO_2 tension; it expands when H^+ secretion is stimulated by CO_2. Shrinkage is brought about by endocytotic retrieval of apical cell membrane; expansion is caused by exocytosis and fusion of membrane vesicles to the apical membrane (Gluck et al 1982, Stetson & Steinmetz 1983, 1986). Hence, for any given cellular acid-base condition the maximal J_H is a function of the total H^+ pump population in place at the apical cell membrane. Andersen et al (1985) have presented several lines of evidence indicating that the J_H response to an imposed ΔpH is not caused by rapid changes in the number of H^+ pumps at the apical membrane, but is determined by the intrinsic properties of a given population of H^+ pumps. The J_H response to a ΔpH is instantaneous and is completed within two minutes. Electron microscopy shows no detectable reduction in the apical area of the intercalated cells when the luminal pH is lowered to 4.5 (D. Stetson & P.R. Steinmetz, unpublished observations). Furthermore, the J_H versus ΔpH relation is preserved in bladders pre-treated with a cytoskeleton-disrupting drug such as colchicine, which inhibits exocy-

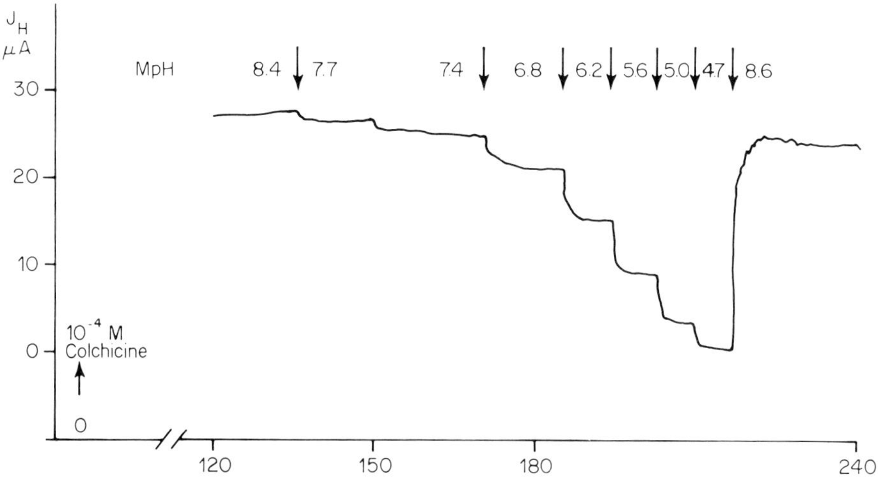

FIG. 2. Tracing of J_H as a function of luminal pH (MpH) in a bladder pre-treated with 10^{-4} M colchicine. The J_H versus ΔpH relation was preserved. Note rapid recovery of J_H after MpH was restored from 4.7 to 8.6. Time scale in minutes. Reproduced from Andersen et al (1985) by copyright permission of The Rockefeller University Press.

tosis in turtle bladder (Andersen et al 1985). In Fig. 2 such an experiment is shown. Lowering luminal pH at constant serosal pH and p_{CO2} caused stepwise inhibition of J_H. The inhibition was promptly reversed when mucosal pH was increased.

The interpretation that the J_H versus ΔpH relation reflects the intrinsic properties of the H^+ pump complex is strengthened by studies in which the effects on J_H of imposed transepithelial electrical differences (Δψ) were compared with pH gradients. These studies show a remarkable equivalence between the effects of imposed electrical and chemical potential differences. The Δψ data follow the ΔpH data of Fig. 1 not only in the near linear region (Al-Awqati et al 1977), but also in the region where J_H plateaus (Andersen et al 1985). Only at high lumen-negative potentials equivalent to luminal pH values of 8.5 and above could a small difference be demonstrated, J_H being slightly higher with a favourable Δψ than with the equivalent ΔpH (Kohn & Steinmetz, unpublished observations).

The near equivalence of the chemical (ΔpH) and electrical (Δψ) contributions to $\Delta\tilde{\mu}_H$ in the control of J_H has several implications: it suggests that the H^+ pump of urinary acidification resembles a number of H^+-ATPases that have been studied in membrane vesicles. Rates of ATP hydrolysis or H^+ translocation are clearly affected by both ΔpH and Δψ in preparations of mitochondrial, bacterial and plasma membrane H^+-ATPases, including preparations from turtle urinary bladder and bovine kidney medulla (Gluck &

Al-Awqati 1984). More specifically, under the simplified conditions of the sheet preparation, the remarkable equivalence of ΔpH and Δψ over the physiological $\Delta\tilde{\mu}_H$ range has implications for the translocation of H^+ across the H^+ pump complex. The similar maximal values for J_H suggest that the rate-limiting step(s) in the overall H^+ translocation are the same in the two experimental situations and either that the rate-limiting step is voltage insensitive or that this step, which is thought to occur in the catalytic unit located in the cytoplasm, escapes most or all of the applied potential difference. Δψ must fall primarily across the channel component of the H^+ pump within the apical cell membrane.

Structure–function relations of intercalated cells

The intercalated cells of tight urinary epithelia share many structural and functional characteristics, whether they occur in turtle or toad bladder or in mammalian collecting duct. They contain carbonic anhydrase and often are rich in mitochondria. Their apical surface is decorated with prominent microplicae or microvilli. Their main function is the transport of H^+ or HCO_3^-.

Recently Stetson & Steinmetz (1985) were able to distinguish between two populations of intercalated or carbonic anhydrase-rich cells in turtle bladder, an α cell and a β cell population. The α-type intercalated cell is characterized by apical microplicae and its apical membrane can undergo large changes in area in response to cellular acid–base stimuli. Membrane vesicles can be retrieved from, or added to, the apical cell membrane. The β cell, which makes up about one third of the intercalated cells in turtle bladder, is characterized by an apical surface that is decorated with individual microvilli and that fails to respond to acid-base stimuli. The basis for the distinction between α and β cells, however, was found in the ultrastructure of their apical and basolateral cell membranes. As shown in Fig. 3, the cytoplasmic face of the apical cell membrane of the α cell is characterized by an extensive coating with studs. The studs resemble stalked spheres and line the microplicae as well as the cytoplasmic face of many of the subapical membrane vesicles. Freeze-fracture studies of the apical surface of the α cells reveal numerous rod-shaped intramembrane particles (RSP) on the P face of the membranes in the same location, i.e. microplicae and subapical vesicles (Fig. 4). The basolateral cell membrane of the α cell is free of studs and RSPs. In contrast to this distribution in the α cell, the β cell of the turtle bladder has both studs and RSPs at its basolateral membrane, but not at its apical pole.

The occurrence of studs and RSPs at either the apical or the basolateral pole of the intercalated cells was interpreted in the light of transport studies which indicated that the processes of H^+ secretion and HCO_3^- secretion were physiologically separate, although both occurred in carbonic anhydrase-rich cells (Stetson et al 1985). As a result of these and other studies, the function

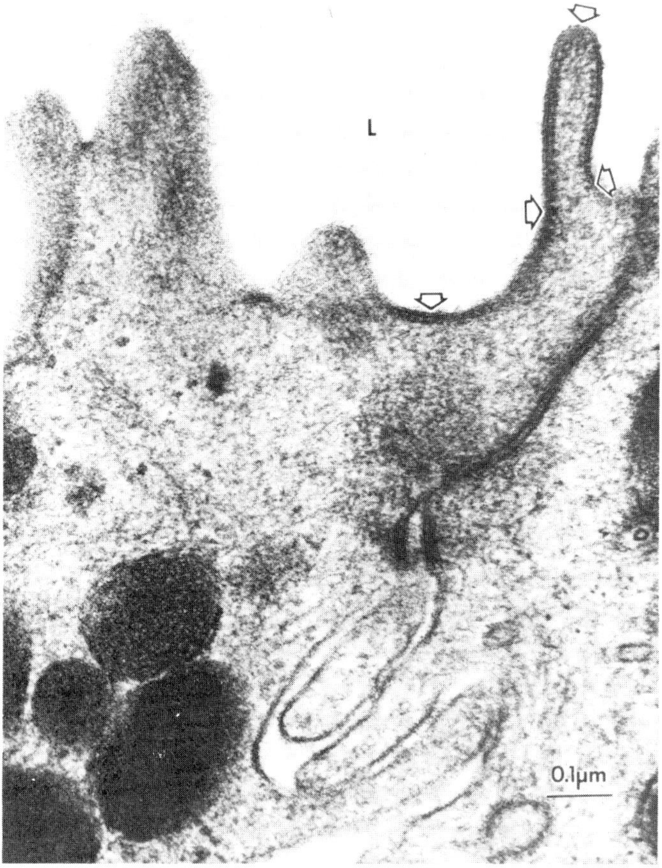

FIG. 3. Electron micrograph of the apical portion of an α-type intercalated cell in turtle bladder. The cytoplasmic face of the apical membrane is coated by a dense array of studs (arrows). L, lumen. Reproduced from Steinmetz & Stetson (1987) by copyright permission of Churchill Livingstone.

of H^+ secretion was assigned to the α cells and the function of HCO_3^- secretion to the β cells.

For this paper I shall focus on the H^+ secreting cells and develop the structure-function correlations for the α intercalated cells. The functional organization of the α intercalated cell has been studied extensively in turtle bladder (Steinmetz 1974, Fischer et al 1983, Andersen et al 1985), as well as in the rabbit outer medullary collecting duct where all intercalated cells are of the α type and their electrophysiology has been studied (Koeppen 1985). In Fig. 5, the current double membrane model for the α cell is shown. The apical cell membrane is densely packed with H^+-translocating ATPase molecules. Studies by Gluck et al (1984, 1987) and Brown et al (1987) indicate that this

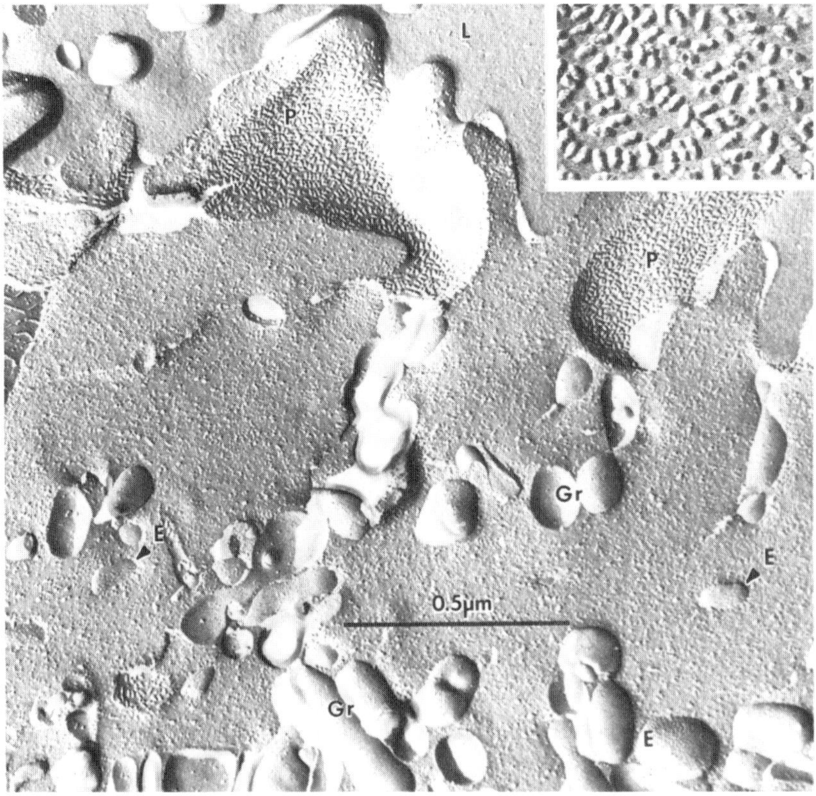

FIG. 4. Freeze-fracture electron micrograph of the apical cell membrane of an α-type intercalated cell in turtle bladder. The P face of the apical membrane (P) contains a dense population of rod-shaped intramembrane particles (RSP). Some intracytoplasmic vesicles also display RSPs on the P face. At high magnification (inset) the RSPs appear to be made up of two or three globular particles. E, E face; L, lumen; Gr, mucin granule. Reproduced from Steinmetz & Stetson (1987) by copyright permission of Churchill Livingstone.

ATPase is of the F_0-F_1 type and that its cytoplasmic domain corresponds to the studded regions of the apical cell membranes and vesicles of α cells. The acid-base permeability and the electrical conductance of the apical membrane are extremely low. In the H^+ secreting cells of the outer medullary collecting duct, Koeppen (1985) found the fractional resistance of the apical cell membrane to be close to unity. No ion conductances for Na^+, K^+ or Cl^- were demonstrable.

The secretion of H^+ leads to formation of OH^- and HCO_3^- in the cytoplasm. The HCO_3^- moves outward across the basolateral cell membrane via a Cl^-/HCO_3^- exchanger which is inhibited by disulphonic stilbenes (Ehrenspeck & Brodsky 1976, Cohen et al 1978) and by Cl^- removal. The basolateral

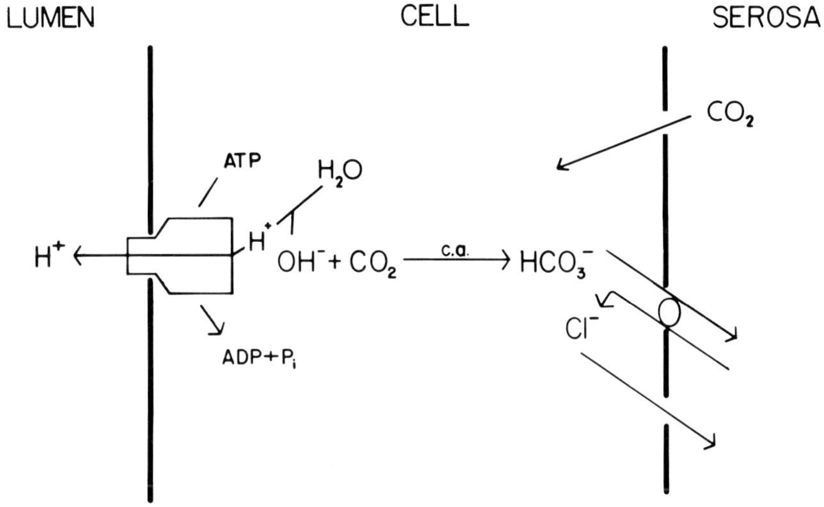

FIG. 5. Cellular organization of H^+ transport in an α-type intercalated cell of a tight urinary epithelium. c.a., carbonic anhydrase. Reproduced from Andersen et al (1985) by copyright permission of The Rockefeller University Press.

membranes of the α intercalated cells show immunoreactivity to the band III protein of the erythrocyte anion exchanger in turtle bladder and in the collecting ducts of rat, rabbit and humans, indicating a close homology between the exchanger proteins of intercalated cells and red cells (Schuster et al 1986, Wagner et al 1987). In parallel to the electroneutral exchanger there is a Cl^- ion conductance in the basolateral cell membrane (Fischer et al 1983, Koeppen 1985). This conductance permits the recycling of Cl^- and allows the overall process of H^+ secretion to be electrogenic.

When the efflux of HCO_3^- is blocked by a disulphonic stilbene, the α cells become transiently alkaline (Cohen et al 1978) and their apical membrane areas shrink by endocytotic membrane retrieval. Conversely, when the cell is acidified by an increase in CO_2 tension, the apical membrane area expands by the fusion of intracytoplasmic membrane vesicles to the apical membrane. The retrieval or addition of membrane containing H^+ pumps tends to correct the initial pH change of the cell. The initiation and control of these events of membrane fusion and retrieval are regulated not only by cell pH but also by the intracellular calcium concentration (van Adelsberg & Al-Awqati 1986).

Since CO_2 addition stimulates the H^+ transport rate and the exocytosis of membrane vesicles in α cells, it is of interest to compare the structural and functional changes in the sheet preparation of turtle bladder. After J_H was measured in paired hemibladders at 0% and 5% CO_2, the bladder sheets were fixed and prepared for thin section and freeze-fracture electron microscopy (EM), as well as for scanning EM. Morphometric studies provided estimates

TABLE 1 Effect of CO_2 on J_H and on apical membrane area of α cells and total number of apical RSPs

	CO_2-free	*5% CO_2*	*Increase (%)*
J_H (μA)	10±1	88±13	880
Apical area (mm^2)	176	693	390
RSP density/μm^2	2424±196	3389±463	not significant
Total RSP (× 10^9)	422	2240	530

RSP, rod-shaped intramembrane particles. Values are given as mean ± SEM. Apical area is given as total area including folds. Summarized from Stetson & Steinmetz 1987.

of the planar and total folded areas of the apical surface of the α-type intercalated cells and freeze-fracture studies permitted a sampling of the density of RSPs on fracture surfaces of the apical membranes of α cells (Stetson & Steinmetz 1986). As summarized in Table 1, CO_2 addition caused an 8.8-fold increase in J_H and an approximately fourfold increase in the total apical membrane area of α cells facing the lumen. The density of RSPs also increased but the sample of fractured surfaces was too small to give statistical significance. In any event, at least half of the increase in transport rate could be accounted for by an increased number of H^+ pumps at the apical surface as judged from the total number of RSPs (see below for structure-function properties of RSPs). These results underscore the importance of the recruitment of H^+ pumps in the J_H response to CO_2, but do not exclude the possibility that kinetic factors also contribute to this response.

Structure and function of the H^+ pump complex

In their analysis of the J_H versus $\Delta\tilde{\mu}_H$ relations in turtle bladder, Andersen et al (1985) concluded that these relations reflect the intrinsic characteristics of the H^+ pumps. They used a simple model for the H^+ pump (Fig. 6) based on the available information on the structure of the F_0-F_1 ATP synthetases (Kagawa et al 1979). The pump was envisaged to consist of two components, a catalytic unit facing the cytoplasm where ATP hydrolysis is coupled to H^+ translocation, and a transmembrane channel. For simplicity, it was assumed that H^+ translocation through each of the two components was independent of the state of the other component and that the two components were linked by an intermediary component (an antechamber) with a high H^+ buffer capacity. They developed a simple kinetic model for active H^+ translocation through the H^+ pump complex which permitted simulation of the J_H versus $\Delta\tilde{\mu}_H$ characteristics shown in Fig. 1.

Recently, biochemical, immunocytochemical and ultrastructural studies have provided new insight into the structure and function of the H^+ pump of urinary acidification.

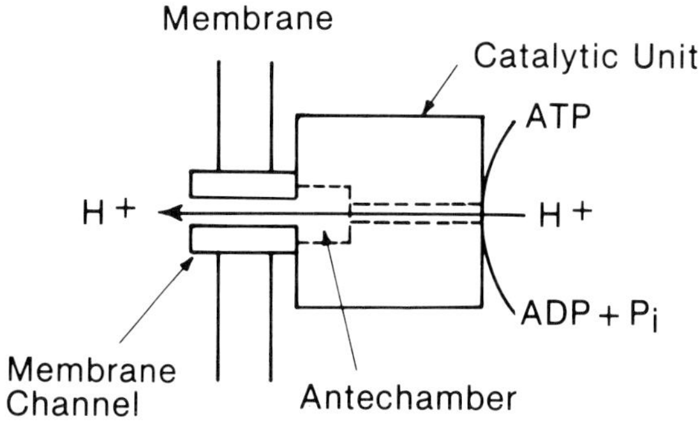

FIG. 6. Schematic representation of the components of an ATP-driven proton pump of F_0-F_1 type. The pump consists of two major components: a membrane channel and a catalytic unit where ATP hydrolysis is coupled to H^+ translocation. The two components are linked by an antechamber that serves as a buffer compartment. Reproduced from Andersen et al (1985) by copyright permission of The Rockefeller University Press.

Studies by Gluck & Al-Awqati (1984) and Gluck & Caldwell (1987) have defined an H^+-translocating ATPase or a family of H^+-ATPases in a fraction of endosomal and epithelial membranes purified from bovine kidney medulla and turtle bladder. This ATPase is inhibited by *N*-ethyl maleimide and dicyclohexyl carbodiimide (DCCD), but not by oligomycin or vanadate. Although it differs from mitochondrial ATPase in its inhibitor characteristics and distribution, it shares the major structural features of a F_0-F_1 type ATPase and it appears to operate without a phosphorylated intermediate. Its subunit structure is complex, as at least 10 different polypeptides have been isolated from endosomal preparations of H^+-ATPase. Since there is some heterogeneity of the various subunits in this group of endosomal H^+-ATPases, the actual H^+-ATPase responsible for H^+ secretion into the urine may contain slightly fewer subunits. The 70 kDa and 56 kDa subunits belong to the cytoplasmic domain of the ATPase and have been used to raise antibodies for immunocytochemical studies. Brown et al (1987) have shown that the studded regions on the cytoplasmic side of the apical membrane are highly immunoreactive to these antibodies. Other subunits of 45 kDa, 42 kDa, 38 kDa, 33 kDa, 31 kDa and 15 kDa have been identified (Gluck & Caldwell 1987). The 31 kDa subunit is hydrophilic and appears to be a component of the catalytic unit of the ATPase, whereas the 15 kDa unit, which binds DCCD, is thought to be part of the transmembrane channel.

As the amino acid sequences of the subunits become known, it is likely that the simple model of Fig. 6 can be improved and refined. In the meantime, progress has been made in several structure-function issues pertaining to the H^+ pump. The nature of the stud-like material on the cytoplasmic side of the H^+ transporting membranes has been clarified. Brown et al (1987) have shown that this material is not clathrin, but is closely related to, and most likely is, an actual part of the cytoplasmic domain of the H^+-ATPase molecule. They were able to examine the underside of apical membrane of toad bladder intercalated cells by rapidly freezing portions of the apical membrane that had been sheared away from the remainder of the cell. This enabled them to visualize regular arrays of studs. Individual studs appeared to be composed of multiple subunits arranged around a central depression and had a diameter of 9.5 nm and a density of about 16 800 per μm^2. They were also able to reconstitute the purified H^+-ATPase from bovine kidney medulla into phospholipid liposomes and demonstrate the presence of studs, as well as ATP-dependent acidification.

Although there is strong evidence that the studded coat material is related to the cytoplasmic domain of the H^+-ATPase, studs are not detectable in any number in most endosomes, lysosomes or clathrin-coated vesicles. They are seen primarily adjacent to the apical cell membranes and subapical membrane vesicles of intercalated cells, membranes capable of high H^+ transport rates to accomplish net transepithelial transport. It is not known whether the H^+-ATPase of urinary acidification differs from other related H^+-ATPases or whether its studs are detectable by EM because of a very high density of H^+ pumps. In the intercalated cells of the turtle bladder, the occurrence of cytoplasmic studs as seen by EM is closely linked to the presence of RSPs on freeze fracture of the adjacent membrane. The two are observed together in the apical pole of the α cells and at the basolateral membrane of the β cells (Stetson & Steinmetz 1985). In the α cell, which has been studied most extensively, CO_2 stimulation clearly increases the macroscopic rate of H^+ secretion and the numbers of RSPs and studs that are in the membrane facing the lumen. Stetson & Steinmetz (1986) were able to determine the total numbers of RSPs in an apical membrane position in the exposed membrane area of Ussing chambers after H^+ secretion had been measured *in vitro* (Table 1). From their morphometric and freeze-fracture studies they calculated an H^+ secretion rate of 196 H^+ ions per second per rod-shaped particle, on the basis of the average values at low and high CO_2. They measured an average RSP density of 2 906 per μm^2 of fracture surface. This RSP density compares with a stud density of 16 800 per μm^2 in the study by Brown et al (1987) in the rapid freeze preparation of α cells in toad bladder.

What is the relation between RSPs and studs? In terms of numbers, studs are about five times more numerous that RSPs. RSPs, however, are larger than studs, since they are composed of two or three globular particles of 12 nm each (Fig. 4), whereas an individual stud has a diameter of 9.5 nm. Hence, each globular particle within a RSP is likely to be composed of several transmembrane polypeptides and could serve as a channel for one or more catalytic units (studs). If one globular particle served as a channel for one catalytic unit of ATPase, each RSP would represent two or three studs or H^+

pumps. The transport rate would be between 65 and 100 H^+ per second per pump, a reasonable number from what is known from other ATPases. The relatively large size of the globular intramembrane particle could also indicate a more complex structure involving additional polypeptides belonging to the membrane skeleton and lipid components.

The precise relationship between the rod-shaped intramembrane particles and the cytoplasmic studs remains to be clarified. Their remarkable coexistence at the same cell membranes of intercalated cells suggest that they are associated with the membrane and cytoplasmic domains of the H^+ pumps of urinary acidification.

Acknowledgements

This work was supported by National Institutes of Health grant #DK-30693.

References

Al-Awqati Q, Mueller A, Steinmetz PR 1977 Transport of H^+ against electrochemical gradients in turtle urinary bladder. Am J Physiol 233:F502–F508

Andersen OS, Silveira JEN, Steinmetz PR 1985 Intrinsic characteristics of the proton pump in the luminal membrane of a tight urinary epithelium. J Gen Physiol 86:215–234

Beauwens R, Al-Awqati Q 1976 Active H^+ transport in the turtle urinary bladder: coupling of transport to glucose oxidation. J Gen Physiol 68:421–439

Brown D, Gluck S, Hartwig J 1987 Structure of the novel membrane-coating material in proton-secreting epithelial cells and identification as an H-ATPase. J Cell Biol 105:1637–1648

Cohen LH, Mueller A, Steinmetz PR 1978 Inhibition of the bicarbonate exit step in urinary acidification by a disulfonic stilbene. J Clin Invest 61:981–986

Ehrenspeck G, Brodsky WA 1976 Effects of 4-acetamido-4′-isothiocyano-2,2′-disulfonic stilbene on ion transport in turtle bladder. Biochim Biophys Acta 419:555–558

Fischer JL, Husted RF, Steinmetz PR 1983 Chloride dependence of the HCO_3^- exit step in urinary acidification by the turtle bladder. Am J Physiol 245:F564–F568

Gluck S, Al-Awqati Q 1984 An electrogenic proton-translocating adenosine triphosphatase from bovine kidney medulla. J Clin Invest 73:1704–1710

Gluck S, Caldwell J 1987 Immunoaffinity purification and characterization of H^+-ATPase from bovine kidney. J Biol Chem 262:15770–15780

Gluck S, Cannon C, Al-Awqati Q 1982 Exocytosis regulates urinary acidification in turtle bladder by rapid insertion of H^+ pumps into the luminal membrane. Proc Natl Acad Sci USA 79:4327–4331

Husted RF, Mueller AL, Kessel RG, Steinmetz PR 1981 Surface characteristics of carbonic anhydrase-rich cells in turtle urinary bladder. Kidney Int 19:491–502

Kagawa Y, Sone N, Hirata H, Yoshida M 1979 Structure and function of H^+ATPase. J Bioenerg Biomembr 11:39–78

Koeppen B 1985 Conductive properties of rabbit outer medullary collecting duct: inner stripe. Am J Physiol 248:F500–F506

Schuster VL, Bonsib SM, Jennings ML 1986 Two types of collecting duct

mitochondria-rich (intercalated) cells: lectin and band 3 cytochemistry. Am J Physiol 251:C347–C355
Steinmetz PR 1974 Cellular mechanisms of urinary acidification. Physiol Rev 54:890–956
Steinmetz PR, Lawson LR 1971 Effect of luminal pH on ion permeability and flows of Na^+ and H^+ in turtle bladder. Am J Physiol 220:1573–1580
Steinmetz PR, Stetson DL 1987 Cellular organization of active proton transport in urinary epithelia. In: Brenner BM, Stein JH (eds) Contemporary Issues in Nephrology. Modern techniques of ion transport. Churchill Livingstone, NY, vol 15:67–89
Steinmetz PR, Omachi RS, Frazier HS 1967 Independence of hydrogen ion secretion and transport of other electrolytes in turtle bladder. J Clin Invest 46:1541–1548
Stetson D, Steinmetz PR 1983 Role of membrane fusion in CO_2 stimulation of proton secretion by turtle bladder. Am J Physiol 245:C113–C120
Stetson D, Steinmetz PR 1985 Alpha and beta types of carbonic anhydrase-rich cells in turtle bladder. Am J Physiol 249:F553–F565
Stetson DL, Steinmetz PR 1986 Correlation between apical intramembrane particles and H^+ secretion rates during CO_2 stimulation in turtle bladder. Plfuegers Arch Eur J Physiol 407:S80–S84
Stetson DL, Beauwens R, Palmisano J, Mitchell PP, Steinmetz PR 1985 A double membrane model for urinary bicarbonate secretion. Am J Physiol 18:F546–F552
Stone DK, Xie XS, Racker E 1983 An ATP-driven proton pump in clathrin-coated vesicles. J Biol Chem 258:4059–4062
van Adelsberg J, Al-Awqati Q 1986 Regulation of cell pH by Ca^{+2} mediated exocytotic insertion of H^+-ATPases. J Cell Biol 102:1638–1645
Wagner S, Vogel R, Lietzke R, Koob R, Drenckhahn D 1987 Immunochemical characterization of a band III-like anion exchanger in collecting duct of human kidney. Am J Physiol 253:F213–F221

DISCUSSION

Machen: You showed that there was an increase in the number of pumps/particles exposed to the lumen but an even larger increase in the rate of H^+ secretion. Does this finding indicate that the increase in acid secretion rate is caused by both an increase in the number of pumps and an increase in the activity of those pumps?

Steinmetz: The results of the morphometric studies can account for only about 60% of the observed increase in H^+ secretion (Table 1). So I think it's quite possible that a given pump can also accelerate its transport rate by a kinetic mechanism.

Machen: How did you measure apical area?

Steinmetz: The planar apical area is obtained from transmission electron micrographs. It is a measure of the degree of folding of a membrane. From these two values a total apical area can be calculated (see Stetson & Steinmetz 1986).

Malnic: You said that the HCO_3^- secretory mechanism is electrically silent.

However, if it is based simply on a reverted α-cell, it should have the same short circuit current but in the opposite direction.

Steinmetz: That is true! The HCO_3^--secreting β-cell, however, is not simply a reverted α-cell. In the model we presented (Stetson et al 1985), only the positions of the H^+ pump and the Cl^-/HCO_3^- exchanger are reversed. The Cl^- channel is at the basolateral membrane in both cell types. Under ordinary conditions, the apical membrane of the β-cell contains only an electroneutral Cl^-/HCO_3^- exchanger, whereas the basolateral membrane has an electrogenic pump in parallel with a conductive Cl^- channel. Each H^+ extruded is matched by a Cl^- exiting via the channel. At the apical membrane of the β-cell, the Cl^- enters in direct exchange for the secreted HCO_3^-.

Machen: How is electroneutrality maintained when H^+ are secreted across the apical membrane of the α-cell at open circuit?

Steinmetz: At open circuit the overall electroneutrality is preserved by the paracellular movement of ions, mainly Cl^-. The apical membrane of the α-cell itself is thought to be extremely tight. Dr Bruce Koeppen (1985) found no appreciable conductances for Cl^- or other ions in the apical membranes of the α-cells of the outer medullary collecting duct. These cells do have a basolateral Cl^- conductance. To return to your question, there is charge separation caused by electrogenic H^+ pumping at the apical membrane. The electron generated on the inside of the cell moves from OH^- to HCO_3^- and ultimately exits the α-cell as a result of the Cl^- recycling across the basolateral membrane in parallel with the Cl^-/HCO_3^- exchanger.

Stone: These rod-shaped particles are intramembranous and 12 nm in diameter.

Steinmetz: No, the rod-shaped particle is made of two or three spherical particles, each of which is 12 nm in diameter as measured by this technique of freeze fracture. The rod-shaped particle is really a linear assembly of intramembrane particles!

Stone: That's very different from F_0 which is a very tiny integral membrane protein. By analogy with the F_1 F_0 type pumps, you could say that the studs that you see are the hydrolytic core. The reason I bring this up is that Dennis Brown (Brown et al 1987) can't find any rod-shaped particles. If he takes the purified enzyme, reconstitutes it into liposomes then looks at those by freeze fracture electron microscopy, he can't detect any rod-shaped particles.

Steinmetz: He does find the rod-shaped particles in ordinary tissue, such as turtle bladder. When he reconstitutes the purified enzyme from bovine kidney in an unusual lipid, soybean phospholipid, then he doesn't see rod-shaped particles but there are studs.

Stone: But he gets proton pumping in the absence of the rod-shaped particles.

Steinmetz: These liposomes do acidify but only at a very low rate. I think the uniqueness of the apical membrane, in contrast to other endosomes of these tissues, is that their function is to transport large amounts of protons transepithelially. I believe the clue to these multiple lined-up particles is that they allow high transport rates. I wouldn't necessarily expect the particles to line up in an artificial lipid, such as soybean phospholipid.

Stone: Why? The studs line up. Do you think that there are latent Cl^- channels? It's now been shown that there is a latent channel in the membrane and it is not expressed until you pull it off.

Steinmetz: If there are latent Cl^- channels, we have never seen them activated or in an open state in the apical membranes of α-cells.

Boron: What is the ratio of Dennis Brown's rod-shaped particles to your studs?

Steinmetz: We counted rod-shaped particles per μm^2, so it would be our particles to Dennis' studs! There would be from 2 – 6 studs per rod-shaped particle, or one or two studs per spherical unit in a rod-shaped particle. So the matching for these estimates is pretty close. If you calculate the rate of proton transport per RSP, it is between 150 in low CO_2 and 250 in the high CO_2 conditions. If you assume one stud per spherical unit, you end up with a reasonable pump rate of between 60 and 100 protons per second per molecule. If you assume two studs per spherical unit or five per RSP, it still gives a reasonable range.

Boron: Do the electron micrographs of Dennis Brown's studs show any symmetry of 5:1?

Steinmetz: Dennis Brown cuts below the level of the membrane, so he cuts through the studs and doesn't see the RSPs simultaneously. Different techniques are used for demonstrating studs and RSPs. If you use regular freeze fracture, you go through the membrane leaflet. If you happen to cut through the cytoplasm you often see what looks like studs. But studs are very difficult to see by this technique, they are best studied by high magnification transmission EM.

Boron: But does the stud show any symmetry that is suggestive of 3:1 or 5:1 ratio?

Steinmetz: The ratio is based entirely on a comparison of our counts of RSPs (Stetson & Steinmetz 1986) with Dennis Brown's independently obtained counts of studs in toad bladder (Brown et al 1987).

Harvey: It's very nice to see optical evidence for a transporter that we have been working on by kinetic means and electrophysiology. Returning to the point raised by Machen about the charge balance for H^+ efflux: we have heard about conditions where proton secretion is matched by Na^+ absorption and this would seem to provide charge balance.

Steinmetz: I think the amiloride-sensitive channels would be in the granular cells and may be not in intercalated cells.

Harvey: Well they seem to be in the mitochondrial rich cells as well. There is evidence that in other tight epithelia, the intercalated cells do have an amiloride-sensitive Na^+ channel and a ouabain-sensitive Na^+/K^+-ATPase (Larsen et al 1987).

Steinmetz: In turtle bladder, if you separate the intercalated cells from the granular cells, they don't respond to ouabain in terms of oxygen consumption. Under open-circuit conditions, we get a higher H^+ transport rate than under short-circuit conditions, and the increment is entirely explained by the lumen-negative potential.

Harvey: In open-circuit conditions, the H^+ efflux requires a balancing movement of charge. In frog skin this counter-current is provided by Na^+ influx. I would contend that the short-circuit state is non-physiological.

Steinmetz: I agree but it's the state that allows you to do experiments.

Morad: Could I ask about the increase in the number of rod-shaped particles as you increase p_{CO2}. How do you see this—are these proteins not present before and being synthesized *de novo*, or are they there but you cannot detect them for some reason?

Steinmetz: Gluck et al (1982) showed that if you acutely increase the p_{CO2} to 40 mm Hg, you get fusion of pre-existing membrane vesicles to the apical membrane in a matter of seconds. So probably most of them were already present within the cell. If you remove CO_2, you get a substantial amount of endocytosis. When CO_2 is then increased from 0 to 5%, the increase in rod-shaped particles is due to fusion of pre-existing vesicles that have the studs and the rod-shaped particles on them. About half of the vesicles that are sub-apical in α-cells have studs and rod-shaped particles.

Morad: When you showed us the non-studded particles, my conclusion was that there was not much difference between the studded and the non-studded ones.

Steinmetz: Fig. 3 shows an extensive coating of the cytoplasmic surface of the apical membrane with studs. I haven't shown the way the various sub-apical membranes look; see Stetson & Steinmetz (1985) for more micrographs. It is true that clathrin-coated vesicles may superficially resemble studded vesicles. However, they can be distinguished on a good electron micrograph or by anti-clathrin antibodies. The studded vesicles fail to react to anti-clathrin antibodies (Brown et al 1987). The clathrin-coated vesicles are surrounded by a basket of clathrin molecules and appear fuzzy; studs, on the other hand, appear as little stalked spheres just below the surface of the inner leaflet of the membrane.

Thomas: So one stud is one proton pump?

Steinmetz: Yes! The stud is the cytoplasmic component of a proton pump. It contains subunits that are cytoplasmic and have been used to raise antibodies, the 70 kDa and 56 kDa subunits (Brown et al 1987, Gluck & Caldwell 1987).

Morad: Is there any reason to believe that these proton pumps are somehow associated with the Na^+ channel itself? Is it a separate entity within membrane, or or is it somehow combined with the five or six subunit structure that you describe?

Steinmetz: No, we believe that the proton pump is separate from the Na^+ channel. We have concentrated on the apical membrane of this subtype of cell, because it is highly specialized in H^+ secretion. The evidence suggests that these cells transport little or no Na^+. The same is true for intercalated cells in the outer medullary collecting duct, where they can be punctured.

Harvey: There is a way to facilitate the puncture of these cells. You can increase their number by chronic acid loading of the animals, and in frogs treated thus the intercalated cells are much bigger and their number matches the number of granular cells in the outer layer.

References

Brown D, Gluck S, Hartwig J 1987 Structure of the novel membrane-coating material in proton-secreting epithelial cells and identification as an H-ATPase. J Cell Biol 105:1637–1648

Gluck S, Caldwell J 1987 Immunoaffinity purification and characterization of H^+-ATPase from bovine kidney. J Biol Chem 262:15770–15780

Gluck S, Cannon C, Al-Awqati Q 1982 Exocytosis regulates urinary acidification in turtle bladder by rapid insertion of H^+ pumps into the luminal membrane. Proc Natl Acad Sci USA 79:4327–4331

Koeppen B 1985 Conductive properties of rabbit outer medullary collecting duct: inner stripe. Am J Physiol 248:F500–F506

Larsen EH, Ussing HH, Spring KR 1987 Ion transport by mitochondria-rich cells in toad skin. J Membr Biol 99:25–40

Stetson DL, Steinmetz PR 1986 Correlation between apical intramembrane particles and H^+ secretion rates during CO_2 stimulation in turtle bladder. Pflugers Arch Eur J Physiol 407:S80-S84

Stetson DL, Beauwens R, Palmisano J, Mitchell PP, Steinmetz PR 1985 A double membrane model for urinary bicarbonate secretion. Am J Physiol 18: F546–552

Epithelial pH and ion transport regulation by proton pumps and exchangers

Brian J. Harvey and Jordi Ehrenfeld

Laboratoire Jean Maetz, Département de Biologie, Commissariat à l'Energie Atomique, Station Marine B.P. 68, F-06230 Villefranche-sur-Mer, France

Abstract. This study reports on the interaction between transepithelial Na^+ transport and H^+ secretory and intracellular pH (pH_i) regulating mechanisms in the model 'tight' epithelium of frog skin. We have used ^{22}Na isotope fluxes and fixed end-point titration to measure undirectional Na^+ fluxes, net Na absorption (J_{net}^{Na}) and proton secretion (J_{net}^{H}), and electrophysiological techniques (double-barrelled ion-sensitive microelectrodes and cell membrane current–voltage relations) to determine intracellular activities of Na^+, Cl^- and H^+ and the conductance of apical membranes to Na^+ (g_{Na}) and of basolateral membranes to K^+ (g_K).

In dilute mucosal solutions or in the absence of a permeant anion (Cl^-) or counter-current (open-circuit conditions) to accompany Na^+ uptake, the J_{net}^{Na} is electrically coupled to J_{net}^{H} via an electrogenic apical H^+-ATPase (located in mitochondria-rich cells). Both fluxes proceed via mitochondria-rich cells and are inhibited by blockers of carbonic anhydrase and H^+-ATPase and stimulated by aldosterone and acid load.

In high NaCl-containing mucosal solutions or in short-circuit conditions, the J_{net}^{Na} becomes uncoupled from J_{net}^{H} and proceeds mainly via the principal cells in the epithelium, in which pH_i is regulated by basolateral Na^+/H^+ and Cl^-/HCO_3^- exchangers. Under these conditions, J_{net}^{Na}, g_{Na} and g_K vary directly and in parallel with pH_i, when pH_i is changed by permeable weak acids or bases. There is also co-variance between g_{Na} and pH_i accompanying spontaneous variations in J_{net}^{Na} and when Na^+ transport is stimulated by aldosterone or inhibited with ouabain.

We conclude that the level of intracellular H^+, modulated by H^+ pump and Na^+/H^+ and Cl^-/HCO_3^- exchangers provides an intrinsic regulation of epithelial Na^+ transport.

1988 Proton passage across cell membranes. Wiley, Chichester (Ciba Foundation Symposium 139) p 139–164

Since the pioneering work of Koefoed-Johnsen and Ussing (1958), the amphibian skin and urinary bladder have proved to be useful and highly successful models of epithelial vectorial ion transport.

Ion transport across an epithelium is different from modes of transport in symmetrical or isolated cells, such as muscle, neurons or blood cells, in that ions must cross two membrane barriers in series, each with different ion selective permeabilities. It is this asymmetry in permeability that allows epithelial structures such as the kidney or amphibian skin and urinary bladder to maintain net sodium absorption and hydrogen ion secretion.

The transepithelial transport of H^+ and Na^+ have, however, been mainly studied separately without serious consideration given to a possible interaction between these two systems. Studies in our laboratory have focused on the coupling existing between J_{net}^{Na} and J_{net}^{H} *in vivo* and *in vitro* (Garcia-Romeu et al 1969, Ehrenfeld & Garcia-Romeu 1977). We have shown that in addition to its involvement in proton secretion, like in turtle urinary bladder (Steinmetz & Andersen 1982), an essential role exists for the electrogenic H^+ pump in the genesis of a favourable electrical driving force for apical Na^+ influx (Ehrenfeld et al 1985, Harvey & Ehrenfeld 1986).

Besides proton secretion via the H^+-ATPase pump, frog skin epithelium also regulates its intracellular pH (pH_i) by electroneutral Na^+/H^+ and Cl^-/HCO_3^- exchangers (Duranti et al 1986, Ehrenfeld et al 1987, Harvey et al 1988). In particular, we observed that the operation of pH_i regulatory mechanisms can have profound effects on the permeabilities to Na^+ and K^+ of the two opposing barriers and consequently on transepithelial Na^+ transport. Thus it will be shown that pH_i can be an intrinsic regulator of epithelial ion transport and mediate the coupling between ion flows across opposing membranes.

Tight epithelia usually contain two main cell types, mitochondria-rich (Mr) cells and principal (Pr) cells (granular cells). There is good evidence that in frog skin epithelium the H^+-ATPase is localized to Mr Cells as in the turtle urinary bladder (Stetson & Steinmetz 1985). In Ussing conditions, Na^+ transport is likely to occur mainly via Pr cells. We have investigated the relative participation of, and interaction between, these two cell types in the overall regulation of transepithelial Na^+ and H^+ transport by exploiting the different effects of pH_i on these transport mechanisms.

Experimental procedures

Our studies reported here were carried out on the isolated whole frog skin ('whole skin') and on sheets of epithelium isolated from whole skin by collagenase treatment. The tissues were mounted in Ussing-type chambers which permit separate superfusion of apical and basolateral sides with saline solutions. Transepithelial and transmembrane unidirectional Na^+ fluxes were measured with the isotope ^{22}Na. Net Na^+ fluxes were calculated by flame photometry; net proton secretion was determined by fixed end-point titration of the apical solutions. Details of these techniques have been described previously (Ehrenfeld & Garcia-Romeu 1977, Ehrenfeld et al 1987).

We used double-barrelled ion-sensitive microelectrodes to determine the intracellular activities of Na^+, Cl^- and H^+; details of the construction, calibration and use of these electrodes are given in Harvey & Kernan 1984, Harvey & Lahlou 1986 and Harvey et al 1988.

Ion transport and pH_i regulation were studied under spontaneous transport conditions (open-circuit) or when the transepithelial potential was voltage clamped to zero (short-circuit) or to any desired voltage in the range ±

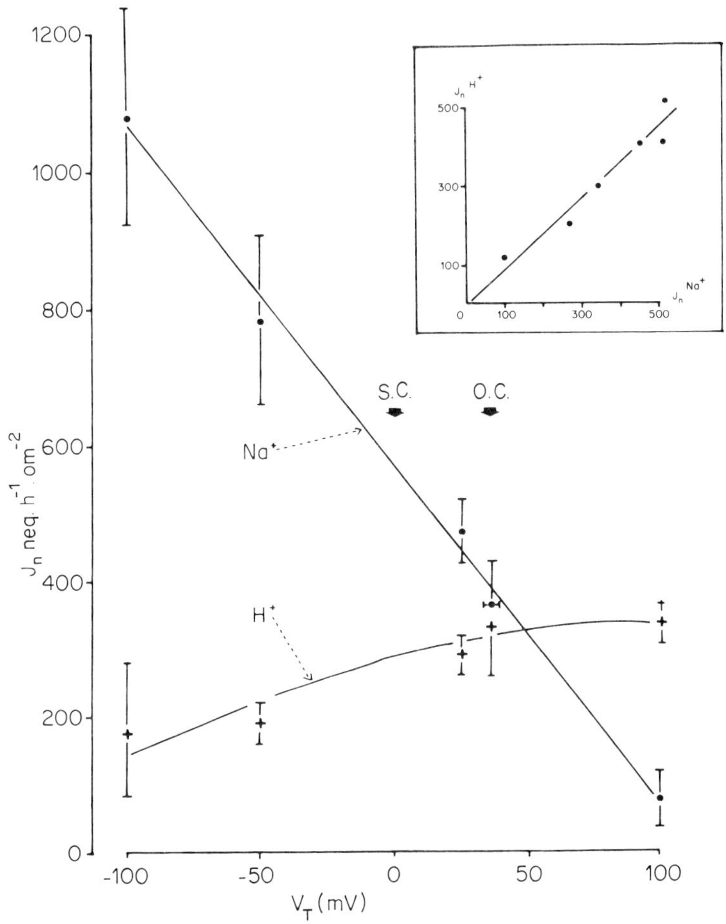

FIG. 1. Evolution of net Na^+ absorption (J_{net}^{Na}) and net proton secretion (J_{net}^{H}) as a function of transepithelial electrical potential (V_t) (referenced to mucosal side) in whole frog skin *in vitro*. The mucosal solution was 1 mM Na_2SO_4 buffered at pH 7.34 with 4 mM imidazole. The serosal solution was normal frog Ringer containing 85 mM NaCl, buffered to pH 7.4 with 5% CO_2, 24 mM $NaHCO_3$. n=8. The V_t was varied by an external e.m.f. Net Na^+ and H^+ fluxes are equal in magnitude (but opposite in direction) only at spontaneous open-circuit potentials (OC). This 1:1 coupling between J_{net}^{Na} and J_{net}^{H} is shown in the inset. Under short-circuit conditions (SC) and at imposed V_t other than OC, the J_{net}^{Na} and J_{net}^{H} are uncoupled and are opposite functions of V_t.

200 mV. From computer-driven current-voltage relations of individual apical and basolateral cell membranes, we determined the conductance and permeability of Na^+, K^+ and H^+ transport pathways. Details of these electrophysiological techniques and data acquisition are given in Harvey et al 1988.

Dependence of Na^+ and H^+ transport on transepithelial electrical potential

Net Na^+ absorption and H^+ secretion are electrogenic processes and are dependent on the magnitude of the transepithelial potential (V_t). This voltage dependence of J_{net}^{Na} and J_{net}^{H} is shown in Fig. 1. J_{net}^{Na} occurs in the apical to serosal direction, whereas J_{net}^{H} has the opposite orientation. These two fluxes vary inversely with V_t and are equal only when the epithelium is in the open-circuit state, i.e. at spontaneous V_t, and are unrelated under short-circuit conditions and at imposed V_t values other than the open-circuit potential. In the presence of dilute mucosal solutions (1 mM Na_2SO_4) under open-circuit conditions, a good correlation exists between J_{net}^{Na} and J_{net}^{H} (see inset of Fig. 1). This one-to-one coupling between J_{net}^{Na} and J_{net}^{H} is observed only in dilute mucosal solutions under open-circuit conditions *in vitro* or *in vivo* (Garcia-Romeu et al 1969, Ehrenfeld & Garcia-Romeu 1977, Ehrenfeld et al 1985).

TABLE 1 The effects of certain drugs on Na^+ and H^+ transport in whole frog skin *in vitro*

Conditions V_T *(mV)*	*Drug used*	$J_{in}^{Na^+}$	$J_{net}^{Na^+}$ *(neq h^{-1} cm^{-2})*	$J_{net}^{H^+}$
Open circuit $V_T = 54$ mV	0 (Control)	285±35	179±54	−179±70
Open circuit $V_T = -43$ mV	Amiloride 5×10^{-5} M	72±13	−43±42	24±13
Clamp $V_T = 54$ mV	Amiloride 5×10^{-5} M	37±13	−19±32	−178±51
Open circuit $V_T = 42$ mV	0 (Control)	290±24	285±40	−232±35
Open circuit $V_T = 75$ mV	Ethoxzolamide (apical) 10^{-4} M	159±12	84±8	−59±10
Clamp $V_T = 100$ mV	0 (Control)	131±20	−41±18	−113±56
	Ethoxzolamide (apical) 10^{-4} M	105±32	−35±33	−22±22
S.C. $V_T = 0$	0 (Control)	—	430±47	−116±23
	Diamox 10^{-3} M	—	440±65	−26±12

Amiloride, Na^+ channel blocker; ethoxzolamide and Diamox, carbonic anhydrase inhibitors. J_{in}^{Na}, unidirectional apical Na^+ uptake; J_{net}^{Na}, net transepithelial Na^+ absorption; J_{net}^{H}, H^+ secretion. V_t, imposed transepithelial voltages. Skins bathed in 1 mM Na_2SO_4 solution on the mucosal side and standard Ringer buffered in 5% CO_2, 24 mM $NaHCO_3$ (pH 7.4) on the serosal side. $n = 6$ in all experiments, except Diamox $n = 10$.

TABLE 2 Effects of inhibition of net H^+ secretion on apical membrane voltage, intracellular pH and Na^+ activity in isolated epithelia

Experimental condition	*% inhibition of $J_{net}^{H^+}$*	V_a	pH_i	a_{Na}^{i} *(mM)*	*n*
Control	—	−47±2	7.20±0.03	6.2±0.5	12
Ethoxzolamide 10^{-4} M	75	−20±3	7.38±0.08	3.4±0.6	12
Amiloride 10^{-5} M	100	−105±4	7.06±0.05	1.2±0.8	6
Na^+ free mucosa	100	−106±5	7.02±0.06	0.8±0.5	6
Voltage clamp (V_a = −100 mV)	100	−100	7.04±0.06	9.5±0.5	6
DCCD 5 × 10^{-5} M	72	—	7.02±0.05	—	6
Vanadate 10^{-3} M	100	—	7.02±0.06	—	6
Oligomycin 14 μg/ml	83	—	—	—	6
O_2 free	100	—	—		6

J_{net}^{H}, net H^+ secretion; V_a, apical membrane voltage; pH_i, intracellular pH; a_{Na}^{i} activity. DCCD, vanadate, oligomycin and anoxia experiments were performed in the presence of apical amiloride (5 × 10^{-6} M) to inhibit Na^+ transport in voltage-clamped frog skins (V_t = 50 mV). (From Harvey & Ehrenfeld 1986 by permission of Springer-Verlag.)

Uncoupling of J_{net}^{Na} and J_{net}^{H}

H^+ secretion independent of Na^+ absorption

Amiloride, when applied from the apical side in micromolar concentrations, blocks Na^+ entry into the cell and by this effect produces a hyperpolarization of the apical cell membrane potential (V_a). Amiloride causes complete inhibition of both J_{net}^{Na} and J_{net}^{H} under open-circuit conditions in dilute mucosal solutions (Tables 1, 2). The inhibitory effect of amiloride on J_{net}^{H} is indirect, since the diuretic has no effect on J_{net}^{H} provided that the voltage changes consequent to inhibition of Na^+ entry are prevented by imposing a transepithelial voltage clamp (Table 1). We have used this property to investigate the characteristics of the H^+ pump in the absence of Na^+ transport. We find that H^+ secretion is dependent on both transepithelial electrical and pH gradients, and J_{net}^{H} is anulled at a mucosal pH of 5.75 (serosal pH 7.4) and a V_t of − 80 mV (serosa negative with respect to mucosa).

Na^+ absorption independent of H^+ secretion

Ethoxzolamide and acetazolamide (Diamox) are powerful inhibitors of carbonic anhydrase, which catalyses the formation of H^+ and HCO_3^- from the hydration of CO_2. Under open-circuit conditions, when ethoxzolamide is

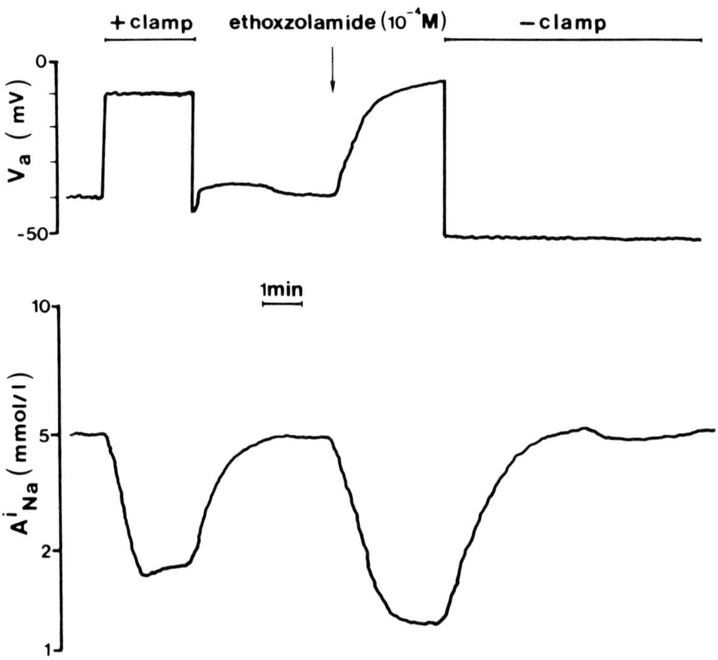

FIG. 2. Recording of intracellular sodium activity (a_{Na}^{i}) and apical membrane potential (V_a) with a double-barrelled Na^+-sensitive microelectrode in whole frog skin *in vitro* under open-circuit conditions and bathed in similar solutions as in Fig. 1. At the bar marked '+ clamp', the apical membrane was depolarized to −10 mV by passing a transepithelial current in the apical to serosal direction. This produced a fall in a_{Na}^{i} from 5 to 1.8 mM, which was reversed on release of the voltage clamp. Thus depolarization alone of V_a can lower the cell Na^+ transport pool. When ethoxzolamide (10^{-4} M) was added to the apical solution, V_a depolarized from −42 mV to −7 mV and a_{Na}^{i} decreased to 1.2 mM. While the drug was still present, the V_a was voltage clamped to − 50 mV (at bar marked ' − clamp') which restored a_{Na}^{i} to 5.2 mM. This experiment shows that the effect of inhibition of the H^+ pump on Na^+ absorption is purely electrical. (From Harvey & Ehrenfeld, 1986 by permission of Springer-Verlag.)

added to the apical or basolateral side of the epithelium both J_{net}^{Na} and J_{net}^{H} decrease by the same amount (Table 1) with an accompanying increase in transepithelial electrical resistance and increase in V_t to more positive values. Under short-circuit conditions or when V_t is clamped to values other than at open-circuit, i.e. in conditions when J_{net}^{Na} is uncoupled from J_{net}^{H}, only J_{net}^{H} is affected by ethoxzolamide (Table 1). Thus the effect of ethoxzolamide on Na^+ absorption in open-circuit is indirect and is a result of the voltage changes induced by ethoxzolamide, as is shown by the experiment in Fig. 2. Ethoxzolamide causes a rapid depolarization of V_a and a fall in intracellular Na^+ activity (a_{Na}^{i}). J_{net}^{H} and J_{net}^{Na} are both inhibited in this situation. When V_a is subsequently repolarized by an external voltage clamp, the a_{Na}^{i} increases to control values and J_{net}^{Na} but not J_{net}^{H} recovers. An analogous effect on a_{Na}^{i} and J_{net}^{Na} is observed when V_a is clamped to values similar to the Nernst potential for Na^+ (E_{Na}) (Fig. 2). This result demonstrates that depolarization of V_a can decrease J_{net}^{Na} and the cellular Na^+ transport pool.

Under open-circuit conditions in dilute mucosal solutions (1 mM Na_2SO_4), the Na^+ chemical gradient across the apical membranes opposes passive Na^+ entry (a_{Na}^{i} = 6 mM). H^+ secretion, however, is a transepithelial rheogenic process and imposes a transepithelial potential which is reflected in hyperpolarization of V_a below E_{Na}, thus facilitating passive electrodiffusion of Na^+ into the cell. The H^+ pump acts as a 'physiological' voltage clamp whose contribution to apical membrane voltage depends on the ratio of apical membrane resistance to total transcellular resistance.

The role of the H^+ pump in the control of the cell Na^+ transport pool and of cell pH under conditions in which Na^+ absorption is linked to H^+ secretion is demonstrated by the use of various inhibitors of these two transport systems (Table 2). Ethoxzolamide increases pH_i, possibly by its inhibitory effect on CO_2 hydration, and decreases a_{Na}^{i} consequent to the decrease in apical membrane potential and electrical driving force for Na^+ entry. Amiloride or a mucosal Na^+-free solution blocks Na^+ entry and reduces a_{Na}^{i}. The accompanying hyperpolarization of V_a appears to be responsible for cell acidification under these conditions. Both pH_i and a_{Na}^{i} are sensitive to apical membrane voltage changes. Voltage clamp of V_a to very negative values (−100 mV) causes a fall in pH_i and an increase in a_{Na}^{i} (Table 2).

Rheogenicity of the H^+ pump

The electrical characteristics of the H^+ pump pathway were obtained from ethoxzolamide-sensitive transcellular current–voltage relations (I_H–V_H). The I_H–V_H relation was a saturable exponential function of the form:

$$I^P = I_{max}^{P} \left(1 - \exp \left(E_R^{P} - V_a\right)/\xi\right)$$

where the maximum pump current I_{max}^{P} = 155 neq h^{-1} cm^{-2} (4.15 μA cm^{-2}) and the pump reversal potential E_R^{P} = − 130 mV. (ξ is an empirical constant and defines the shape of the I_H–V_H curve.) The ethoxzolamide-sensitive proton current is practically constant for V_a between ± 50 mV and thus the H^+ pump acts as a constant current source over this voltage range, as reflected in J_{net}^{H} (see Fig. 1). That amiloride inhibits J_{net}^{H} can now be understood, since amiloride causes V_a to hyperpolarize to levels approaching the reversal potential of the H^+ pump.

When Na^+ transport is inhibited by amiloride and the changes in V_t are prevented by a voltage clamp, we find that drug- or acid load-induced variations in J_{net}^{H} are reflected by equal changes in transepithelial current (I_t). Comparing simultaneous variations in I_t and J_{net}^{H}, we find a good linear relation between both parameters (r = 0.90 ± 0.09, n = 27) when H^+

secretion is inhibited by oligomycin, diocyclohexyl carbodiimide (DCCD) or acidic mucosal pH. This is evidence that H^+ secretion is mediated solely by a rheogenic H^+-ATPase pump.

Coupling of H^+ secretion to metabolism

The energy source for proton secretion is oxidative metabolism, since $J_{net}{}^{H}$ is completely abolished in anoxia (Table 2). In this respect the H^+ pump in frog skin epithelium differs from that in turtle urinary bladder, which can use either aerobic or anaerobic metabolism to drive active H^+ transport (Schwartz & Steinmetz 1977). DCCD and oligomycin are powerful inhibitors of mitochondrial ATPases (Bowman & Slayman 1977) and it has been suggested that DCCD exerts a primary inhibitory action on the H^+ pump in turtle urinary bladder (Steinmetz et al 1981). These agents drastically reduce $J_{net}{}^{H}$ in frog skin, as does vanadate (Table 2), which does not inhibit mitochondrial ATPase (Bowman et al 1978). Vanadate may thus inhibit the H^+-ATPase pump directly without affecting mitochondrial respiration. Under spontaneous transport conditions, the calculated free energy of the H^+ pump is 4.7 kcal mol^{-1} and the molar free energy for ATP hydrolysis in frog skin is 15 kcal mol^{-1} (Civan et al 1983). If ATP is the sole energy source for H^+ secretion, this gives a stoichiometry of 3 H^+ per ATP hydrolysed.

In summary, we envisage two essential physiological roles for the proton pump:

(1) The high rate of H^+ secretion indicates that the pump plays a major role in the acid–base balance of the whole animal.

(2) As a result of its electrogenicity, transepithelial H^+ secretion acts as a physiological voltage clamp which provides an electrical driving force for passive Na^+ uptake from dilute mucosal solutions or in the absence of a permeant anion or counter-current to Na^+ absorption. Under these conditions the H^+ pump participates in the regulation of the cellular Na^+ transport pool.

HCO_3^- reabsorption

Secretion of protons via the H^+ pump will leave excess HCO_3^- (or OH^-) in the cell which must be extruded across basolateral membranes to the blood side in order to keep a stable cell pH and explain the transepithelial nature of H^+ secretion (reflected in I_H). We have found a 1:1 correlation between the rate of acidification of the mucosal Ringer and the rate of alkalinization of the serosal solution ($J_{net}{}^{HCO3-}$). $J_{net}{}^{H}$, $J_{net}{}^{HCO3-}$ and net Na^+ absorption (linked to H^+ secretion) are simultaneously and equally inhibited by ethoxzolamide (83% of control) or by removal of Cl^- from the serosal Ringer (67% of control) (Duranti et al 1986). Inhibitors of Cl^-/HCO_3^- exchange reduce bicarbonate reabsorption: DIDS (4,4′-diisothiocyanatostilbene-2,2′-disulphonic acid), furosemide and meclofenamate inhibit J_{net}^{HCO3-} by 35%, 21% and 53%, respectively. Depolarization of the basolateral membrane does not inhibit HCO_3^- reabsorption, indicating an electrically silent HCO_3^- efflux mechanism. We conclude that H^+ secretion via the apical proton pump and HCO_3^- reabsorption via an electroneutral basolateral Cl^-/HCO_3^- exchanger are linked in series via cellular production of H^+ and HCO_3^-. An electrogenic Cl^- efflux in parallel with the Cl^-/HCO_3^- exchanger may provide the associated basolateral membrane current in series with the electrogenic apical H^+ pump.

Na^+/H^+ exchange and intracellular pH regulation

In acidosis, the amphibian skin and urinary bladder respond by increasing the capacity of the proton pump (Cohen & Steinmetz 1980, Ehrenfeld & Garcia-Romeu 1977, Frazier 1984). Cell pH must also be regulated and in this case the H^+ pump and its associated Cl^-/HCO_3^- exchanger may not be the unique pH_i regulating mechanisms, especially if they are localized solely to mitochondria-rich cells. We have studied pH_i regulation using analogues of amiloride which specifically block Na^+ channels (phenamil) or Na^+/H^+ exchangers (ethyl isopropyl amiloride, EIPA). We found that a Na^+/H^+ exchanger is present only at the basolateral cell membrane (Ehrenfeld et al 1987), as

TABLE 3 Evidence for basolateral Na^+/H^+ exchange in frog skin epithelia

	Pre-treatment of the epithelium		J_{32}^{Na} *(neq h^{-1} cm^{-2})*	*Difference*
Acid stimulation of the basolateral J_{32}^{Na}	control	(n = 11)	617±105	908±150
	CO_2		1525±156	
	control	(n = 6)	468±62	499±105
	NH_4^+		967±164	
EIPA inhibition of the basolateral J_{32}^{Na}	Non-acid loaded epithelia (n = 5)	control	796±210	271±79
		EIPA	525±140	
	Acid-loaded epithelia (n = 6)	control	1525±156	621±126
		EIPA	904±137	

Upper panel: stimulation of unidirectional ^{22}Na influx across the basolateral cell membrane (J_{32}^{Na}) by intracellular acid loads provoked by pre-treatment of the epithelia with CO_2 (5%) or NH_4Cl (15 mM). Lower panel: sensitivity of J_{32}^{Na} in non-acid loaded and acid loaded epithelia to ethyl isopropyl amiloride (EIPA), a specific inhibitor of the Na/H exchanger. J_{32}^{Na} was measured in the presence of the Na^+ channel blocker, prenamil, in standard Ringer equilibrated in air. (From Ehrenfeld et al 1987 by permission of Springer-Verlag.)

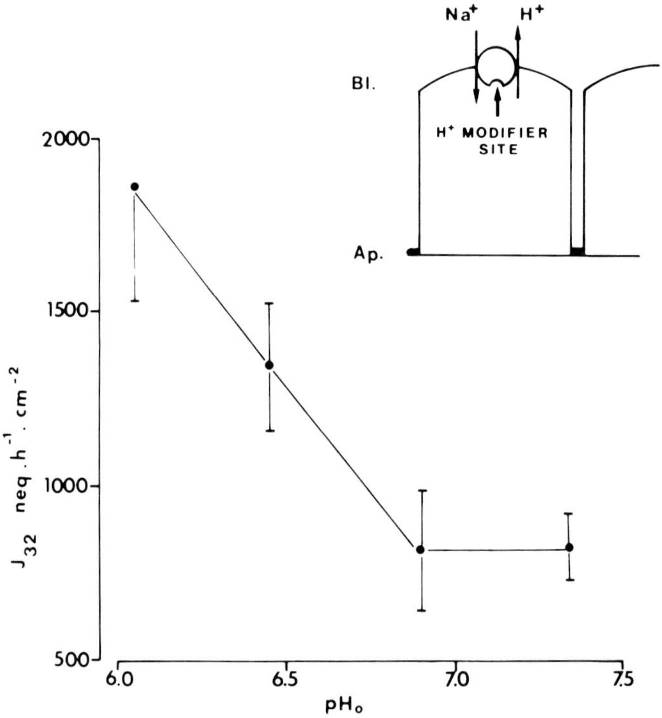

FIG. 3. The effect of cell acidification on unidirectional ^{22}Na uptake across the basolateral cell membranes of isolated epithelia from frog skin. Sheets of epithelia were pre-incubated in frog Ringer solutions containing different amounts of bicarbonate gassed with 5% CO_2 and having the different pH values given on the abscissa. n= 5. The activity of the Na^+/H^+ exchanger may be positively modulated by intracellular protons, thus explaining its low activity at alkaline pH. (From Ehrenfeld et al 1987.)

demonstrated by the stimulation of ^{22}Na influxes across the basolateral cell membrane (J_{32}^{Na}) after an intracellular acid load (Table 3, Fig. 3). The acid-stimulated J_{32}^{Na} was inhibited by amiloride ($K_i = 7 \times 10^{-6}$ M) or EIPA ($K_i = 5 \times 10^{-7}$ M) (Table 3), whereas phenamil had no effect on these basolateral Na^+ fluxes, although it completely blocks Na^+ conductive entry ($K_i = 4 \times 10^{-7}$ M).

The Na^+/H^+ exchanger is not quiescent, since EIPA or amiloride inhibit basal J_{32}^{Na} under spontaneous transport conditions (Table 3). The antiporter is also reversible in Na^+ loaded epithelia (ouabain-treated) in which $^{22}Na^+$ efflux across the basolateral membranes is stimulated by an inwardly directed H^+ gradient (serosa to cell) and blocked by EIPA or amiloride (Ehrenfeld et al 1987).

The operation of the Na^+/H^+ exchanger is essential for the recovery of pH_i

from an intracellular acid load (Harvey & Ehrenfeld 1988). Inhibition of the antiporter by EIPA, amiloride or Na^+ free serosal solutions greatly slows down the rate of pH_i recovery (Fig. 4a). We also found that the Na^+/H^+ exchanger is active over the physiological range of pH_i (7.1 to 7.4) and thus participates in the maintenance of steady-state pH_i.

Na^+/H^+ exchange is electroneutral and proton efflux is dependent on the inwardly directed Na^+ chemical gradient across the basolateral membrane. This relationship can be described by simple Michaelis–Menten kinetics with a V_{max} for H^+ extrusion of 680±50 neq h^{-1} cm^{-2} and a K_m for Na^+ of 18±2 mM (n = 8) (Harvey & Ehrenfeld 1988).

The Na^+/H^+ exchanger is not driven to thermodynamic equilibrium, shown by the fact that there is sufficient energy in the Na^+ gradient to drive pH_i to 8.4, whereas the normal pH_i is 7.3. The antiporter is activated by an increasing acid load (Fig. 3) but appears inactivated or quiescent at pH_i more alkaline than 7.5 (Harvey & Ehrenfeld 1988). These results agree with a postulated modifier role for cytoplasmic protons in controlling the activity of the Na^+/H^+ exchanger (Aronson et al 1982).

Cl^-/HCO_3^- exchange and pH_i regulation

We recently investigated the possible participation of Cl^-/HCO_3^- exchange in the control of cell pH. The evidence for such an exchange is that:
(1) alkalinization stimulates basolateral ^{36}Cl uptake.
(2) Cl^- free serosal solutions produce an intracellular alkalinization.
(3) Recovery of pH_i from an alkali load is prevented in Cl^- free solutions (Fig. 4b).

In summary, frog skin epithelium possesses classical pH_i regulating mechanisms as reported in most animal cells (Na^+/H^+ and Cl^-/HCO_3^- exchangers). The localization of these exchangers to the basolateral cell membranes allows the maximum use of the imposed physiological thermodynamic conditions to operate these modes of ion transport.

pH_i regulates Na^+ transport

Experimental manoeuvres designed to produce an intracellular acid load have been shown to decrease transepithelial Na^+ transport rate in frog skin (Funder et al 1967, Mandel 1978) and to decrease apical Na^+ permeability in toad urinary bladder (Palmer 1985). Furthermore, an intracellular acidosis reduces K^+ transfer in *Necturus* (salamander) proximal tubule (Kubota et al 1983) and in cultured bovine retinal pigment cells (Keller et al 1986).

We find that in frog skin epithelium the conductance of Na^+ at the apical membrane (g_{Na}) and of K^+ at the basolateral membrane (g_K) (measured from cell membrane I–V relations under short-circuit conditions) vary

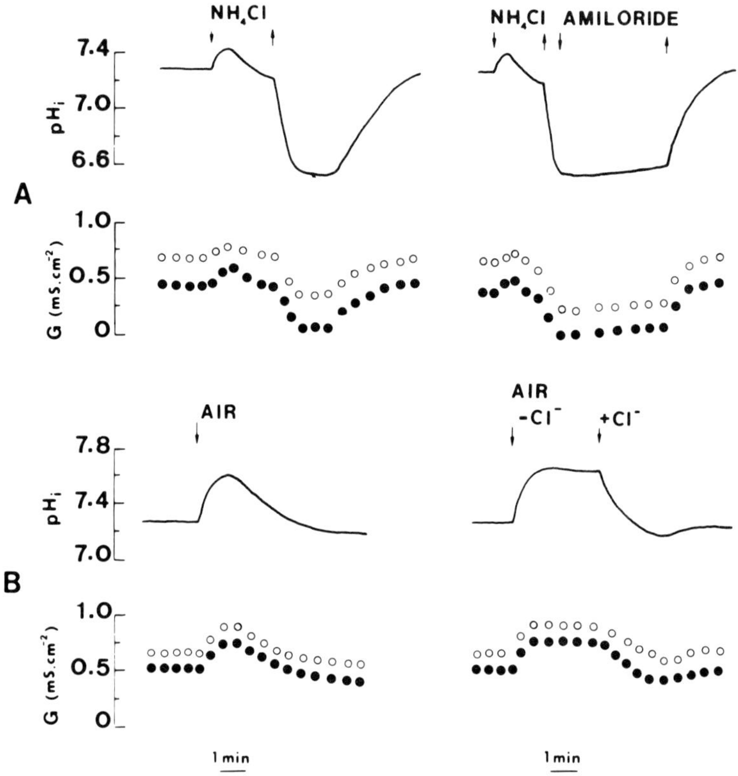

FIG. 4. Recording of the effects of variations in pH_i (measured by double-barrelled H^+-sensitive microelectrodes) on apical membrane Na^+ conductance (g_{Na}●) and basolateral membrane K^+ conductance (g_K o). Measured from current–voltage relations of these membranes in the isolated epithelium of frog skin under short-circuit conditions with both sides bathed in Ussing Ringer (115 mM NaCl), equilibrated in air and buffered in BES (N,N-bis[2-hydroxyethyl]-2-aminoethane-sulphonic acid) (10 mM) at pH 7.4. a) When 15 mM NH_4Cl was added to the basolateral Ringer solution the cell pH alkalinized transiently due to entry of NH_3 and capture of cell protons. Subsequent removal of NH_4Cl produced a 'rebound' acidification caused by dissociation of cell NH_4^+ and efflux of NH_3. These variations in pH_i were accompanied by simultaneous and parallel changes in g_{Na} and g_K. When this experiment was repeated (right-hand side of figure) with amiloride (5×10^{-4} M) present in the basolateral Ringer following washout of NH_4Cl, the recovery in pH_i and conductances was inhibited. Subsequent washout of amiloride permitted the restoration of these parameters. Thus, recovery of pH_i, g_{Na} and g_K from an intracellular acid load requires the operation of a basolateral Na^+/H^+ exchanger. b) In these experiments an intracellular alkali load was produced by pre-incubating isolated epithelia for 10 min in CO_2/HCO_3^- buffered Ringer on the basolateral side and then rapidly replacing this solution with 'Ussing' Ringer (HCO_3^- free and equilibrated in air), at first arrow. The rapid loss from the cell of CO_2 relative to HCO_3^- produced a transient alkaline shift in pH_i and an associated increase in both g_{Na} and g_K. On the right-hand side of the figure, the recovery of pH_i from the alkali load was inhibited when the basolateral Ringer solution was chloride-free (1st arrow) and g_{Na} and g_K were maintained at elevated values. The subsequent return to standard chloride-containing Ringer solution (2nd arrow) produced a rapid recovery in pH_i and membrane conductances.

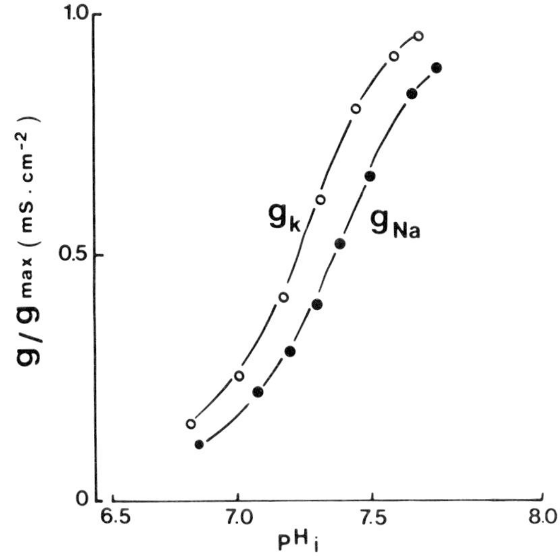

FIG. 5. Dependence of apical membrane Na^+ conductance g_{Na} (•) and basolateral membrane K^+ conductance g_K (o) on intracellular pH. pH_i was varied by incubating epithelia in weak acid (CO_2) or base (NH_4Cl) under short-circuit 'Ussing' conditions at constant external pH. g_{Na} and g_K values were 'normalized' by dividing each value by the maximum measured conductance (g_{max}) for the particular ion. The relations g/g_{max} for Na^+ and K^+ were fit by the equation $g/g_{max} = K^n/(K^n + [H^+]_i^n)$, where $K = 10^{-pK}$, $[H^+]_i$ is the intracellular concentration of hydrogen ions and 'n' is the Hill coefficient.

simultaneously and in parallel with changes in pH_i (Fig. 4a, b). Both g_{Na} and g_K are extremely sensitive to pH_i variations over the physiological range (pH_i 7.1–7.4) (Fig. 5). The relationship between Na^+ and K^+ conductances and pH_i shown in Fig. 5 could be described by titration curves with a Hill coefficient of two and a pK of 7.25 and 7.1 for the titration sites at the Na^+ channel and K^+ channel, respectively, indicating a highly cooperative binding between H^+ and channel proteins, possibly histidine residues. Thus Na^+ and K^+ channels are gated by intracellular protons. This appears to be a direct effect of cell pH, since it is voltage independent (Harvey et al 1988) and does not require changes in cell calcium (Harvey & Thomas 1987). Under normal transport conditions the g_{Na} and g_K are poised on the most sensitive part of the titration curve. Since pH_i is controlled by Na^+/H^+ and Cl^-/HCO_3^- exchangers, these will influence Na^+ transport indirectly via their modulation of the level of cytoplasmic H^+ concentration. Inhibition of Na^+/H^+ or Cl^-/HCO_3^- exchangers prevents the recovery of g_{Na} and g_K from an intracellular acid or alkali load, respectively (Fig. 4a, b). A similar inhibitory effect on K^+ conductance was found in frog proximal tubule (Oberleithner et al 1986, and this volume) when Na^+/H^+ exchange was blocked by amiloride. Intracellular protons can, therefore, provide an intrinsic regulation of transepithelial Na^+ transport, since the overall Na^+ efflux via Na^+/K^+ ATPase is similarly sensitive to pH_i

(Eaton et al 1984). This important role for cell protons is given further support from our recent experiments in which we find co-variance between pH_i variations and g_{Na} and g_K associated with changes in Na^+ transport induced spontaneously or following aldosterone or ouabain treatment (Harvey et al 1988, Harvey & Ehrenfeld 1988).

Cellular localization of H^+ pumps and exchangers and their relative importance in controlling Na^+ transport

The H^+ ATPase appears to be localized to one cell type: the mitochondria-rich cell (or the intercalated cell of urinary bladder as discussed by Professor Steinmetz, this volume). In fact, net proton secretion is blocked by inhibitors of carbonic anhydrase, which in frog skin epithelium is found only in Mr cells. Apical amiloride-sensitive Na^+ channels and Na^+/K^+ ATPase exist in both Pr and Mr cells (Larsen et al 1987), thus both cell types have the ability to engage in transepithelial Na^+ transport.

The increase in the number of Mr cells in freshwater (NaCl < 100 μM) (Ehrenfeld et al 1976) to equal the population of granular cells, and the correlation demonstrated between J_{net}^{Na} and J_{net}^{H}, could implicate Mr cells in Na^+ transport in 'physiological' conditions. However, the granular cells are known to transport Na^+ in 'Ussing' conditions (high [NaCl] bathing solutions and short-circuit conditions). We have attempted to determine the relative Na^+ transport function of Mr and Pr cells, under 'physiological' and 'Ussing' conditions, by exploiting the different effects of an intracellular acid load on transepithelial Na^+ transport, g_{Na}, and fractional resistance, $F(R_a)$ (ratio of apical to transcellular electrical resistance). We record from Pr cells, since the probability of penetrating Mr cells with microelectrodes is very low because of the shape and depth of the body of these cells.

An inhibitory effect of an intracellular acid load on Na^+ transport is observed under short-circuit conditions, whether Cl^- or SO_4^{2-} is the main anion in the mucosal bath (Table 4). Under these conditions no coupling exists between J_{net}^{Na} and J_{net}^{H}, and Na^+ transport proceeds via Pr cells as deduced from measurements of $F(R_a)$ and transcellular Na^+ current (I_{Na}).

Changing from short-circuit to open-circuit conditions with 57.5 mM Na_2SO_4 in the mucosal bath caused a 70% reduction in Na^+ absorption from 1679±281 to 478±52 neq h^{-1} cm^{-2} (n = 4), together with an increase in the $F(R_a)$ of Pr cells towards unity (0.93±0.05, n = 37), indicating a reduction of I_{Na} through these cells. Subsequent addition of amiloride (apical side) caused a further 10-fold decrease in total transepithelial Na^+ transport rate but had only slight effects on I_{Na} and $F(R_a)$ in Pr cells. In fact, with 57.5 mM Na_2SO_4 mucosal solutions under open-circuit conditions the J_{net}^{Na} is strictly coupled to J_{net}^{H} and both fluxes are inhibited by amiloride or ethoxzolamide (Ehrenfeld & Garcia-Romeu 1977). Thus Na^+ absorption under these conditions must

TABLE 4 Effects of an intracellular acid load on Na^+ and H^+ transport in frog skin *in vitro*

		J_{in}^{Na+}	J_{net}^{Na+}
OCC	control	181±25	108±24
1 mM Na_2SO_4	acid load	222±30	200±40
(n = 6)	(NH_4^+)		
	reversal	184±26	149±32
OCC	control	304±25	170±35
2 mM NaCl	acid load	438±23	400±44
(n = 6)	(CO_2)		
	reversal	357±38	272±47
SCC	control	1664±378	1448±263
57.5 mM Na_2SO_4	acid load	997±192	533±191
(n = 10)	(CO_2)		
	reversal	1279±304	1306±280
SCC	control	1365±519	1881±652
R_{air}/R_{air}	acid load	1002±403	1034±389
(n = 4)	(NH_4^+)		
	reversal	388±115	416±146
	(first 15 min)		

The intracellular acid load was at constant external pH. J_{in}^{Na}, unidirectional apical Na^+ uptake; J_{net}^{Na}, transepithelial net Na^+ absorption; J_{net}^{H}, H^+ secretion. Upper panel: tissues under open-circuit conditions (OCC) bathed on the mucosal side with either 1 mM Na_2SO_4 or 2 mM NaCl solutions buffered at pH 7.34 with 4 mM imidazole. The serosal solution was standard Ringer buffered at pH 7.34 with 5% CO_2, 24 mM $NaHCO_3$. Lower panel: tissues under short-circuit conditions (SCC) bathed on the mucosal side either in 57.5 mM Na_2SO_4 solution buffered at pH 7.34 with 4 mM imidazole or in standard Ringer buffered at pH 7.34 with 10 mM BES (*N,N*-bis-[2-hydroxyethyl]-2-aminoethane sulphonic acid) and equilibrated in air (R_{air}). Standard Ringer also present on serosal side. Acid loads were created by application of CO_2/HCO_3 to the apical side or NH_4Cl to the basolateral side (N.B. 'rebound' acidification after removal of NH_4Cl in SCC). Fluxes are given in neq h^{-1} cm^{-2}.

occur via a transcellular pathway other than Pr cells, i.e. via Mr cells.

An intracellular acid load in this situation causes a stimulation of J_{net}^{Na} by 96% with a corresponding increase in J_{net}^{H} from 147±32 to 284±263 neq h^{-1} cm^{-2} (n = 6). Similar results were found under open-circuit conditions in dilute (2 mM Na^+) mucosal solutions with either Cl^- or SO_4^{2-} as the main anion (Table 4). Since these latter conditions are close to those found *in vivo* and practically all epithelial ion transport studies are performed in 'Ussing' conditions (short-circuit and high external Na^+ solutions), the participation of Mr cells in Na^+ transport has, up until now, been missed. The ion transport functions of Pr and Mr cells are summarized in Fig. 6.

We conclude that transepithelial Na^+ transport occurs mainly via principal cells under Ussing conditions, in the presence of high external Na^+ and a

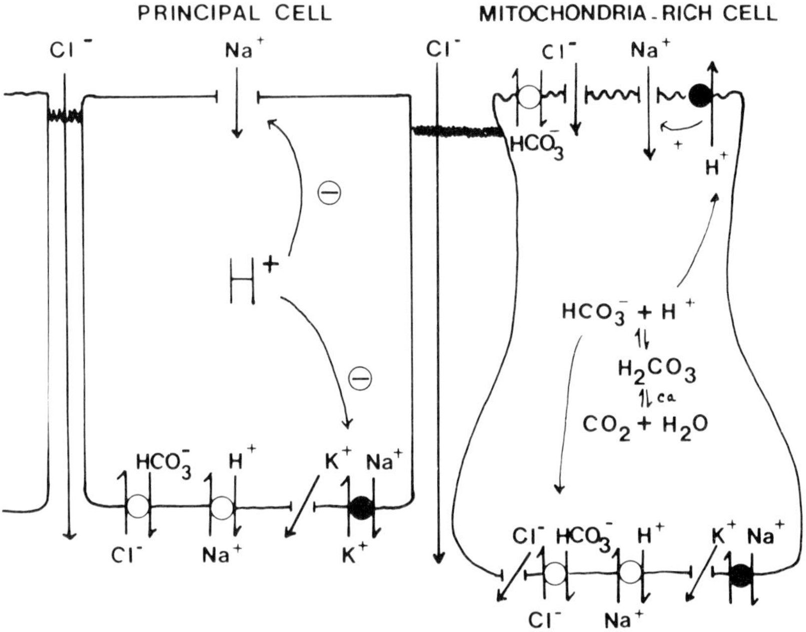

FIG. 6. Schema of the sodium and proton transport functions of principal (granular) cells and mitochondria-rich cells in the epithelium of frog skin. Passive Na^+ entry across the apical membranes occurs through amiloride-blockable channels in both Pr and Mr cells, with the active transport step (Na^+/K^+ ATPase) situated at the basolateral membranes. Regulation of intracellular pH is provided by basolateral Na^+/H^+ and Cl^-/HCO_3^- exchangers. Transepithelial H^+ secretion is assured by an electrogenic apical H^+-ATPase, in series with a basolateral Cl^-/HCO_3^- exchanger, localized in Mr cells. Na^+ absorption occurs mainly via Pr cells in 'Ussing' conditions (high NaCl mucosal solutions or short-circuit conditions), which have a high V_{max} and K_m for apical Na^+ entry. Na^+ absorption via Mr cells occurs preferentially in dilute NaCl or Na_2SO_4 solutions under open-circuit conditions. Coupling of Na^+ absorption to H^+ secretion in Mr cells releases Na^+ entry from dependence on a co-permeant anion (Cl^-) or counter-current (from external e.m.f). Apical Na^+ entry and basolateral Na^+/K^+ ATPase and K^+ conductance are inhibited by increasing intracellular H^+ in Pr cells, whereas an intracellular acid load increases H^+ secretion and associated Na^+ absorption in Mr cells. Na^+ channels in Mr cells may not be gated by cell H^+ or else pH_i may be invariable due to the dynamic buffering power of the carbonic anhydrase (c.a.) catalysed CO_2+H_2O/HCO_3^-+H^+ buffer system and rapid secretion of H^+ by the pump. H^+ secretion via Mr cells is associated with a transepithelial current which can voltage clamp the apical membranes of Mr cells and neighbouring Pr cells (indirectly via effects on transepithelial voltage), thereby facilitating Na^+ entry via these cells. Transepithelial Cl^- fluxes which accompany Na^+ transport may occur through Mr cells via concentration- and voltage-gated Cl^- channels (Larsen et al 1987) and Cl^-/HCO_3^- exchangers or via paracellular shunt pathways.

permeant anion (Cl^-) or balancing current (short-circuit current), and predominantly via mitochondria-rich cells when Na^+ uptake is coupled to H^+ secretion (open-circuit with SO_4^{2-} as the main anion, or in dilute mucosal solutions [Na^+] < 4 mM). This is supported by the different K_m and V_{max} for Na^+ absorption under these conditions (Ehrenfeld & Garcia-Romeu 1980). When Na^+ transport is coupled to H^+ secretion, K_m = 0.22 mM and V_{max} = 331 neq h^{-1} cm^{-2}; this implies that Mr cells have a high affinity but a low relative capacity for Na^+ absorption. Under high external NaCl concentrations (> 4 mM), J_{net}^{Na} and J_{net}^{H} are to a large extent uncoupled and K_m = 13 mM and V_{max} = 904 neq h^{-1} cm^{-2}, indicating that Pr cells have a low affinity but high capacity for Na^+ transport.

The relative participation of Mr and Pr cells in Na^+ transport, depending on the mucosal Na^+ concentration and on the available counter ions or counter current (Cl^-, SO_4^{2-}, H^+, or electrons from external e.m.f), can serve to explain the apparent conflicts in modes of Na^+ transport reported in *in vivo* and *in vitro* studies.

Acknowledgements

The technical assistance of Corinne Raschi and Anny Giovagnoli is gratefully acknowledged. The authors work is supported by grants from the Centre National de la Recherche Scientifique (Physiologie des Membranes UA 638 associée au C.E.A.). B.J.H. is Chargé de Recherche of the C.N.R.S. and J.E. is Maître de Conférence at the University of Nice (Laboratorie de Physiologie Cellulaire).

References

Aronson PS, Nee J, Shum MA 1982 Modifier role of internal H^+ in activating the Na^+/H^+ exchanger in renal microvillus membrane vesicles. Nature (Lond) 299: 161–163

Bowman B, Slayman CW 1977 Characterization of plasma membrane adenosine triphosphatase of Neurospora crassa. J Biol Chem 252:3357–3363

Bowman B, Mainzer S, Allen K, Slayman CW 1978 Effects of inhibitors on the plasma membrane and mitochondrial adenosine triphosphatases of Neurospora crassa. Biophys Acta 512:13–28

Civan MM, Peterson-Yantano K, Di Bona DR, Wilson DF, Erecinska M 1983 Bioenergetics of Na^+ transport across frog skin : chemical and electrical measurements. Am J Physiol 245:F621–F700

Cohen HL, Steinmetz PR 1980 Control of active proton transport in turtle urinary bladder by cell pH. J Gen Physiol 76:381–393

Duranti E, Ehrenfeld J, Harvey BJ 1986 Acid secretion through Rana esculenta skin : involvement of an anion-exchange mechanism at the basolateral membrane. J Physiol (Lond) 378:195–211

Eaton DC, Hamilton KL, Johnson KE 1984 Intracellular acidosis blocks the basolateral Na-K in rabbit urinary bladder. Am J Physiol 247:F946–F954

Ehrenfeld J, Masoni A, Garcia-Romeu F 1976 Mitochondria-rich cells of frog skin in transport mechanisms : morphological and kinetic studies on transepithelial excretion of methylene blue. Am J Physiol 231:120–126

Ehrenfeld J, Garcia-Romeu F 1977 Active hydrogen excretion and sodium absorption through isolated frog skin. Am J Physiol 233:F46–F54

Ehrenfeld J, Garcia-Romeu F 1980 Kinetics of ion transport across frog skin : two concentration dependent processes. J Membr Biol 56:139–147

Ehrenfeld J, Garcia-Romeu F, Harvey BJ 1985 Electrogenic active proton pump in Rana esculenta skin and its role in sodium transport. J Physiol (Lond) 359:331–355

Ehrenfeld J, Cragoe EJ, Harvey BJ 1987 Evidence for a Na^+/H^+ exchanger at the basolateral membranes of the isolated frog skin epithelium : Effect of amiloride analogues. Pfluegers Arch Eur J Physiol 409:200–207

Frazier LW 1984 Characteristics of proton excretion in normal and acidotic toad urinary bladder. Biochim Biophys Acta 817:75–84

Funder T, Ussing HH, Wieth TO 1967 The effects of CO_2 and hydrogen ions on active Na transport in the isolated frog skin. Acta Physiol Scand 71:65–76

Garcia-Romeu F, Salibian A, Pezzani-Hernandez S 1969 The nature of in vivo sodium and chloride uptake mechanisms through the epithelium of the Chilean frog Calyptocephalella gayi. J Gen Physiol 53:816–835

Harvey BJ, Kernan RP 1984 Intracellular ionic activities in relation to external sodium and effects of amiloride and/or ouabain. J Physiol (Lond) 349:501–517

Harvey BJ, Lahlou B 1986 Ion selective microelectrode studies of the electrochemical potentials in trout urinary bladder. J Physiol (Lond) 370:467–488

Harvey BJ, Ehrenfeld J 1986 Regulation of intracellular sodium and pH by the electrogenic H^+ pump in frog skin. Pfluegers Arch Eur J Physiol 406:362–366

Harvey BJ, Thomas RC 1987 Intracellular pH and calcium effects on sodium conductance and transport in isolated frog skin epithelium. J Physiol (Lond) 394:92

Harvey BJ, Ehrenfeld J 1988 Role of Na^+/H^+ exchange in the control of intracellular pH and cell membrane conductances in frog skin epithelium. J Gen Physiol (in press)

Harvey BJ, Thomas SR, Ehrenfeld J 1988 Intracellular pH controls cell membrane Na^+ and K^+ conductances and transport in frog skin epithelium. J Gen Physiol (in press)

Keller SK, Jentsch TJ, Koch M, Wiederholt M 1986 Interactions of pH and K^+ conductance in cultured bovine retinal pigment epithelial cells. Am J Physiol 250:C124–C137

Koefoed-Johnsen V, Ussing H 1958 The nature of the frog skin potential. Acta Physiol Scand 42:298–398

Kubota T, Biagi BA, Giebisch G 1983 Effects of acid base disturbances on basolateral membrane potential and intracellular potassium activity in the proximal tubule of Necturus. J Membr Biol 73:61–68

Larsen EH, Ussing HH, Spring KR 1987 Ion transport by mitochondria-rich cells in toad skin. J Membr Biol 99:25–40

Mandel LJ 1978 Effects of pH, Ca, ADH and theophylline in kinetics of Na entry in frog skin. Am J Physiol 235:C35–C48

Oberleithner H, Münich G, Schwab A, Dietl P 1986 Amiloride reduces potassium conductance in frog kidney via inhibition of Na/H exchange. Am J Physiol 251:F66–F73

Oberleithner H, Kersting U, Gaßner B 1988 Aldosterone-controlled linkage between Na^+/H^+ exchange and K^+ channels in fused renal epithelial cells. In: Proton passage across cell membranes. Wiley, Chichester (Ciba Found Symp 139) p 201–219

Palmer LG 1985 Modulation of apical Na permeability of the toad urinary bladder by intracellular Na, Ca and H. J Membr Biol 83:57–69

Schwartz JH, Steinmetz PR 1977 Metabolic energy and PCO_2 as determinants of H^+ secretion by turtle urinary bladder. Am J Physiol 233:F145–F149

Steinmetz PR 1988 Electrogenic proton transport by intercalated cells of 'tight' urinary epithelia. In: Proton passage across cell membranes. Wiley, Chichester (Ciba Found Symp 139) p 122–138

Steinmetz PR, Husted RF, Mueller A, Beauwens R 1981 Coupling between H^+

transport and anaerobic glycolysis in turtle urinary bladder : effects of inhibitors of H^+ ATPase. J Membr Biol 59:27–34
Steinmetz PR, Andersen OS 1982 Electrogenic proton transport in epithelial membranes. J Membr Biol 65:155–174
Stetson DL, Steinmetz PR 1985 α and β types of carbonic anhydrase rich cells in turtle bladder. Am J Physiol 249:F553–F565

DISCUSSION

Frömter: Is the Na^+ uptake mechanism at low external Na^+ concentration identical to that under Ussing conditions? Do they have the same inhibition kinetics with respect to amiloride? What is known about fluctuation analysis under both conditions?

Harvey: The kinetics of Na^+ uptake differs under these two conditions. We are planning to do some fluctuation analysis in collaboration with Van Driesshe and Zeiske to see if there is a different K_i for amiloride, for example, between the two transport systems.

Steinmetz: How do you distinguish between Na^+ entry in the two cell types?

Harvey: The microelectrode recordings are from principal cells and we can switch very rapidly from 'Ussing' conditions (short-circuit) to 'physiological' conditions (open-circuit). In the latter, Na^+ uptake is dependent on the movement of a counter ion. If we remove the counter ion, which is Cl^-, and replace it with sulphate, Na^+ uptake becomes dependent on H^+ secretion. Similarly, if we rapidly change from short-circuit conditions to the open circuit (in sulphate solutions), we have removed the counter current provided by the external e.m.f. and Na^+ entry is now dependent on proton secretion. In this situation, the fractional resistance of principal cells goes towards unity, indicating block of Na^+ entry into these cells. However, we still have an amiloride-sensitive ^{22}Na absorption, so Na^+ is now passing through a transcellular pathway but through cells other than the principal cells. Since this flux is inhibited by ethoxzolamide, a carbonic anhydrase inhibitor, and this enzyme is present only in the mitochondria-rich cells, Na^+ must be going through these mitochondria-rich cells. In addition, increasing intracellular H^+ concentration stimulates Na^+ influx in 'physiological' conditions but blocks it in 'Ussing' conditions.

Steinmetz: We agree that the coupling is indirect—there is no magic transporter that is capable of primary active transport of both H^+ and Na^+.

Harvey: The relation between Na^+ influx and H^+ efflux is one of indirect electrical coupling. The proton pump creates a current balance and a charge balance for Na^+ absorption.

Steinmetz: I think it's interesting but it is difficult to judge the carbonic anhydrase cell by assuming that the Na^+ transport in a much larger cell number is abolished.

Harvey: Passing from a large to a small population of Na^+ transport sites agrees with our kinetic data. We go from a very high capacity Na^+ transporter in Ussing conditions, 3000 nanoequivalents per hour, to 300 nanoequivalents per hour of Na^+ absorption in physiological conditions. In the latter situation, we also find 300 nanoequivalents of protons being secreted.

Morad: In the open-circuit condition, what is the signal to keep these events coupled?

Harvey: It's membrane potential.

Morad: Is there a kinetic match or is there a lag between the two transport systems?

Harvey: Both fluxes are obligatorily coupled. There has to be charge balance and it must be instantaneous. Na^+ uptake is limited by the capacity of the proton pump to generate a counter current, thus Na^+ influx indirectly reflects the kinetics of the H^+ efflux mechanism.

Boron: You propose that there is a basolateral Na^+/H^+ exchanger in your granular cells?

Harvey: We know it's in the granular cells because that's where the pH-sensitive microelectrode is. We can't eliminate its presence from mitochondria-rich cells.

Boron: And the Cl^-/HCO_3^- exchanger?

Harvey: That is in the granular cells. All the microelectrode work was done in the principal cells.

Thomas: Is that because the other cells are too small to penetrate?

Harvey: There are many fewer mitochondria-rich cells and they are very difficult to penetrate with microelectrodes, probably because of their small volume and non-syncytial contact with neighbouring cells.

Sabolic: What do you mean by inhibitor of H^+-ATPase? You mentioned DCCD and vanadate, these are also potent inhibitors of the Na^+/K^+-ATPase.

Harvey: There are, as yet, no specific inhibitors of the H^+-ATPase, although DCCD, oligomycin and vanadate inhibit proton secretion in frog skin (Ehrenfeld et al 1985).

Sabolic: The same inhibitors, except oligomycin, strongly inhibit the Na^+/K^+-ATPase and thus could change Na^+ gradients.

Harvey: The inhibition of Na^+ influx by these drugs is an indirect electrical effect following inhibition of proton secretion. The membrane potential has depolarized towards the E_{Na} because the rheogenic component of proton secretion has been lost. If the membrane potential is re-clamped to control levels, Na^+ absorption re-occurs since the counter charge to Na^+ influx is now provided by the clamping current; so the coupling that existed in open circuit was indirect, it was electrical.

Russell: What is the pharmacology of these channels when they are open? Are they sensitive to barium?

Harvey: The apical Na^+ channel is pH sensitive; it is blocked by amiloride or phenamil. The basolateral K^+ channel is blocked by barium and is also pH sensitive.

Sabolic: In renal endosomes a charge balance can be maintained by opening and closing an anion channel, which seems to be quite specific. Is there any such anion channel here?

Harvey: Apparently, in principal cells there is no anion conductance in the apical membranes. In the mitochondria-rich cells there is some evidence for a voltage-activated Cl^- conductance but only with high external Cl^-.

Sabolic: Our endosomes also have no significant Cl^- conductance until we add ATP and the proton pumps are operative. So it seems as though an electrogenic proton pump opens up this Cl^- conductance and after that there is a charge balance, but not before.

Harvey: Proton secretion can occur normally in the absence of external Cl^-. The H^+ pump has the same capacity in Cl^- or sulphate apical solutions. In open circuit, however, the Na^+ uptake is dependent on H^+ secretion, if no permeant anion (Cl^-) is present; if the concentration of external Cl^- exceeds 4 mM, the Cl^- influx through a cellular or paracellular pathway provides the charge balance for Na^+ entry. In that case, Na^+ transport is uncoupled from proton secretion.

Steels: In which cells are the effects of aldosterone and insulin observed?

Harvey: Aldosterone and insulin stimulate Na^+ transport in both conditions. In the Ussing conditions, it's probably a sum of effects on intercalated cells and principal cells. However, in physiological conditions, where Na^+ transport is coupled to proton secretion, aldosterone stimulates proton secretion and this would cause an obligatory increase in Na^+ absorption, because these fluxes remain electrically coupled.

Grinstein: You tried to explain the difference between the mitochondria-rich cells and the granular cells towards an acid load on the basis of carbonic anhydrase. Can you give any more detail? Why do they behave differently?

Harvey: An acid load blocks Na^+ entry through granular cells but stimulates Na^+ influx when it is coupled to H^+ efflux, i.e. the situation in which we think Na^+ is passing through the mitochondria-rich cells. These may be protected from a fall in pH by the dynamic buffering power of carbonic anhydrase. The mitochondria-rich cells contain carbonic anhydrase, the granular cells do not. In the physiological conditions, inhibitors of carbonic anhydrase, such as ethoxzolamide, cause parallel inhibition of Na^+ uptake and proton secretion. This is indirect inhibition of Na^+ absorption because ethoxzolamide, by inhibiting H^+ secretion, causes the membrane potential to depolarize and thus reduces the electrical driving force for Na^+ uptake. If these voltage changes are prevented by using a voltage clamp, Na^+ absorption is unaffected by ethoxzolamide (Fig. 2, Harvey & Ehrenfeld 1986). In Ussing conditions, ethoxzolamide has little effect on Na^+ absorption because the fraction of Na^+ influx which is uncoupled from H^+ secretion amounts to more that 90% of the total.

Steinmetz: What chamber did you use for the physiological conditions?

Harvey: We have used Ussing and Foskett chambers.

Steinmetz: Ussing has also studied transport under open-circuit conditions.

Harvey: The electrical coupling of Na^+ uptake and H^+ secretion is the mechanism for Na^+ absorption *in vivo*. Garcia-Romeu and colleagues have shown that in the living frog, proton secretion and Na^+ absorption are obligatorily coupled with a 1:1 stoichiometry (Garcia-Romeu & Ehrenfeld 1975, Garcia- Romeu et al 1969). This is why we call it the physiological condition.

Aickin: Do your experiments in physiological conditions mean that the principal cells don't work normally and that Na^+ absorption is taken over by the mitochondria-rich cells?

Harvey: The Na^+ transport function of the principal cells in Ussing conditions is well known but the role of the mitochondria-rich cell was unknown except for proton secretion. Now we can ask, what are the principal cells doing under physiological conditions, if only the Mr cell is providing the necessary electrical driving force for Na^+ absorption.

Thomas: They keep the frog in one piece!

Harvey: We think that the principle cells are working but at a much lower rate than in 'Ussing' conditions; from conductance measurements the Na^+ current passing through these cells is not sufficient to account for the net transepithelial ^{22}Na absorption (J_{Na})—a sizeable fraction (>60%) of J_{Na} must also be going via the Mr cells. The proton secretion function of the Mr cells generates a transepithelial current, which imposes a transepithelial potential. This is reflected in the apical membrane of all cells, depending on their fractional resistance. So although there may be no direct electrical coupling between neighbouring Mr cells and principal cells, the H^+ current acts as a physiological voltage clamp, which drives the apical membrane potential of principal cells to values more negative than the E_{Na}, so that Na^+ may now enter the cell (Fig. 2). This could explain why under these conditions the fractional resistance is not unity and why Na^+ entry into principal cells is dependent on H^+ secretion even though the H^+ pump is not localized to these cells.

Morad: So the Na^+ channels are not gated? What regulates the current?

Harvey: The Na^+ channels in the principal cells are voltage dependent. Conductance and Na^+ transfer via the channels increase as membrane potential becomes more negative than E_{Na}. Either a short-circuit current or the H^+ pump can achieve this.

Frömter: Recent model calculations by Larsen & Rasmussen (1985) have indicated that, apart from cellular pathways, frog skin also has a distinct paracellular conductance pathway for Cl^-.

Harvey: But we were working without Cl^-, in sulphate we still have the same transport.

Frömter: OK, but if the frog sits in the pond, would not some Cl^- have a chance to pass across the paracellular conductance and to shunt some of the Na^+ absorption, which would result in a coupling ratio of Na^+:H^+ different from 1:1?

Harvey: Garcia-Romeu & Ehrenfeld (1975) showed that *in vivo*, in low external NaCl, Cl^- uptake is electroneutral across the apical membrane via a

Cl^-/HCO_3^- exchanger. So we still need charge balance, irrespective of the anion.

Frömter: But even if you have only some backflux of Cl^-, the charge balance of Na^+/H^+ can no longer be exactly 1:1.

Harvey: Probably the answer is that back-leakage of Cl^- into dilute bathing solutions is extremely low.

Oberleithner: Wolfram Nagel showed that stimulation of transport increases the paracellular Cl^- flux.

Harvey: Nagel has also activated a transepithelial Cl^- conductance with a voltage clamp and with procaine. The Cl^- shunt, whether it is paracellular or going via cells other than the granular cells, is induced in non-physiological conditions.

Steels: What's your evidence that the pH sensitivity of the Na^+ and K^+ conductance does not operate via Ca^{2+}?

Harvey: Roger Thomas and I did some experiments which addressed this question (Harvey & Thomas 1987). It's very difficult to dissociate effects of intracellular pH on channels from subsequent effects on Ca^{2+}. One way in which we have approached this problem is to look at the effects of pH_i on Na^+ conductance in the presence of intracellular Ca^{2+} chelators or ionophores. Intracellular acid loads in the presence of MAPTAM still gave the same percentage inhibition of Na^+ conductance. So it appears that the pH effect does not require changes in intracellular Ca^{2+}. We then asked whether Ca^{2+} can change the Na^+ conductance in the absence of pH changes. We have used manoeuvres designed to change intracellular Ca^{2+} in the presence of ionomycin, and high or low external Ca^{2+} in the presence of the intracellular pH buffer, procaine. This situation should show pure Ca^{2+} effects on the Na^+ channel but we don't get any. If we remove Ca^{2+} using EGTA in the presence of ionomycin, Na^+ conductance remains stable for about five minutes. This is in the presence of procaine so pH_i is stabilized. This result agrees with the patch-clamp work of Palmer & Frindt (1987), which showed that in isolated cortical collecting tubule patches it's pH and not Ca^{2+} which gates the channels.

Al-Awqati: I think he found that there is an effect with Ca^{2+}.

Harvey: Not in the isolated patch, Ca^{2+} had a delayed (>10 minutes) and possibly indirect effect only in the whole cell.

Frömter: Ca^{2+} may stimulate, for example, a protein kinase that phosphorylates the channel.

Thomas: It's easy enough to show that pH has all these effects but it's very hard to prove it's not linked with Ca^{2+} because Ca^{2+} competes with H^+ for binding sites and all sorts of things. But the link with the pH that Brian (Harvey) and I showed was very close and surely it would not be as tight if it was really an indirect effect (Harvey & Thomas 1987).

Machen: Another question about the physiological condition—Foskett and

Ussing did experiments on frog skin with high Cl^- on both sides of the tissue and the transepithelial potential clamped to +50 mV. Under these conditions, they showed, using a vibrating electrode, that there was a large negative current (mucosa to serosa) through the mitochondria-rich cells. Also, measurements of cell volume with an optical sectioning technique showed that the mitochondria-rich cells swelled quite dramatically.

Harvey: Yes, from cell swelling and vibrating probe experiments they showed that the Cl^- flux was close to or above an Mr cell.

Machen: Since the cells swelled, the data seem very convincing in showing that Cl^- was going through the cells and not around them, e.g. through the paracellular pathway. Plus 50 mV is fairly physiological so the only thing that's not physiological is the high Cl^- concentration. Do you think that this is somehow inducing a Cl^- conductance?

Harvey: The Cl^- conductance is activated by increasing external Cl^- concentration and by voltage clamping the transepithelial potential to positive (serosal side) values. This Cl^- conductance could provide a counter current for Na^+ absorption under these conditions.

Morad: Have measurements been made with any voltage- sensitive dyes?

Harvey: In frog skin and in all multicellular epithelia it's practically impossible to use dyes for membrane potential or for ionic measurements. First because of the heterogeneous nature of the epithelium and second because it is a multi-layered structure. The fluorescent signal will not be from a single cell because of the whole stack of cells underneath.

de Hemptinne: With the microelectrode work, how do you know where the electrode is sitting, that it is in a principal cell and not in a mitochondria-rich cell?

Harvey: Without staining we can see the Mr cells, they are so optically dense that we can identify them in the living tissue and we can see that the microelectrode is five cells away from the nearest Mr cell.

Grinstein: Are there gap junctions between Mr and granular cells?

Harvey: We have not seen gap junctions between these cells. The Mr cells are spatially isolated from their neighbouring principal cells, which would seem to eliminate electrical coupling and chemical coupling between these cells.

Al-Awqati: What's the evidence that there are no gap junctions between Mr cells and principal cells?

Harvey: Nobody has ever found them and Ussing's osmotic shock work shows that the Mr cells will swell independently of the principal cells.

Grinstein: If you inject fluorescent dye into the granular cells, do the Mr cells stain?

Harvey: Graber et al (1987) loaded fura-2 and BCECF dyes into granular cells in the toad and turtle urinary bladder. The dyes seemed not to spread into adjacent Mr cells or into unloaded neighbouring granular cells.

Morad: The problem may be that the molecule is too big. Gap junctions may allow the passage of Na^+ but not of fluorescein, for example.

Harvey: That's right, it is well known that the granular cells are a functional syncytium for Na^+ transport, therefore the dye results were a little surprising. However, the lack of dye transport does not necessarily indicate the absence of ionic coupling between cells.

Boron: It is thought that there is a pH difference between intercalated cells and principal cells in the rabbit cortical collecting tubule. This would imply that if cytoplasmic communication between cell types exists, it is overwhelmed by other factors.

Al-Awqati: We have developed a method where the cells are exposed to 6-carboxy fluorescein. Because carbonic anhydrase, which is a very potent esterase, is present only in the mitochondria-rich cells, only these are labelled, if the exposure is for only a minute or so. We watch the dye diffuse into adjacent cells. Fluorescein should go through the gap junction, hence this method is a measure of junctional permeability. We found that the fluorescein stays in the cells. However, what one would like to do is to put an electrode in the mitochondria-rich cell and another electrode in the granular cell, pass current and see what happens, to get a more definitive result.

Frömter: The question of ionic coupling between mitochondria-rich cells and granular cells has already been answered by Dörge and collaborators working on frog skin (Rick et al 1978). Inorganic ion concentrations were found to be virtually equal in different granular cells, but those of mitochondria-rich cells were distinctly different. This suggests that the latter cells do not participate in the syncytial property of the frog skin epithelium and hence cannot be coupled ionically with granular cells.

Harvey: There is another possible way to identify whether Na^+ can move between Mr cells and principal cells, that is by autoradiography with ^{22}Na under Ussing and physiological conditions. In physiological conditions we would expect the label to show up in the Mr cell and not in the granular cell.

Steels: What happens to the frog when it gets into metabolic acidosis? What happens to the Na^+ transport?

Harvey: We have pre-adapted frogs which are in metabolic acidosis. They are kept in high concentrations of KCl for about a week. These frogs have an increased number of mitochondria-rich cells, almost the same number as of granular cells. They also have an extremely high rate of proton secretion, almost threefold greater than that of frogs in normal tap water. Na^+ transport is obligatorily coupled to this proton secretion and is correspondingly increased threefold. So these frogs are absorbing more Na^+ and secreting more protons.

Oberleithner: But that's the aldosterone effect.

Thomas: But frogs spend a lot of their time in real life in air, how do they get rid of their acid then?

Harvey: They have a very thin film of mucus on the skin. If Cl^- is present in this solution, it will exchange for cellular bicarbonate, which in turn will buffer the secreted H^+. The CO_2 formed can then be lost as volatile acidity.

References

Ehrenfeld J, Garcia-Romeu F, Harvey BJ 1985 Kinetics of ion transport across frog skin: two concentration dependent processes. J Membr Biol 56:139–147

Garcia-Romeu F, Ehrenfeld J 1975 *In vivo* Na^+ and Cl^- independent transport across the skin of *Rana esculenta*. Am J Physiol 228:839–844

Garcia-Romeu F, Salibian A, Pezzani-Hernandez S 1969 The nature of *in vivo* sodium and chloride uptake mechanisms through the epithelium of the Chilean frog *Calyptocephalella gayi*. J Gen Physiol 53:816–835

Graber M, Brink PR, DiLillo D, Devine P, Pastoriza-Munoz E 1987 Permeabilizing the granular cell of toad and turtle bladder: lack of cell coupling. Am J Physiol 253:F588–F594

Harvey BJ, Ehrenfeld J 1986 Regulation of intracellular sodium and pH by the electrogenic H^+ pump in frog skin. Pfluegers Arch Eur J Physiol 406:362–366

Harvey BJ, Thomas RC 1987 Intracellular pH and calcium effects on sodium conductance and transport in isolated frog skin epithelium. J Physiol (Lond) 394:92P

Larsen EH, Rasmussen BE 1985 A mathematical model of amphibian skin epithelium with two types of transporting cellular units. Pfluegers Arch Eur J Physiol 405:S50-S58

Palmer LG, Frindt G 1987 Effects of cell Ca and pH on Na channels from rat cortical collecting tubule. Am J Physiol 253:F333–339

Rick R, Dörge A, Arnim EV, Thurau K 1987 Electron microprobe analysis of frog skin epithelium: Evidence for a syncytial sodium transport compartment. J Membr Biol 39:313–331

General discussion II

Cl^-/HCO_3^- exchange in the oxyntic cell of the stomach and in the intercalated cell of the kidney

Machen: I would like to say a few words about the oxyntic cell of the stomach. On the luminal side of the oxyntic cell there is a H^+/K^+-ATPase that exchanges K^+ in the lumen for a proton. This pump secretes large amounts of highly concentrated HCl; the stomach lumen can reach pH 1. The cell has to eliminate the base that's left behind and indirect physiological evidence has indicated that this is accomplished by Cl^-/HCO_3^- exchange at the serosal membrane. I would like to present both physiological and biochemical evidence that the oxyntic cell Cl^-/HCO_3^- exchanger is localized to one cell type.

Anthony Paradiso, Roger Tsien and I did experiments using a pH-sensitive fluorescent dye (BCECF) and digital image processing to look for HCO_3^-/Cl^- exchange. We changed the Cl^- concentration in the bathing medium and looked for predictable changes in intracellular pH. In a bright field micrograph of an isolated rabbit gastric gland, the oxyntic cells can be identified as those that bulge away from the perpendicular axis of the gland. In between the oxyntic cells are the chief cells, which secrete enzymes. Cl^--free solutions cause the oxyntic cells but not the chief cells to become alkaline by about 0.4 pH units (from pH 7.1 to 7.5). When Cl^- was returned to the bathing solution, the oxyntic cells returned to the control pH and, again, the pH of the chief cells was unaffected. This effect can be blocked by 200 μM SITS or DIDS or dihydro DIDS, and is Na^+ independent. We concluded that the oxyntic cells have a Cl^-/HCO_3^- exchanger but the chief cells do not.

We wanted to know whether this anion exchanger is similar to the one in the red blood cell. Ron Kopito has generated an antibody against the 12 amino acids of the C-terminal region of the Band 3 anion exchanger. Holly Thomas, Ron Kopito and I used this antibody to localize Band-3 like molecules. Gastric glands were isolated, fixed and permeabilized. To identify the oxyntic cells we labelled the H^+/K^+-ATPase with a monoclonal antibody obtained from Adam Smolka. The anti-Band 3 antibody labelled all the oxyntic cells and none of the chief cells. Immunoelectron microscopy by Roland Baron showed that the anion exchanger was localized to the basolateral membrane of oxyntic cells. Western blots indicated that the molecular weights of the two proteins that we can identify are 180 kDa and 140 kDa. These molecular weights are similar to those of proteins observed in the cortical collecting duct of the kidney but different from the red cell Band 3 protein, which has an apparent molecular weight of 95–100 kDa.

Vaughan-Jones: Is there any Na^+ sensitivity of that anion exchanger? I appreciate that it looks like the Band 3 one.

Machen: We can't detect any Na^+ dependence. We have done experiments where we take away all the Na^+ and measure rates of change of pH in response to a Cl^- gradient, and the rates are the same in Na^+-containing and Na^+-free solutions. If there is a Na^+ dependent anion exchanger, it's a fairly minor component in these cells.

Grinstein: Are these Western blots of whole glands just ground up? What do you make of the three bands that you get?

Machen: The Western blots were done using whole glands. The bands may be distinct forms of the same protein that were processed differently, but it is impossible to tell at this time.

Al-Awqati: We have done some similar work on the anion exchanger in the kidney. We recently found that the intercalated cell of the kidney collecting tubule exists in two forms: one secretes acid, the other secretes HCO_3^-. The acid-secreting cell has a Cl^-/HCO_3^- exchanger on the basolateral side, the HCO_3^--secreting cell has an apical exchanger. The question now arises: are these two exchangers composed of the same protein or of different proteins? In order to be properly targeted to the correct membrane, they must have some differences. Recently, we and others have found that the basolateral Cl^-/HCO_3^- exchanger is a form of the red cell Band 3 protein. To test the apical exchanger, we have now purified the HCO_3^--secreting cell and can grow it in culture. These monolayers have high resistance and exhibit Cl^--dependent transepithelial HCO_3^- secretion. Using an antibody raised against the C-terminal 12 amino acids of the Band 3 protein, we were able to demonstrate apical labelling of the intercalated cells.

We prepared poly$(A)^+$ mRNA from these cells and probed it with a DNA fragment that codes for the membrane spanning form of Band 3. Northern blot analysis showed that the intercalated cells contain message for Band 3. Hence, we suggest that both apical and basolateral types of Cl^-/HCO_3^- exchanger represent forms of the Band 3 protein. We are now studying whether they are encoded by the same gene or are the products of two different genes. Harvey Lodish and his group, and Peter Curtis and his group, have already identified a non-erythroid Band 3 gene. There may therefore be a gene family that codes for Cl^-/HCO_3^- exchangers.

Oberleithner: Do you have an antibody for the Band 3 protein and does it functionally interfere with the exchange system?

Al-Awqati: We have three antibodies. One labels the apical membrane and the HCO_3^- secreting cells. Most antibodies to Band 3 do not label this apical membrane *in situ*, but we don't know why that is. In cultured cells, it seems to work. Other antibodies to Band 3 label the basolateral membrane of the acid-secreting cells. But I don't think there is any antibody to Band 3 that inhibits function.

Thomas: So the antibodies stick to the protein without inhibiting its action?

Al-Awqati: Yes. A lot of the binding sites for the antibodies are on the cytoplasmic domain of the protein. That's the most immunogenic part of the protein and that's the non-transporting domain. We have made anti-peptide antibodies. One binds to the extreme C terminal, which is not in the transporting domain. Other antibodies, raised against the region that we think may be the active site, don't even bind to the protein.

Steinmetz: Qais, with respect to the apical exchanger, in which tissue can you show that it is present in the apical cell membrane?

Al-Awqati: This is in the cultured HCO_3^--secreting cells from the rabbit cortical collecting duct.

Steinmetz: That's the first time that the exchanger has been demonstrated by immunocytochemistry in an apical membrane.

Boron: How did you raise those antibodies?

Al-Awqati: They are made in rabbits but they are functionally monoclonal because the epitope is so limited; the peptide was only 12 amino acids long.

Steinmetz: Are these apical antigens different from the ones you can show in the other cells?

Al-Awqati: No, they are present on both kinds of cells. The C-terminal peptide against which these antibodies were raised is conserved in the two Band 3 genes. There is a lot of homology between the erythroid and the non-erythroid forms of the protein and the C-terminal peptide that we have is identical in both of them.

Grinstein: The C terminal sticks out into the cytoplasm?

Al-Awqati: Probably.

Proton channels in snail neurons studied with surface pH glass microelectrodes

Roger C. Thomas

Physiology Department, School of Medical Sciences, Bristol University, Bristol BS8 1TD, UK

Abstract. Surface pH was recorded on voltage-clamped snail neurons with Hinke-type glass microelectrodes. During stepwise depolarization from − 50 mV to +40 mV the surface pH usually increased at first and then fell rapidly during each 7.5 sec voltage step. The pH changes appeared to be due to proton passage through voltage-sensitive channels. The pH changes were inhibited by cadmium and zinc, and were not as sensitive to external pH as to internal pH. Changes in internal pH lagged behind surface pH. Changes in external calcium had little effect on the surface pH increases seen with small depolarizations. These results confirm earlier conclusions that depolarization opens proton-permeable channels.

1988 Proton passage across cell membranes. Wiley, Chichester (Ciba Foundation Symposium 139) p 168–183

From the papers in this volume it is abundantly clear that the great majority of protons pass across cell membranes on a carrier of some sort. This carrier may be a special protein in the cell membrane or a weak acid, such as CO_2, or a weak base. Free protons are at such a low concentration on both sides of the membrane that at one time it seemed impossible that H^+ passage through a channel could carry a measurable current or cause a significant pH change. But in 1979 I found that pH_i recovered rapidly after HCl injection into snail neurons depolarized by isotonic KCl (Thomas 1979). I therefore suggested that the rapid recovery was due to K^+/H^+ exchange. I failed to see any signs that the protons left as a current. But R.W. Meech (personal communication 1980) then discovered that voltage-clamped snail neurons also exhibited rapid intracellular pH (pH_i) changes when depolarized in normal snail Ringer solution. Together we later showed that depolarized snail neurons somehow became highly permeable to H^+ (Thomas & Meech 1982, Meech & Thomas 1987). At normal pH levels the H^+ leave a cell when the membrane potential (E_m) is positive.

To confirm that H^+ do leave depolarized snail neurons, I have recently recorded surface pH (pH_s) changes with a liquid-sensor pH microelectrode (Thomas 1987, 1988). Such microelectrodes were first used to record ion activity changes on the surface of nerve cells by Neher & Lux (1973). They used K^+-sensitive liquid ion-exchanger microelectrodes to record the K^+ leaving snail neurons during action potentials or voltage clamp pulses. The pH_s electrodes I used had a tip diameter of 1–2 μm and had a neutral ligand pH sensor. I have now used glass pH microelectrodes to confirm my earlier observations and establish that pH_s changes occur over a relatively large area.

Surface pH was first recorded as such with microelectrodes by de Hemptinne (1980, see also de Hemptinne et al 1987) using an elegant design I was unable to duplicate. Instead I have used a Hinke-style microelectrode with a 30–40 μm tip placed sideways against the cell surface. Such electrodes work very well, and are much more reliable than the ligand type I used earlier. I have continued to use ligand electrodes intracellularly for their unrivalled response time. In spite of this, I have found that the pH_i changes seem to lag behind the pH_s changes on depolarization. Both pH changes are inhibited by cadmium and zinc.

Methods

Experiments were done on exposed 70–150 μm diameter cells in the brain of the common snail, *Helix aspersa*, dissected as previously described (Thomas 1988).

Solutions. The normal CO_2-free snail Ringer solution contained NaCl 80 mM, KCl 4 mM, $CaCl_2$ 7 mM, $MgCl_2$ 5 mM, Hepes 20 mM, the pH being adjusted by the addition of NaOH. The normal CO_2 solution was bubbled with 2.7% CO_2 in air, and contained 20 mM $NaHCO_3$ but no Hepes. Solutions with lower concentrations of Hepes or $NaHCO_3$ had additional NaCl to compensate; the pH electrodes were calibrated in a Ringer solution buffered to pH 6.5 with 20 mM Pipes. This solution was led from its control valve direct to the bath, but the others were led there via a manifold and pump. Some experiments were done with solutions equilibrated with 0.5% CO_2 to give a higher pH with 20 mM $NaHCO_3$.

Microelectrodes. All pH_i measurements were done using eccentric double-barrelled microelectrodes. These were made as previously described (Thomas 1988) from aluminosilicate glass. They were usually used the same day as they were filled, and always required careful tip breakage before the reference or pH side had a reasonable resistance or pH response, respectively. An acceptable pH_i electrode had a reference side resistance of 20–40 megohms and a pH response of over 50 mV/pH unit.

All pH_s measurements were made with Hinke-design glass pH microelectrodes, made as described in Thomas (1978). The exposed length of pH glass protruding from the insulating glass tip was 30–40 μm. These all-glass pH_s electrodes lasted for several weeks and were much more reliable than the ligand pH_s electrodes used earlier (Thomas 1988). Experiments in which the voltage baseline of any electrode shifted by more than 3 mV were discarded.

Microelectrodes for voltage clamping and ion injection were usually made from the same aluminosilicate blanks as the pH microelectrodes, but without silanization. The outer micropipette was filled with 3 M CsCl for voltage clamping, while the inner was filled with 1M HCl. In most experiments HCl was not injected, and a single CsCl-filled micropipette was used.

Electrical arrangements. The way the micropipettes were positioned and connected electrically is illustrated diagramatically in Fig. 1. A simple voltage clamp was used to control the E_m via the CsCl microelectrode, and was set manually at −50 mV. At appropriate intervals, commands were provided by an Apple IIe microcomputer to change the set voltage in 10 mV steps. In some experiments HCl was injected by a symmetrical double-current clamp.

Voltages from the amplifiers connected to the various electrodes and current and voltage clamp outputs were displayed on an oscilloscope and potentiometric pen recorder, and stored on floppy discs by the Apple IIe microcomputer (Rickard & Thomas 1985) for later analysis. The measured values of pH_i and pH_s quoted in the text and used to plot graphs were all taken from the data sorted on disc rather than from the pen-recordings.

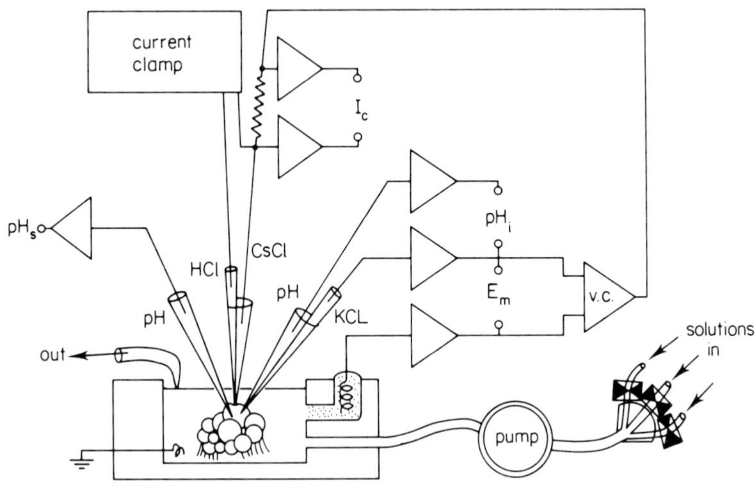

FIG. 1. Diagram of the experimental set up of snail brain, microelectrodes, electrical equipment and solution superfusion system.

General procedure

Once the dissection of the snail brain was finished, the microelectrodes were mounted with their tips in the bath and tested for current passing and pH response. Often the ligand microelectrode tip needed careful breaking. Once the electrode voltages had stabilized, a well-exposed cell was selected and penetrated first with the eccentric E_m and pH_i electrode, and then with the current-passing one. The voltage clamp was then switched on and set to −50 mV. If all was stable, the surface pH electrode was then pushed sideways onto the cell so that the midpoint of the exposed pH glass was level with the centre of the cell body. The superfusate was then switched to a lower buffering power and the computer set to apply a series of voltage clamp steps. The usual protocol was for an initial hyperpolarization to −60 mV, followed by 10 mV steps to +40 mV. Each step usually lasted for 7.5 sec.

Results

In spite of their much larger tip sizes, the glass pH_s microelectrodes recorded falls in surface pH similar to those seen with the ligand microelectrodes

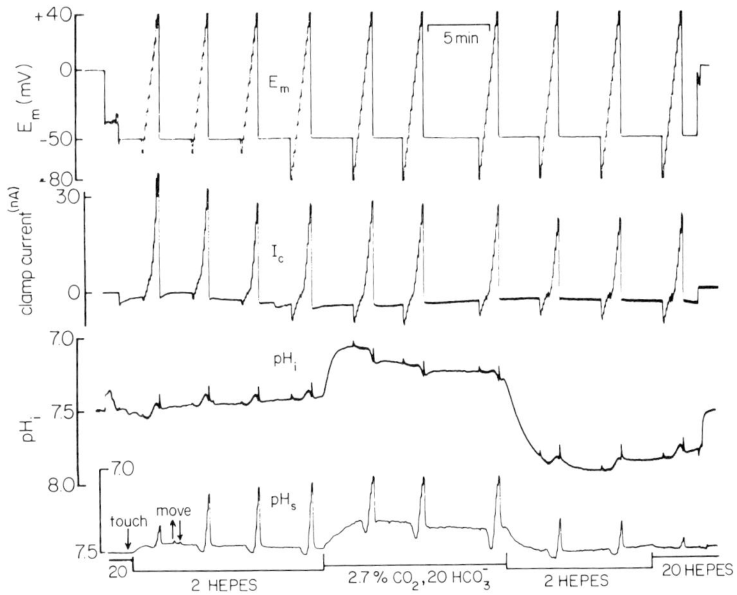

FIG. 2. Pen-recording from an experiment to show the surface pH responses of a snail neuron to depolarization in different solutions, all at pH 7.5. The four records show the membrane potential (E_m), the clamp current (I_c), intracellular pH (pH_i) and surface pH (pH_s). The buffer content of the superfusing saline is indicated at the bottom of the figure.

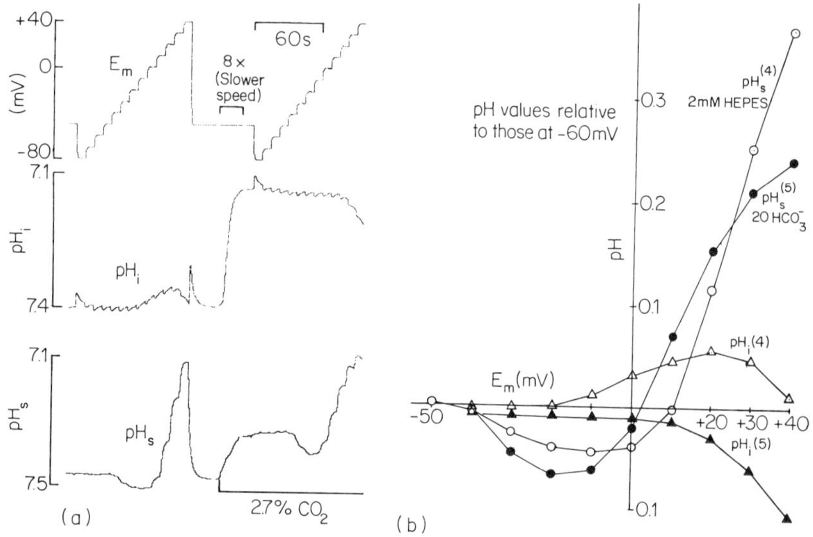

FIG. 3. Analysis of the fourth and fifth series of voltage clamp steps in the experiment shown in Fig. 2. (a) Print-outs after calibration of the data points recorded on the microcomputer disc. (b) Graphs of the relative values of pH_i and pH_s at the end of each period of stepwise depolarization.

(Thomas 1988). Indeed some of the pH_s changes seen with Hinke electrodes were larger than any recorded under comparable conditions in the earlier experiments. A total of 38 cells were tested with Hinke pH_s electrodes during depolarization to +40 mV. The average fall in pH_s for the first depolarization was 0.14 pH units, ranging from a barely detectable 0.02 to over half a pH unit. It was often possible to increase the size of the recorded fall in pH_s by relocating the Hinke electrode, as shown in the experiment of Fig. 2. Between the first and second series of depolarizing steps the Hinke electrode was moved from one side of the cell to the other: the fall in pH_s increased from 0.11 to 0.30 units.

This experiment was designed to show the effect on the changes in pH_s of switching the external solution from one buffered with 2 mM Hepes to one equilibrated with 2.7% CO_2 and containing 20 mM $NaHCO_3$. No attempt was made to block pH_i regulation, so during the exposure to CO_2 pH_i partially recovered from the CO_2-induced decrease. On removal of CO_2 pH_i increased well beyond its original level.

These changes in pH_i and the change in external solutions had relatively small effects on the pH_s and on the changes in pH_s caused by the series of depolarizations imposed every four to six minutes. As E_m decreased from the holding potential (having been stepped initially to − 60 or − 80 mV), pH_s at first increased, then more rapidly decreased.

The fourth and fifth series of depolarizations are shown in more detail in Fig. 3a. This shows a printout (screen dump) of part of the data recorded during the experiment by the Apple II computer. Fig. 3b shows a graphical analysis of these two series of depolarizations, in which the pH values at the end of the 7.5 sec period at each E_m are plotted against E_m. The open symbols show values for the fourth series, the closed symbols those for the fifth.

As E_m was decreased in 10 mV steps from −80 mV, there was no change in pH_s until −40 mV. With each further 10 mV step pH_s then increased to a peak, decreased, and beyond +10 mV fell rapidly. The changes in pH_i were opposite to, much smaller than, and lagged behind, those of pH_s. Thus pH_i only started to fall once E_m had reached −20 mV. The minimum pH_i was at an E_m of +20 mV, and it never increased beyond its starting value. If the changes in pH_s were caused by H^+ moving first into, and later out of, the cell, the minimum pH_i should occur at the same E_m as that at which the changes in pH_s reverse. The 10 mV difference suggests that the pH_i at the recording site is lower than that near the inside of the cell membrane. The pH_i microelectrode responded at least as rapidly to the calibration solution as the Hinke electrode (not shown), so I believe the recorded differences are real.

The pH_s and pH_i values for the fifth series of depolarizations are similar to those for the fourth, but the pH_s increases more at its maximum, and decreases less at +40 mV. The pH_i is somewhat distorted by the pH_i recovery from the exposure to CO_2. The larger increase in pH_s may be due to a lower buffering power outside the cell or increased H^+ mobility inside and outside the cell, but cannot be due to an increased gradient for H^+ entry. Since pH_i decreased more than pH_s, the H^+ reversal potential (E_H) was in fact more negative, reducing the gradient for passive H^+ influx.

The importance of external buffering power is confirmed by the last series of depolarizations shown in Fig. 2, which were done with the cell superfused with 20 mM Hepes solution. The change in pH_s was reduced from 0.15 units in 2 mM Hepes to only 0.05 units in 20 mM Hepes. There is a clear difference between the two pH_i records as well, with pH_i increasing more at +30 and +40 mV in 20 mM than in 2 mM Hepes. This suggests that more H^+ leave the cell when it is depolarized in 20 mM rather than 2mM Hepes. Perhaps there is also a greater H^+ flux away from the cell surface in the 20 mM Hepes solution. There is clearly a smaller H^+ gradient near the cell surface, as shown by the smaller changes in pH_s.

The changes in pH_s and pH_i are consequences of the same basic (or to be more exact, acidic!) events. This is confirmed by the experiment shown in Fig. 4, in which cadmium and zinc inhibit both pH changes. Cadmium was first shown to inhibit the pH_i changes in depolarized snail neurons by Thomas & Meech (1982), while the more potent blocking action of zinc was described later by Mahaut-Smith (1987). Both agents themselves lowered pH_s. Cadmium reduced the pH_s response to depolarization, while zinc virtually

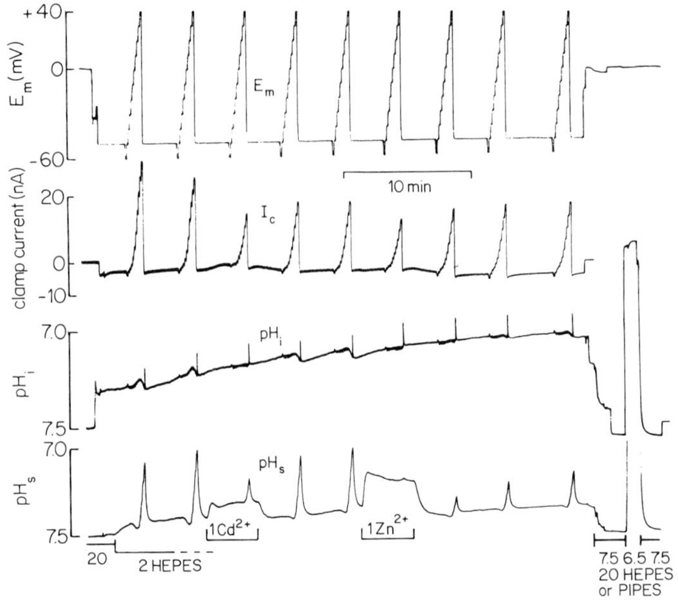

FIG. 4. Experiment to show the effect of 1 mM $CdCl_2$ and 1 mM $ZnCl_2$ on the pH_i and pH_s responses to depolarization. The cell superfusate was buffered with 2 mM Hepes pH 7.5 except where indicated.

abolished it. Both agents apparently blocked any change in pH_i.

In other experiments I found that 0.1 mM zinc was about as effective as 1 mM cadmium, although the recovery from cadmium was faster.

It is notable that both agents also reduced the clamp current needed to depolarize E_m. This is presumably partly because some of the outward current is normally carried by the H^+ leaving the cell. The zinc-inhibited current was therefore calculated and plotted against E_m (not shown) but it did not correspond very well with the pH_s changes. Presumably zinc inhibits more than H^+ channels.

If the pH_s changes are caused by the passive movement of H^+ through channels, the E_m at which there is no change should equal E_H, the equilibrium potential as given by the Nernst equation. In earlier experiments with ligand surface electrodes, I found good agreement between E_H (calculated from $E_H = 58\ [pH_i\text{-}pH_s]$) and the E_m at which pH_s starts to decrease when the external pH was kept at 7.5 and pH_i was varied (Thomas 1988). I have now examined the effect of changing pH_o while leaving pH_i essentially unchanged, as shown in Fig. 5.

In this experiment I changed external pH from 7.5 to a nominal 8 and then 7, adjusting the Hepes concentration to keep the external buffering power constant. As can be seen from the number of steps in pH_s as E_m was stepped

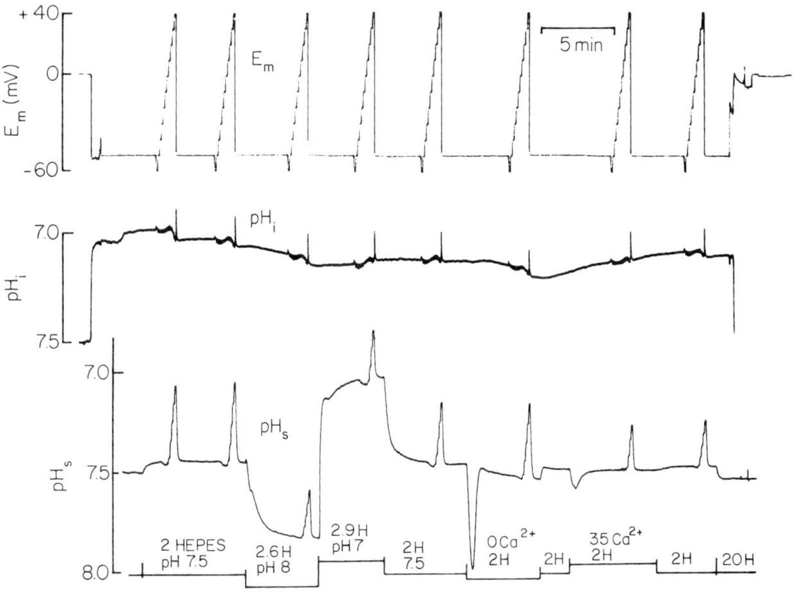

FIG. 5. The effect of changing extracellular pH and calcium on the pH_i and pH_s responses to depolarization. In the pH 7 and 8 solutions the Hepes concentration was chosen to give the same buffering power as 2 mM Hepes at pH 7.5. (The large pH_s changes seen on Ca^{2+} removal and increase were caused by contamination of the solutions with carbonate deposits in the tubing between aspirator and manifold.)

down to +40 mV, the potential at which pH_s started to fall did not change much. These values for the 2nd, 3rd, 4th and 5th depolarizations are indicated by the filled symbols in Fig. 6, which shows points from four other experiments as well. It appears that external pH over the range 7–8 has less effect on the E_m at which H^+ start to leave the cell than does pH_i. This fits in with the observations on H^+ currents by Byerly et al (1984). It would be worth repeating these experiments in well-buffered external solutions in case poor control of pH_s has distorted the results.

An alternative explanation for the increases in pH_s is that they are caused by Ca^{2+} entry through Ca^{2+} channels rather than directly by H^+ entry. The depletion of Ca^{2+} from the cell surface might increase H^+ binding and cause a transient rise in pH_s. I have seen such a rise in pH_s when external Ca^{2+} is removed from the superfusate, as shown in Fig. 5, and indeed when external Ca^{2+} is increased, but this proved to be due to carbonate deposits in the tubing. The removal of Ca^{2+} only rarely blocked a subsequent pH_s increase on stepwise depolarization.

Raising external Ca^{2+} usually shifted the E_m threshold in a depolarized direction, but removal had little effect. One reason may be that I did not add

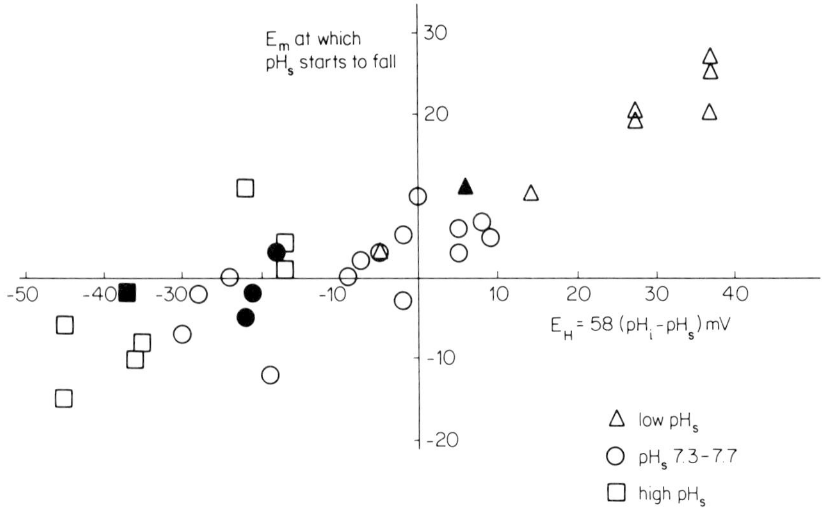

FIG. 6. Graph of the E_m value at which pH_s starts to decrease against E_H calculated from the corresponding measured values of pH_i and pH_s. Data points from five experiments in which pH_o was varied between 7 and 8 but pH_i was close to normal. Filled symbols show points from the experiment illustrated in Fig. 5.

any Ca^{2+} chelator, so external Ca^{2+} would still have been much greater than internal. More experiments are needed to clarify the role of calcium in the pH_s changes I observe.

Conclusions

The above results confirm that the surface pH changes previously recorded with ligand microelectrodes are real, and not some artifact caused by ligand leakage from the electrode tip. They also suggest that pH_i away from the cell membrane may well differ from that closer to the membrane when the cell is depolarized for some seconds. The changes in pH_s and pH_i are otherwise closely correlated and both are blocked by cadmium and zinc ions.

The simple picture of a channel opening in the cell membrane on depolarization and allowing H^+ to move down their electrochemical gradient is not enough to explain the effects of external pH. Either the channel itself has a complicated sensitivity to pH or the local pH at each end is often quite different from the pH_i and pH_s I measure. The stability and sensitivity of the glass pH microelectrode should allow me to use much higher external buffering powers in the hope of making the pH_s I measure closer to the pH at the outer end of the supposed channel.

The size and possible physiological significance of the pH_s decrease likely to occur as H^+ leave hyperactive neurons are unknown, but it may be that the H^+ are modifying Ca^{2+} channels (see Morad, this volume).

Acknowledgements

I am grateful to Michael Rickard for training the Apple IIe computer and general technical assistance, Sue Maskell for secretarial help, and the M.R.C. for money.

References

Byerly L, Meech RW, Moody W 1984 Rapidly activating hydrogen ion currents in perfused neurones of the snail *Lymnaea stagnalis*. J Physiol (Lond) 351:199–216

de Hemptinne A 1980 Intracellular pH and surface pH in skeletal and cardiac muscle measured with a double-barrelled pH microelectrode. Pfluegers Arch Eur J Physiol 386:121–126

de Hemptinne A, Morranne R, Vanheel B 1987 Surface pH and the control of intracellular pH in cardiac and skeletal muscle. Can J Physiol Pharmacol 65: 970–977

Mahaut-Smith M 1987 The effect of zinc on calcium and hydrogen ion currents in snail neurones. J Physiol (Lond) 382:129P

Meech RW, Thomas RC 1987 Voltage-dependent intracellular pH in *Helix aspersa* neurones. J Physiol (Lond) 390:433–452

Morad M 1988 Proton-induced transformation in gating and selectivity of the calcium channel in neurons. In: Proton passage across cell membranes. Wiley, Chichester (Ciba Found Symp 139) p 187–200

Neher E, Lux HD 1973 Rapid changes of potassium concentration at the outer surface of exposed single neurons during membrane current flow. J Gen Physiol 61:385–399

Rickard MJ, Thomas RC 1985 Microcomputer system for simultaneous solution switching and data collection for experiments with ion-sensitive microelectrodes. J Physiol (Lond) 364:8P

Thomas RC 1978 Ion-sensitive intracellular microelectrodes: how to make and use them. Academic Press, London

Thomas RC 1979 Recovery of pH_i in snail neurones exposed to high external potassium. J Physiol (Lond) 296:77P

Thomas RC 1987 Extracellular acidification at the surface of depolarised voltage-clamped snail neurones detected with eccentric combination pH microelectrodes. Can J Physiol Pharmacol 65:1001–1005

Thomas RC 1988 Changes in the surface pH of voltage-clamped snail neurones apparently caused by H^+ fluxes through a channel. J Physiol (Lond) 398:313–327

Thomas RC, Meech RW 1982 Hydrogen ion currents and intracellular pH in depolarized voltage-clamped snail neurones. Nature (Lond) 299:826–828

DISCUSSION

Vaughan-Jones: You showed that there is a surface alkalosis when you make E_m negative with respect to the holding potential. Does that suggest that this channel is open at rest?

Thomas: No, surface alkalinization does not occur until E_m reaches -30 or -40 mV.

Vaughan-Jones: What's producing the alkalinization then?

Thomas: It could be protons going into the cell through these hypothetical channels which would happen to open at that potential. This alkalinization is very variable, in a lot of experiments you don't see it at all. If it's not protons, it could be that some cells have a lot of Ca^{2+} channels which open at that sort of potential. The entry of Ca^{2+} could expose H^+ binding sites, which could cause the observed alkalinization.

Vaughan-Jones: Is the alkalosis removed in Ca^{2+}-free solutions?

Thomas: It's sometimes reduced but it's not very often removed. I have not got a nice clear result. I need a reliable Ca^{2+} channel blocker that has no other effect. It's very hard to prove a negative effect.

Aickin: The alkalosis did seem to be knocked out by zinc.

Thomas: Everything was knocked out by zinc, but zinc also blocks Ca^{2+} channels.

Morad: What happens if you make the membrane potential more negative, e.g. -80 mV? Does the surface pH change? Is there a cross-over point again?

Thomas: I have only taken E_m as far as -100 mV and nothing happens.

Morad: If the relation becomes flatter, you would need a lot of depolarization before you could detect much current.

Thomas: It depends on the intracellular pH. If the cells have been acidified, then surface pH acidification occurs with quite modest depolarizations. The potential at which the surface pH starts to decrease is very sensitive to intracellular pH, that's the major determinant. In these experiments I have been careful not to work on cells that are particularly acifidied, but most of my earlier experiments (Thomas 1987) were done on rather acidified cells.

Morad: Can you do the experiment where you can keep the H^+ concentration high only on the outside of the cell without worrying about inside?

Thomas: Yes, I can.

Morad: And then can you drive H^+ backwards?

Thomas: No, I never got that to work at all well. I did try to do that, to exaggerate the surface alkalinization effect. But, I get the impression that these channels are rather pH sensitive, as naturally they would be.

de Hemptinne: A technical question—isn't it advisable to use double-barrelled electrodes for surface pH measurement when transmembrane current is applied?

Thomas: I did in those earlier experiments I just mentioned. I used an eccentric electrode which was double-barrelled and the reference barrel was filled with an unbuffered Ringer solution. But I saw so little surface potential change that I am now quite happy not having a reference electrode on the cell surface.

de Hemptinne: When you apply transmembrane current you have a voltage drop between the surface of the cells and the ground. Is the voltage drop so small that you don't see it?

Thomas: Yes, I think the proof is in the zinc experiment, which shows no surface pH changes.

de Hemptinne: But after applying zinc, the clamping current is also smaller.

Thomas: No, only 10% less or 20% less, not enough to account for the abolition of the surface pH effects.

Kaila: Are you sure that your surface pH electrode is completely immune to zinc and cadmium?

Thomas: Yes, I use a Hinke-type glass electrode. I haven't actually added zinc to my calibration solutions and proved that, but I have never heard of any problem with such ions.

Aickin: Your surface pH jumped quite a long way when you added zinc.

Thomas: I agree, I presume zinc and cadmium have all sorts of effects on surface charges. I have also studied it with lower levels of zinc.

Aickin: That means you ought to have a reference electrode at the same place.

Thomas: No, because I don't care about the absolute pH. In none of these experiments have I claimed that the actual values of pH I am recording are really significant. I am interested in the changes of surface pH and you see I have big changes before adding zinc when I depolarize and none in the presence of zinc.

Syková: Can you get such biphasic alkaline-acid changes by stimulating the afferent input of those neurons?

Thomas: I don't know. I haven't done anything in the way of natural stimulation, or even unnatural stimulation!

Syková: Would it be possible in your preparation to stimulate input pathways by repetitive electrical stimuli?

Thomas: It must be possible, yes. I don't know anything about the input pathways to these neurons at all, I am afraid.

Boron: You said that Bob Meech is enamoured of the Ca^{2+} hypothesis.

Thomas: He has measured these proton currents and they do not appear to flow into the cell, they only come out. He did this with *Lymnaea* neurons (Byerly et al 1984). He feels that these channels must be rectifying—I have an open mind.

Boron: But when you start depolarizing in your step pulses, as you go from -80 mV to +40 mV, once you get beyond some threshold at which the channel presumably opens, the surface pH starts getting alkaline and then it becomes acid. How does that jive with the Ca^{2+} story? The Ca^{2+} equilibrium potential can't be anywhere near the normal membrane potential.

Thomas: I agree, the Ca^{2+} equilibrium potential must be over +100 mV. My results could be explained by a combination of Ca^{2+} influx at all voltages between -30 and +40 mV, and an H^+ efflux at positive potentials. Initially, the Ca^{2+} influx causes a small increase in surface pH, but this is then swamped by the large exit of H^+ ions at positive voltage.

Boron: So Ca^{2+} is going into the cell and exposing H^+ binding sites?

Thomas: Yes, there are probably a lot of binding sites for Ca^{2+} on the outside of the cell. These sites could be being stripped of Ca^{2+} as this enters the cell when these channels open. These sites then bind H^+ and cause the surface pH increase. I agree, the more I think about it, the less I like it.

Oberleithner: My impression is that it's quite critical that your pH electrode is very close to the surface. Does that mean that the closer the electrode is, the higher the signal?

Thomas: Yes, you can certainly see pH changes if you hold the electrode close to the cell but not touching it, but these changes are much smaller.

Oberleithner: One way to maximize the signal is to move the electrode towards the cell. Another way would be that the cell moves towards the electrode. For example, depolarization induces cell swelling because of the Cl^- influx, that causes the surface of the cell to approach your electrode and then your signal gets bigger. Can you rule out this possibility?

Thomas: In the experiment shown in Fig. 4, everything about Cl^- is going to be the same, zinc doesn't affect Cl^- movement. I don't think it affects anything other than these Ca^{2+} and H^+ channels. In this case, the surface pH response is completely wiped out and it's quite reversible. Don't forget that the corresponding intracellular pH change is again wiped out by zinc. There are some cells in which you don't see much pH change, however much you move the electrode. I think the density of channels does vary from cell to cell.

Harvey: I have trouble understanding the zinc effect on surface pH. How is it changing the surface pH at a membrane potential at which these H^+ channels are not open?

Thomas: I don't know; I don't think this effect has anything to do with channels.

Kaila: It looks like a *de novo* generation of protons. There is an acidosis both intracellularly and extracellularly.

Thomas: The effect is purely extracellular. It could be a surface charge effect—whatever that means! Zinc ions are binding to the surface charges on the cell surface and somehow changing the pH detected by the surface electrode.

Vaughan-Jones: Or the pH of your solution is wrong?

Thomas: That's another possibility. I added a small amount of 1M zinc to the 2 mM Hepes-buffered solution. The key thing is that zinc blocks the channels—I don't honestly think it matters what the pH is.

de Hemptinne: Why are you using the Hinke-type electrode for the measurement of surface pH? With such an electrode a relatively large surface of the pH-sensitive glass is not in direct contact with the surface of the preparation. Why don't you use another electrode shape and have the whole surface of the pH-sensitive glass facing the cells?

Thomas: There is a very elegant design of electrode invented by you but I have tried making those electrodes and technically I cannot! The Hinke design is quite easy to make, it works for weeks, it's reliable, doesn't drift. I may only be seeing half the pH change but since I am not concerned about the absolute value of the pH change, it doesn't matter too much. I have a plan to insinuate this Hinke electrode between two cells in an undisturbed brain. Then stimulate one cell in a pseudophysiological way and see if I actually record the kind of pH change that might occur in a real cleft.

Boron: It seems to me that when you apply a voltage clamp and see a change in pH_i, you should be able to calculate from the rate of intracellular pH change, the buffering power and the surface area:volume ratio, the permeability of the membrane under these conditions to H^+, and whether or not it's reasonable, given the physiological proton concentrations.

Thomas: My fear is that it would be unreasonable and everyone will laugh!

Vaughan-Jones: You will certainly get an extremely high proton permeability coefficient, if you apply Constant Field theory. But this surely assumes that the movement of protons through the channel is similar to the movement of other ions. It's documented, however, that protons can move rapidly along lipid–aqueous interfaces, rather like electrons in a metal conductor. So it's probably not relevant to use the Constant Field equation to estimate proton permeability.

Boron: If the answer was that the permeability is greater than the free diffusion of protons in water (or ice), we should start getting worried.

Thomas: It is quite likely to be greater because of all the buffers and H^+-binding sites.

Grinstein: I think the movement of protons in ice and in other similar semi-conducting water chains is pH insensitive.

Oberleithner: What is the physiological relevance of this channel?

Thomas: I haven't the remotest idea. Martin Morad has discovered these crazy effects of pH on channels (Morad, this volume). May be it's an advantage that the protons come out of some cells and turn round and attack the channels either in the same cell or in a neighbouring cell and convert them from Ca^{2+} to Na^+ channels. His discovery of this effect gives me a marvellous mechanism which these protons might be involved in.

Syková: How can you exclude that these surface pH changes are not due to changes in strong ion concentration differences? These can influence extracellular pH by affecting the dissociation of water. We can explain extracellular pH changes in our experiments by differences in strong ion concentration. Do you think that the relative amount of protons actually leaving the cells could explain large extracellular pH changes? I am very interested to know whether the surface pH changes are comparable to the extracellular pH changes which occur *in vivo* in the brain and whether there is evidence that the pH_e changes are due to protons leaving the cell.

Thomas: Meech and I (1987) have shown that you can get intracellular pH changes after all ions that might be relevant have been removed—that is Na^+, K^+, Ca^{2+}, Mg^{2+}, Cl^-. The superfusate had only N-methyl glucamine and gluconate—are they strong ions or not?

Boron: Stewart would define those as being strong ions.

Vaughan-Jones: Surely you can just measure possible ion changes at the surface using ion-selective microelectrodes?

Thomas: Yes, there are ion changes at the cell surface. For example, Neher & Lux (1973) long ago recorded increases in surface K^+. In my experiments, I also see large changes in intracellular pH. During long depolarizations, large numbers of H^+ must have left the cell and I can see them coming out. Alternatively, OH^- or HCO_3^- could be entering the cell, but I have not seen any effect of changing OH^- or HCO_3^- levels. I cannot see how both surface and intracellular pH could be altered by any change in strong ion levels.

Malnic: A few years ago we did some studies in renal tubules, injecting current into proximal tubules and observing what happened to the pH. Making the lumen of the tubule negative causes flow of H^+ into the lumen. When we reversed this current, the direction of H^+ movement did not reverse. This is similar to what you are showing. At the time we interpreted this as a passive H^+ permeability coupled with buffer properties (Malnic & Cassola 1977).

Thomas: Yes it could be, there are one or two other possible examples. So far, proton channels have only been found in snail neurons and axolotl oocytes.

Oberleithner: Could it be that you drive HCO_3^- out of the tubule lumen when you make the lumen electrically negative? That would acidify the luminal pH.

Malnic: But why then couldn't we reverse it? HCO_3^- flow through the paracellular pathway should be symmetrical.

Boron: Roger, you described an earlier experiment (Thomas 1979), in which you removed Na^+ by itself and the pH oscillated slightly but not much happened. However, when you removed Na^+ and at the same time raised the K^+, there was a huge decrease in the pH_i of the cell. Does that relate to this channel story?

Thomas: The K^+ causes a big depolarization but given that Ca^{2+} was present in the Ringer solution it's hard to rule out Ca^{2+} entry. Ca^{2+} could enter the cell through Ca^{2+} channels which should be opened by the depolarization, this would cause a large pH_i change (Meech & Thomas 1977). Or it could have been H^+ going in. Obviously the depolarization will open up not only proton channels but also Ca^{2+} channels, either of which could cause an intracellular pH change like the one we see. There is some evidence that you don't get such a big change if there is no Ca^{2+} around. It's a traumatic thing to do, to depolarize a cell like that in high K^+ medium.

References

Byerly L, Meech RW, Moody W 1984 Rapidly activating hydrogen ion currents in

perfused neurones of the snail *Lymnaea stagnalis*. J Physiol (Lond) 351:199–216
Malnic G, Cassola AC 1977 Effect of current-induced transepithelial PD changes on H ion movements in renal tubules. In: Anagnostopoulos (ed) Electrophysiology of the nephron. Coll, INSERM 67:53–70
Meech RW, Thomas RC 1977 The effect of calcium injection on the intracellular sodium and pH of snail neurones. J Physiol (Lond) 265:867–879
Meech RW, Thomas RC 1987 Voltage-dependent intracellular pH in *Helix aspersa* neurones. J Physiol (Lond) 390:433–452
Morad M 1988 Proton-induced transformation in gating and selectivity of the calcium channel in neurons. In: Proton passage across cell membranes. Wiley, Chichester (Ciba Found Symp 139) p 187–200
Thomas RC 1979 Recovery of pH_i in snail neurones exposed to high external potassium. J Physiol (Lond) 296:77P
Thomas RC 1987 Extracellular acidification at the surface of depolarised voltage-clamped snail neurones detected with eccentric combination pH microelectrodes. Can J Physiol Pharmacol 65:1001–1005

General discussion III

GABA-activated movements of formate and acetate: influence on intracellular pH and surface pH in crayfish skeletal muscle fibres

Kaila: Proton channels (Thomas, this volume) are not the only conductive pathways capable of mediating movements of acid equivalents across cell membranes. Together with J Voipio, I have recently shown that, in the presence of CO_2/HCO_3^-, the inhibitory transmitter compound γ-aminobutyric acid (GABA) induces a fall in intracellular pH (pH_i) and an increase in extracellular surface pH (pH_s) in crayfish muscle fibres (Kaila & Voipio 1987). The mechanism underlying these changes is a GABA-activated HCO_c^- conduct-

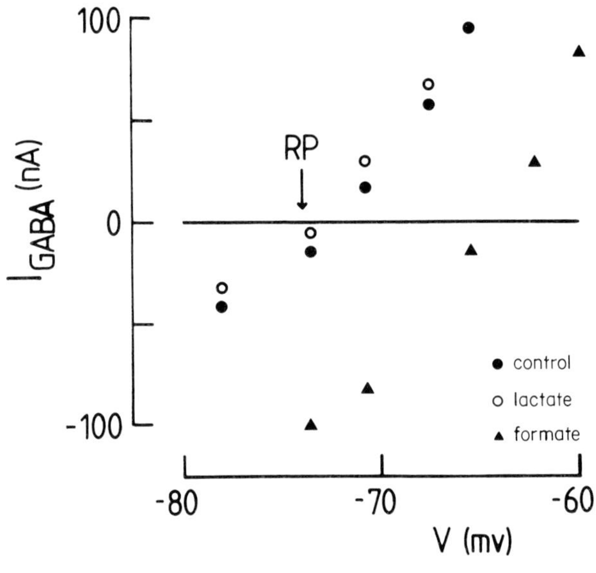

FIG. 1. Influence of lactate and formate on E_{GABA}. A two microelectrode voltage clamp was used to measure GABA-activated currents (vertical axis) evoked at different holding potentials by brief pulses of transmitter added to the bath. Under control conditions and in the presence of 30 mM lactate, the GABA-induced current changes polarity at potentials close to the resting membrane potential, which, due to the high background Cl^- permeability, is virtually identical to E_{Cl}. In 30 mM formate, there is a positive shift of about 8 mV in E_{GABA}. The control solution contained: NaCl 200 mM, KCl 5.4 mM, $CaCl_2$ 7.0 mM, $MgCl_2$ 2.6 mM, Hepes 10 mM pH 7.4. In the lactate and formate solutions, 30 mM NaCl was replaced by an equal amount of the sodium salt of the appropriate weak acid anion.

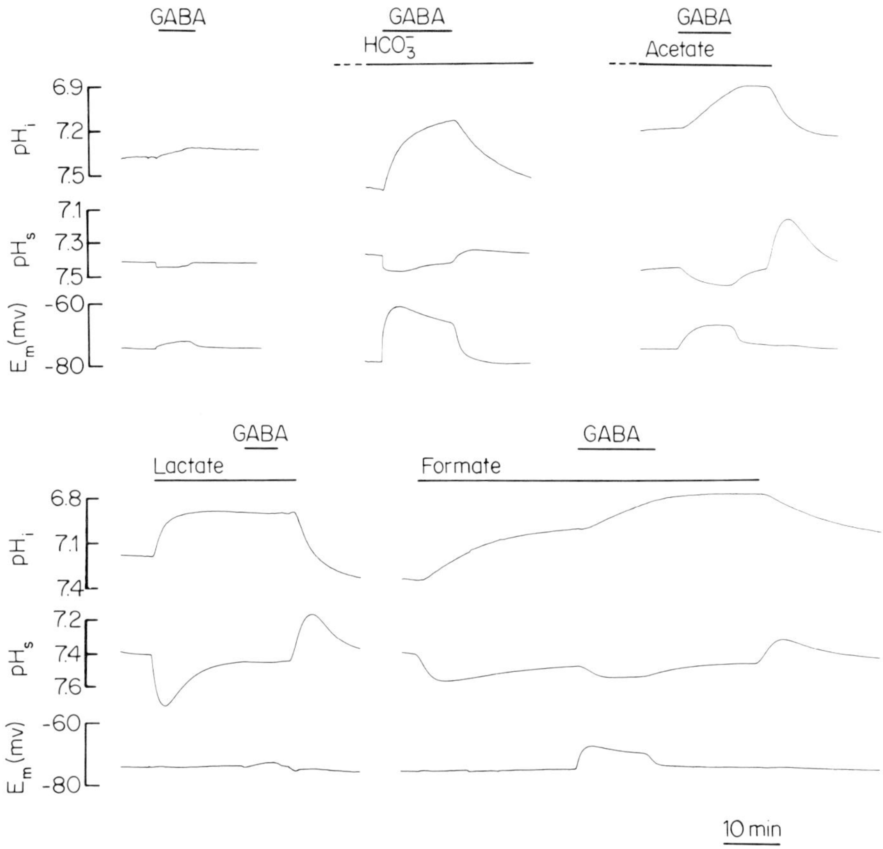

FIG. 2. Influence of bicarbonate, acetate, lactate and formate on the actions of GABA on resting membrane potential (E_m), pH_s and pH_i. In the presence of HCO_3^- (5% CO_2), acetate or formate, application of GABA leads to a substantial depolarization and to a fall in pH_i which is coupled to an increase in pH_s. Such effects are not observed in the presence of lactate. In the absence of HCO_3^-, the GABA-induced intracellular acidoses show little recovery (cf Galler & Moser 1986). All recordings shown are from a single, continuous experiment. Solutions as in Fig. 1 (concentrations of all weak acid anions 30 mM). pH_i and pH_s measured using liquid-sensor H^+-selective microelectrodes.

ance which promotes a passive efflux of HCO_3^- down its electrochemical gradient.

If a cell is exposed to a weak acid which in its non-dissociated form (HA) readily permeates the plasmalemma, then by thermodynamics, after equilibration of HA, the equilibrium potential of the anionic form (E_A) equals that of protons (E_H). Due to active regulation of pH_i (see Roos & Boron 1981), E_H is more positive than the resting membrane potential, and, consequently, activation of a conductance for the anionic species A^- brings about a depolarizing current, a fall in pH_i and an increase in pH_s. Furthermore, if in a crayfish muscle fibre such a conductance is activated by GABA, there should be a shift to more positive potentials in the reversal potential of the GABA-gated current (E_{GABA}) such that $E_{GABA} > E_{Cl}$.

My colleagues MJ Mason, K Mattsson, J Voipio and I have examined the influence of formate, acetate and lactate on the GABA reversal potential and found that both formate and acetate, but not lactate, induced a positive shift in E_{GABA} (Fig. 1). These results indicate a significant permeability of the inhibitory postsynaptic channels to formate and acetate anions but not to lactate (cf Takeuchi & Takeuchi 1967, Bormann et al 1987). A positive shift in E_{GABA} to values higher than E_{Cl} is also observed in the presence of HCO_3^- (Kaila & Voipio 1987).

In agreement with their effects on E_{GABA}, both formate and acetate promoted a large GABA-induced fall in pH_i which was linked to an increase in pH_s and to a depolarization of the membrane potential (Fig.2). In contrast to this, lactate did not produce significant GABA-induced changes in pH_i, pH_s or membrane potential.

These results show that in the presence of permeant weak acid anions, such as bicarbonate, acetate and formate, activation of GABA-gated channels can have a significant effect on postsynaptic pH and may thereby have a modulatory effect on synaptic inhibition. In this context, it is noteworthy that changes in pH_i and pH_s can exert a strong influence on the properties of ion channels in excitable cells (Moody 1984, Morad 1988). On the other hand, the fact that the equilibrium potential of any permeant weak acid anion depends on the plasmalemmal proton gradient (see above) suggests that *in vivo*, E_{GABA} may be sensitive to physiological and pathophysiological alterations of intracellular and extracellular pH.

References

Bormann J, Hamill O P, Sakmann B 1987 Mechanism of anion permeation through channels gated by glycine and γ-aminobutyric acid in mouse cultured spinal neurones. J Physiol (Lond) 385:243–286

Galler S, Moser HJ 1986 The ionic mechanism of intracellular pH regulation in crayfish muscle fibres. J Physiol (Lond) 374:137–151

Kaila K, Voipio J 1987 Postsynaptic fall in intracellular pH induced by GABA-activated bicarbonate conductance. Nature (Lond) 330:163–165

Moody W Jr 1984 Effects of intracellular H^+ on the electrical properties of excitable cells. Annu Rev Neurosci 7:257–278

Morad M 1988 Proton-induced transformation in gating and selectivity of the calcium channel in neurons. In: Proton passage across cell membranes. Wiley, Chichester (Ciba Foundation Symp 139) p 187–200

Roos A, Boron WF 1981 Intracellular pH. Physiol Rev 61:296–433

Takeuchi A, Takeuchi N 1967 Anion permeability of the inhibitory post-synaptic membrane of the crayfish neuromuscular junction. J Physiol (Lond) 191:575–590

Thomas RC 1988 Proton channels in snail neurons studied with surface pH glass microelectrodes. In: Proton passage across cell membranes (Ciba Found Symp 139) p 168–183

Proton-induced transformation in gating and selectivity of the calcium channel in neurons

Martin Morad

University of Pennsylvania, Department of Physiology, Philadelphia, PA 19104, USA

Abstract. Steady-state elevation of $[H^+]_o$ inhibits both Ca^{2+} and Na^+ currents in neurons. Step changes in $[H^+]_o$, however, induce a transient inward Na^+ current ($I_{Na(H)}$) in isolated neurons. In outside-out patches of dorsal root ganglion cells, $I_{Na(H)}$ was fully activated within 2 ms at pH 6.7, and inactivated with a time constant of 300 ms. Deactivation $t_{1/2}$ was 3 ms at pH 7.9. $I_{Na(H)}$ was blocked by Ca^{2+} channel blockers. This observation, and the finding that the voltage-gated I_{Ca} disappeared rapidly during activation of $I_{Na(H)}$ and reappeared with inactivation of $I_{Na(H)}$, suggested that $I_{Na(H)}$ occurs through a transformed Ca^{2+} channel. The proton-sensitive site was located at the external mouth of Ca^{2+} channel. The single channel conductance of $I_{Na(H)}$ was 28 ps in symmetrical 120 mM NaCl solutions. Increase of $[H^+]_o$ during the activation of I_{Ca} suppressed I_{Ca} within 2 ms. Our studies suggest that the Ca^{2+} channel exists in two conformational states; a voltage-gated Ca^{2+}-transporting state, and a proton-gated Na^+-transporting state. The dominance of the proton-gated state over the voltage-gated state suggests that proton modificiation of the Ca^{2+} channel may be extremely important in neurophysiological and neurosecretory function.

1988 Proton passage across cell membranes. Wiley, Chichester (Ciba Foundation Symposium 139) p 187–200

Proton concentrations, though highly regulated both in the intracellular and extracellular environments, are known to change transiently under a variety of physiological and pathological conditions. Transient increases in proton concentrations occur, for instance, in the extracellular space during epileptic episodes (Cragg et al 1977) and rapid stimulation periods (Kraig et al 1983). Intracellular alkalinizations are also known to accompany Ca^{2+} release signal in muscle (Baylor et al 1987), while fatigued or ischaemic muscles have increased myoplasmic proton concentrations. Changes in $[H^+]_o$ also have marked effects on neuronal ionic channels. While large increases in $[H^+]_o$ are required to suppress the Na^+ current (Hille 1968), the Ca^{2+} current is suppressed with moderate decreases in the extracellular pH (Iljima et al 1986). Interestingly, rapid increases in $[H^+]_o$ around a variety of different peripheral and central neuronal cells have been shown to activate a large tetrodotoxin

(TTX)-insensitive Na^+ current lasting 1–2 seconds, which may cause a transient depolarization of about 20–30 mV (Krishtal & Pidoplichko 1980, 1981, Konnerth et al 1987, Davies et al 1988).

In this paper, I show that such a proton-gated Na^+ current is carried through transformed Ca^{2+} channels which have transiently lost their voltage-gating and divalent selectivity. The proton-sensitive site appears to be associated only with the external mouth of the Ca^{2+} channel. The single channel conductance of $i_{Na(H)}$, the similar pharmacological sensitivities of $i_{Na(H)}$ and i_{Ca}, and the temporal exclusiveness of $i_{Na(H)}$ and i_{Ca} suggest that $i_{Na(H)}$ is transmitted by a transformed state of the Ca^{2+} channel.

Methods

Two- to three-day-old chick cultured dorsal root ganglion (DRG) neurons or mouse neuroblastoma cells (Type NIE-115) were studied using the patch-clamp technique, in whole cells or in isolated membrane patches (Hamill et al 1981). Cells were dialysed with Cs^+ and tetraethyl ammonium (TEA) to suppress K^+ current and in addition the solution usually contained 11 mM EGTA (ethyleneglycol-bis-[β-aminoethyl ether]-N,N,N′,N′-tetracetic acid), 20 mM Hepes, 5 mM Mg^{2+}ATP, 0.2 mM cAMP. In some experiments 25 mM BAPTA (a pH-insensitive Ca^{2+} buffer) and 90 mM Hepes were included, to increase the buffering capacity for Ca^{2+} and H^+. Extracellular solutions were modifications of standard Tyrode solutions, usually containing high Na^+, no added K^+, 10–20 mM TEA to block K^+ channels, TTX (3×10^{-6} M) to block i_{Na}, and zero to 5 mM Ca^{2+}. Extracellular solutions around the neurons were exchanged rapidly using a multi-barrelled pipette with a common outlet: see Konnerth et al (1987) and Davies et al (1988) for a description of the rapid flow technique and other procedural details.

Results

Protons gate a slowly inactivating Na^+ current in neurons

A step increase in proton concentrations around neurons, from pH 7.9 to 6.7, induces a large inward current in DRG neurons (Fig. 1). The current can be activated rapidly in whole cells (Fig. 1A) and outside-out patches (Fig. 1B), but not in inside-out patches (not shown). The activation of the current was rapid, but it relaxed slowly with a time constant of about 300 ms. The current was rapidly deactivated, $t_{1/2}$ 1.2–3 ms, when proton concentrations were rapidly reduced (Fig. 1A, light trace, and Fig. 1B). The activation half-time of the current was proton dependent, ranging from 1–20 ms for changes from pH 6.0 to 7.2 (Davies et al 1988). Neither the activation nor the inactivation kinetics was significantly voltage dependent.

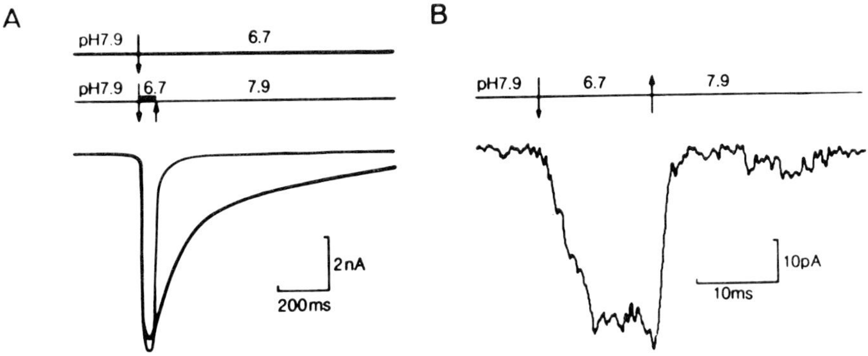

FIG. 1. Activation and deactivation of $i_{Na(H)}$ in a whole cell and an isolated outside-out patch. (A) Heavy trace shows a recording of $i_{Na(H)}$ induced in a whole cell following a step in pH from 7.9 to 6.7, as indicated. In the continued presence of pH 6.7, the response inactivated completely within 1–2 s. If, following the pH step from 7.9 to 6.7, the pH was returned after a brief period to 7.9 (lower protocol), then the response was rapidly deactivated as shown by the light trace. (B) Activation and deactivation of $i_{Na(H)}$ in an outside-out patch following the sequence described above the recording. Note that the time course of the events was much faster in the isolated patch compared with that in the whole cell. In this example the half-time of activation was 2.9 ms; following a prompt return to pH 7.9, deactivation to a residual level occurred with a half-time of 1.3 ms. Holding potential was −80 and −50 mV; $[Ca^{2+}]_o$ was 2 mM and 0.1 mM in A and B, respectively. (Modified from Davies et al 1988.)

Small proton concentrations, which by themselves failed to activate the proton-induced current in outside-out and whole cells, significantly reduced the current accompanying a larger increase in $[H^+]_o$. In outside-out patches, inactivation did not require that dispersed channel openings accompany the conditioning proton concentrations. Thus, by analogy with the voltage-gated currents, inactivation was considered to represent an independent state, which could be reached directly from the closed or open state of the channel. Since this current appears to activate, inactivate and deactivate in response to step increase or decrease of extracellular proton concentration, in a manner similar to the response of the voltage-gated channels to voltage, we concluded that the channel was 'proton-gated.'

Fig. 2A shows the time course of proton-induced current in the presence and absence of Na^+ in an outside-out patch. When extracellular Na^+ was replaced by choline or TEA, the current was completely and reversibly suppressed. In the absence of Na^+, 5 mM Ca^{2+} failed to carry any current through the channel (Fig. 2A, middle trace). Examination of the voltage dependence of unitary events in outside-out patches also showed that this current reversed at the Na^+ equilibrium potential. Fig. 2B shows that the reversal potential of the unitary current is, as expected, near 0 mV in symmetrical

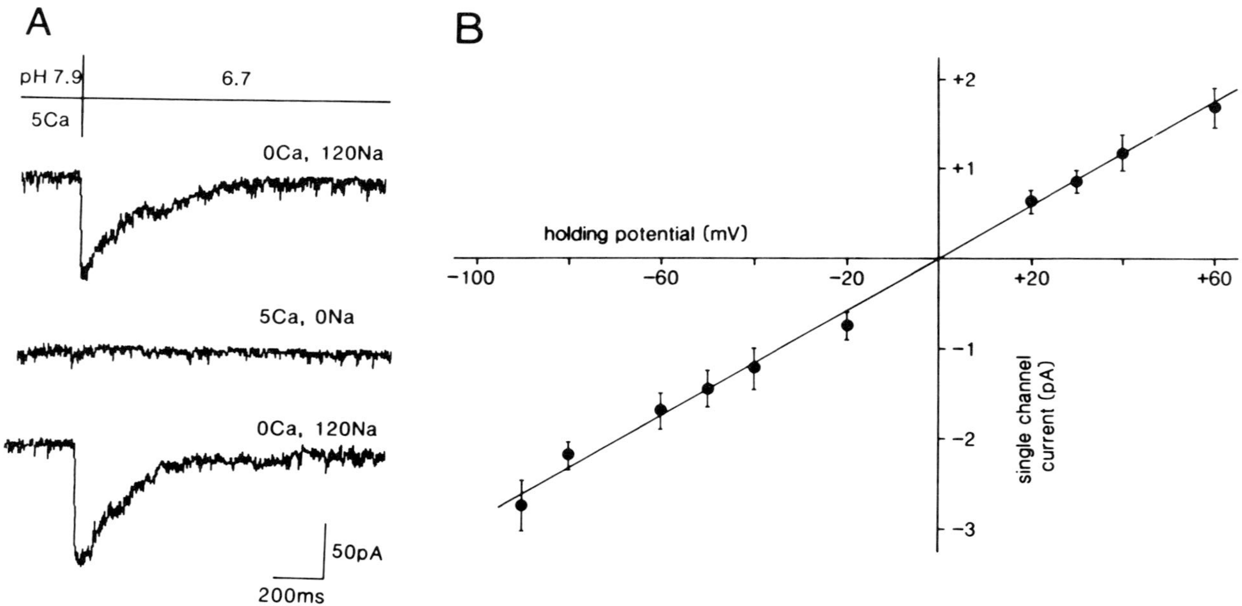

FIG. 2.(A) In the absence of internal Na^+, the replacement of external Na^+ by choline abolished $i_{Na(H)}$ in outside-out patches. This effect was reversible. The mM concentrations of external Ca^{2+} and Na^+ are indicated. Note that no current was induced when 5 mM Ca^{2+} was present in the zero Na^+ activating solution. The holding potential was −80 mV. Identical results were obtained in four other patches. (B) Current–voltage relation of unitary events associated with $i_{Na(H)}$ in an outside-out patch with $[Na^+]_o = 120$ mM $[Na^+]_i$. The line drawn through the data has a slope of 28 pS and a reversal potential near zero mV, close to the calculated Nernst potential for Na^+. Each point is the mean ± *SD* of up to twenty measurements. (Modified from Davies et al 1988.)

(120 mM) Na^+ solutions. The single channel conductance was about 28 ps in symmetrical Na^+ solution, but was slightly lower (20 ps) when $[Na^+]_i$ was reduced to 10 mM. The plot of different ratios of Na_o/Na_i in different cells versus reversal potential of the proton-induced current gave a straight line with the slope of 58 mV/decade (Konnerth et al 1987), suggesting that the channel was highly selective for Na^+. Therefore, the proton-induced current is carried by Na^+ and is abbreviated as $i_{Na(H)}$.

$i_{Na(H)}$ *is blocked by* Ca^{2+} *channel blockers*

Since $i_{Na(H)}$ was not blocked by high concentrations of TTX (10 μM), it was concluded that $i_{Na(H)}$ did not flow through the TTX-sensitive Na^+ channel. On the other hand, $i_{Na(H)}$ was highly modified by the concentrations of divalent cations in the bathing solution. Cd^{2+}, Co^{2+} and Ni^{2+} blocked or strongly suppressed $i_{Na(H)}$. However, both the K_d and the stoichiometry of block were somewhat different for $i_{Na(H)}$ than for i_{Ca} (Davies et al 1988). For instance, Cd^{2+} half block was achieved at 10 μM for i_{Ca}, but 100 μM was required for half maximal block of $i_{Na(H)}$. The stoichiometry of Cd^{2+} or Ni^{2+} block was 1:1 for i_{Ca} and 2:1 for $i_{Na(H)}$. Organic Ca^{2+} channel blockers, with the exception of dihydropyridines, also strongly suppressed $i_{Na(H)}$ and i_{Ca}. In DRG neurons, we were unsuccessful in blocking $i_{Na(H)}$ or i_{Ca} by nifedipine (10 μM). Thus, our results suggest either that $i_{Na(H)}$ and i_{Ca} flow through the same channel or that Ca^{2+} channel blockers also block $i_{Na(H)}$. It should be remembered, however, that if Ca^{2+} channel is the conduit for $i_{Na(H)}$, the channel is not in its usual voltage-gated state, but in a novel, voltage-sensitive 'proton-gated' state. This possibility, however, requires the channel to exist in two different conformational states: a voltage-gated, Ca^{2+}-permeable state, and a proton-gated, Na^+-selective state.

$i_{Na(H)}$ *is carried by a modified* Ca^{2+} *channel*

The ionic independence hypothesis requires not only that discrete and pharmacologically distinct channels transport the different ionic currents, but also that channels function simultaneously and independently of each other (Hodgkin & Huxley 1952). For example, voltage-gated Na^+ and K^+ currents can be separated temporally and pharmacologically even though they co-exist simultaneously. Since $i_{Na(H)}$ lasts about 1–2 seconds and could be activated at a wide range of membrane potentials by step elevation of $[H^+]_o$, it was possible to measure the magnitude of i_{Ca} during the time course of $i_{Na(H)}$ by applying short (30–50 ms) pulses that depolarized the transmembrane voltage to zero mV. Fig. 3 shows that simultaneously with activation of $i_{Na(H)}$, i_{Ca} decreases markedly and rapidly. As $i_{Na(H)}$ inactivates slowly in the presence of high proton concentrations (pH 6.7), i_{Ca} recovers slowly. Thus i_{Ca} and $i_{Na(H)}$ appear

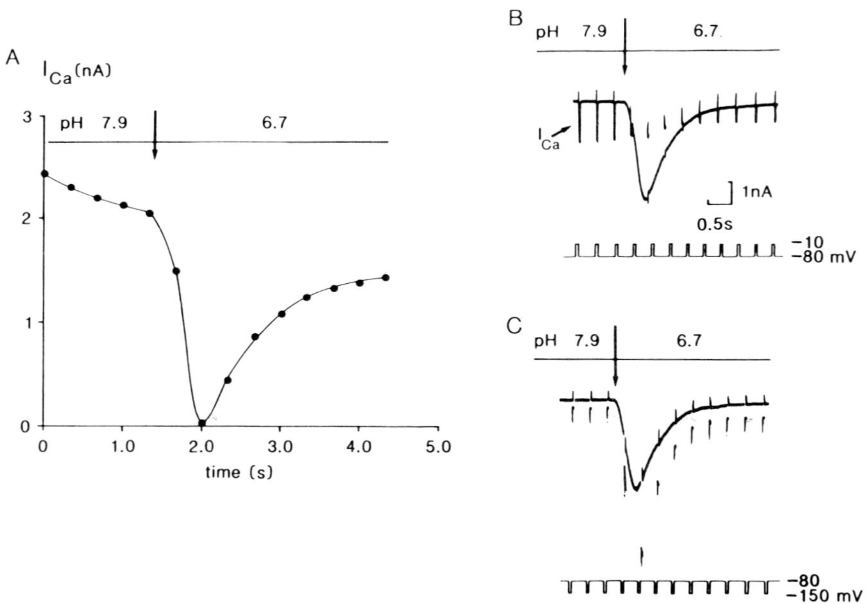

FIG. 3. Transient transformation of Ca^{2+} channel from a voltage-gated to a non-gated state. (B) Inward i_{Ca} activated at a frequency of 3 Hz by short, 30 ms depolarizing pulses before and during the time course of activation of $i_{Na(H)}$. Prior to application of the pH step, i_{Ca} was corrected by substracting the 'leak' current measured during hyperpolarizing pulses (C). i_{Ca} thus determined was plotted prior to and during the application of the pH step (A). Note that when the pH step was applied (arrow) there was a rapid decrease in the amplitude of i_{Ca} as $i_{Na(H)}$ was activated. Slow relaxation of $i_{Na(H)}$ occurred simultaneously with recovery of the voltage-gated i_{Ca}. The decrease in magnitude of i_{Ca} in solutions buffered at pH 6.7 represents the depressive effect of low pH solutions on the voltage-gated Ca^{2+} channel. $[Ca^{2+}]_o$ = 5 mM; TTX, 2 μM; room temperature; pH_i 7.3; $[Ca^{2+}]_i = 10^{-9}$ M. (From Konnerth et al 1987.)

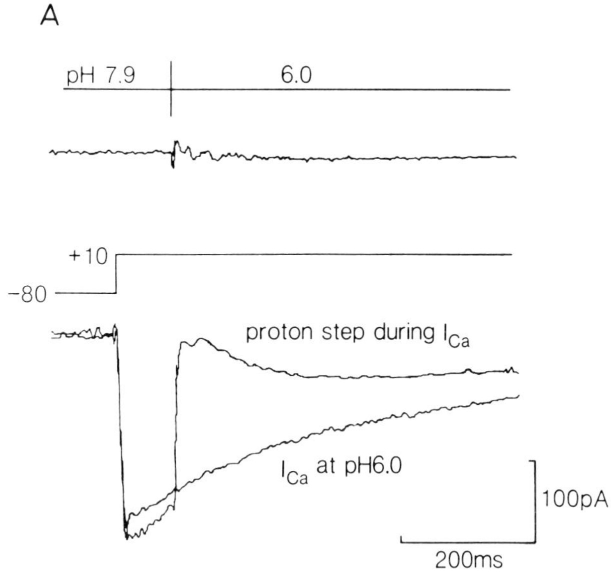

FIG. 4. (A) Application of a proton step during the voltage-gated i_{Ca} in whole cells. The upper trace shows the response obtained following a proton step from pH 7.9 to 6.0 at a holding potential of +10 mV, the reversal potential of $i_{Na(H)}$ under these conditions (symmetrical 100 mM Na^+, 5 min after breaking into the cell). (B) The same proton step applied during the time course of the voltage-gated i_{Ca} (produced by a pulse from −80 mV to +10 mV) rapidly and markedly reduced this current. Note that i_{Ca} recovers during the inactivation of the proton-gated state. i_{Ca} at steady-state pH 6.0 is also shown for comparison. $[Ca^{2+}]_o$ = 5 mM. (From Davies et al 1988.)

to be mutually exclusive, supporting the idea that both currents are carried by the same channel. This finding suggests that rapid elevation of protons at the external mouth of the Ca^{2+} channel leads to its transformation from a Ca^{2+} permeable to a Na^+ permeable state.

The possibility that a similar transformation of Ca^{2+} channel could take place when the Ca^{2+} channel was in the voltage-gated open state was tested by applying the proton step during the activation phase of i_{Ca}. We chose symmetrical Na^+ solutions such that activation of $i_{Na(H)}$ around zero to +10 mV (its reversal potential) was accompanied by little or no net current flow. Fig. 4 shows that step elevation of protons rapidly (1–2 ms) and strongly suppresses the Ca^{2+} current activated at +10 mV. This effect was caused by a mechanism other than simple decrease of i_{Ca} at high proton concentrations, since i_{Ca} activated at pH 6.0 under steady-state conditions was only slightly smaller than that recorded at pH 7.9 at +10 mV (Davies et al 1988, Iljima et al 1986). This transient suppression of voltage-gated Ca^{2+} current did not require the presence of Na^+, and appeared to be an inherent property of the channel. Thus it appears that whether the Ca^{2+} channel is in the closed or open state, changes in proton concentrations can modify its current and gating.

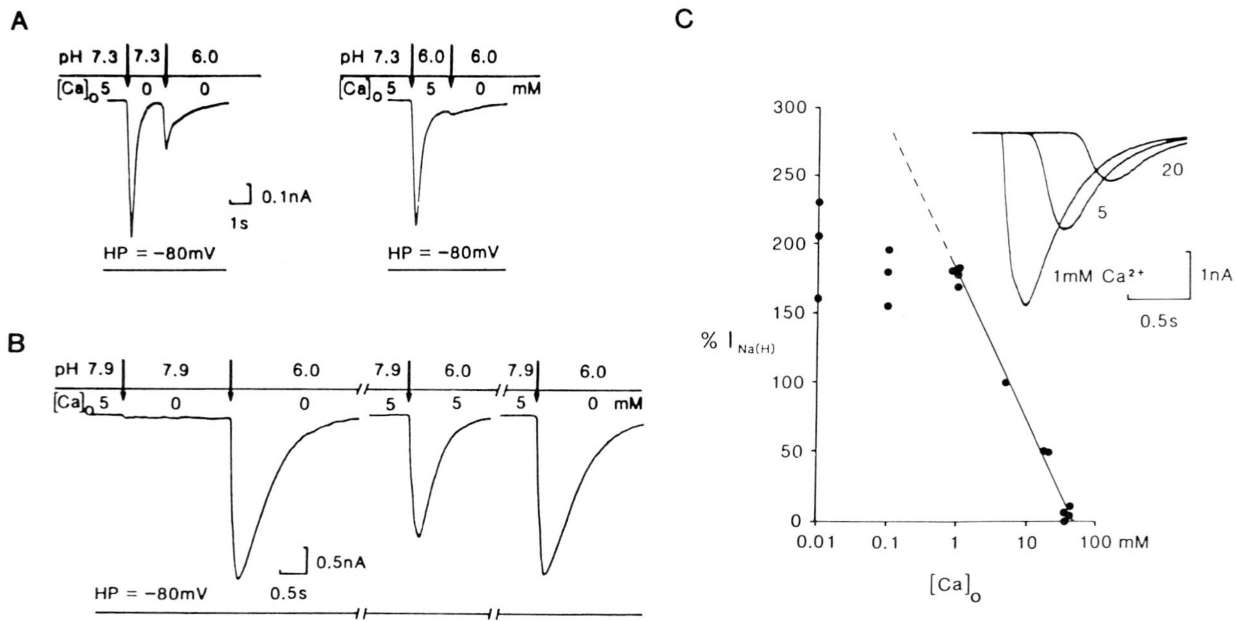

FIG. 5. Ca^{2+} and H^+ interact at a site associated with the Ca^{2+} channel to induce or prevent transformation of Ca^{2+} channel. (A) Time course of activation of zero Ca^{2+}-induced Na^+ current followed by a step increase in $[H^+]_o$. If the current was activated first by application of protons, the zero Ca^{2+} solution (Ca^{2+} omitted plus 1 mM EGTA) failed to activate any inward current. (B) Zero Ca^{2+}-induced $i_{Na(H)}$ failed to activate when $[H^+]_o$ was reduced to pH 7.9. Comparison of the simultaneous application of zero Ca^{2+} and the proton step with that where the cell was incubated first for a short time in a zero Ca^{2+} solution followed by elevation of $[H^+]_o$ showed no difference in the time course or the magnitude of $i_{Na(H)}$. Room temperature; $[Na^+]_i$ = 20 mM; TTX, 2 μM; pH_i 7.3. (C) Time course of $i_{Na(H)}$ at 1.0 mM; 5.0 mM and 20 mM $[Ca^{2+}]_o$ in response to a step change in pH from pH 7.9 to 6.7. $i_{Na(H)}$ is completely blocked by 35–40 mM $[Ca^{2+}]_o$. The graph shows considerable scatter at concentrations below 1 mM. The results from five cells were normalized with respect to $i_{Na(H)}$ measured at 5 mM Ca^{2+}. At least three different Ca^{2+} concentrations were used in each cell. TTX, 2 μM; room temperature; HP, holding potential. (Modified from Konnerth et al 1987.)

How is the voltage-gated Ca^{2+} channel transformed?

Activation of $i_{Na(H)}$ was proton dependent. The data relating the degree of activation of $i_{Na(H)}$ to proton concentration had a Hill coefficient of four (Konnerth et al 1987), suggesting that the binding of at least four protons was required for activation of $i_{Na(H)}$. The activation of $i_{Na(H)}$ shifted to higher proton concentrations when $[Ca^{2+}]_o$ was increased. Fig 5C shows that increases in the $[Ca^{2+}]_o$ suppressed $i_{Na(H)}$ and completely blocked it at concentrations greater than 25 mM. The half maximum block was often at around 5 mM. In fact, simultaneous step elevation of H^+ and Ca^{2+} prevented transformation of the Ca^{2+} channel, as if removal of Ca^{2+} from a site closely associated with the channel was necessary to transform the channel. Consistent with this hypothesis, step reduction of Ca^{2+} to $<10^{-6}$ M activated $i_{Na(H)}$ in normal solutions (pH 7.3, Fig. 5A), but not when proton concentrations were further reduced to pH 7.9 (Fig. 5B). Thus it appears that Ca^{2+} and H^+ compete for a site associated with the Ca^{2+} channel which determines the gating and selectivity of the channel.

Extracellular, but not intracellular, Ca^{2+} also had a significant effect on the kinetics of inactivation of $i_{Na(H)}$. The time constant of inactivation was increased from 300 to 600ms by reduction of Ca^{2+} from 10^{-3} M to 10^{-6} M. Very low Ca^{2+} concentrations ($<10^{-8}$ M) in some outside-out patches arrested the inactivation process. Large $[Ca^{2+}]_i$, up to 1 mM tested, had no effect on either the activation or inactivation of $i_{Na(H)}$, suggesting that both the activation and the inactiyation sites associated with $i_{Na(H)}$ are located at the extracellular side of the membrane. As the proton-induced transformation changes the stoichiometry of Ca^{2+} channel block by inorganic blockers (Cd^{2+} and Ni^{2+}) from 1:1 to 2:1, it is likely that proton-induced transformation of the Ca^{2+} channel is accompanied by exposure of a second divalent binding site. It is intriguing to speculate that the binding of Ca^{2+} to this second site initiates the inactivation process.

Conclusions and possible physiological implications

We found that a large, slowly inactivating Na^+ current can be activated in whole cells or isolated membrane of neuronal cells in response to a moderate step elevation of $[H^+]_o$. This current could be blocked by organic and inorganic Ca^{2+} channel blockers, and its activation was mutually exclusive from the activation of the voltage-gated Ca^{2+} channel. Voltage-gated K^+ and Na^+ channels could be measured simultaneously with activation of $i_{Na(H)}$. Thus it appeared that proton-induced transformation was an exclusive property of the neuronal Ca^{2+} channels. The single channel amplitude (−1.6 pA at −30 mV) was comparable to that measured for the Na^+-permeable low threshold Ca^{2+} channel in DRG neurons (Carbone & Lux 1987). Thus Ca^{2+} channels may exist in two mutually exclusive states: a voltage-gated, Ca^{2+}-permeable state and a proton-gated, Na^+-transporting state. The transition from the voltage-gated to the proton-gated state occurs rapidly, and takes place whether the voltage-gated channel is closed or open.

Proton modification of Ca^{2+} channel may be involved in regulation of neurosecretion, release of Ca^{2+} from the sarcoplasmic reticulum, and repetitive neuronal activity. It has been recently shown that $i_{Na(H)}$ may be activated in a variety of isolated central and peripheral neurons (Davies et al 1988). Although changes in $[H^+]_o$ are measured in the extracellular space in response to repetitive neuronal activity, it may be argued that such changes are moderate and slow. It should be remembered, however, that pH is measured in a space markedly distorted and enlarged by the penetrating H^+-selective electrodes, which have very slow response times (Kline & Morad 1978). Considering the normal proximity of the adjacent cell membrane (10–20 nm), the increase in proton concentration may be sufficiently large and fast to modulate or transform the neighbouring Ca^{2+} channels.

We have recently tested whether Ca^{2+} channels in neurosecretory cells are regulated by localized increases in proton concentration secondary to the release of neurotransmitter from the secretory vesicles, which also contain high proton concentrations. In chromaffin cells, for instance, the adrenergic transmitter is stored in vesicles with very high proton concentrations (pH 5.3; Johnson & Scarpa 1976). In clusters of cultured bovine chromaffin cells, we were able to patch individual cells and measure the Ca^{2+} current before, during, and after secretory activity. We found that voltage-gated Ca^{2+} currents measured from individual cells in the cluster were strongly suppressed immediately following perfusion of the cluster with solutions containing either 30 mM KCl or 10^{-6} M nicotine, which induce secretion of vesicles containing adrenaline and H^+. Step elevation of protons (to pH 6.5), but not of ATP (10 mM) or adrenaline (10^{-5} M), also found in high concentrations in the secretory vesicles, mimicked the suppression of i_{Ca} (G. Callewaert, R. Johnson & M. Morad, unpublished observations). These results suggest that protons may indeed play an important role in presynaptic regulation of the Ca^{2+} channel. It is likely that such proton feedback regulatory mechanisms will be found more often as we start to look for them more frequently.

Acknowledgements

This work was supported by NIH grant #RO1-HL-16152 and A. von Homboldt Senior Scientist Award. In particular I thank Professor H.D. Lux for his help with many of the published experiments cited above.

References

Baylor SM, Hollingworth S, Pape P 1987 Myoplasmic pH transients monitored with indicator dyes in frog skeletal muscle fibers. Biophys J 51:549a

Carbone E, Lux HD 1987 Single low-voltage-activated calcium channels in chick and rat sensory neurones. J Physiol (Lond) 386:571–601
Cragg P, Patterson L, Purves MJ 1977 The pH of brain extracellular fluid in the cat. J Physiol (Lond) 272:137–166
Davies NW, Lux HD, Morad M 1988 Site and mechanism of activation of proton-induced Na^+ current in chick dorsal root ganglion neurons. J Physiol (Lond) 400:159–187
Hamill OP, Marty A, Neher E, Sakmann B, Sigworth FJ 1981 Improved patch-clamp techniques for high-resolution current recording from cells and cell-free membrane patches. Pfluegers Arch Eur J Physiol 391:85–100
Hille B 1968 Charges and potentials at the nerve surface: Divalent ions and pH. J Gen Physiol 51:221–236
Hodgkin AL, Huxley AF 1952 A quantitative description of membrane current and its application to conduction and excitation in nerve. J Physiol (Lond) 117:500–544
Iljima T, Ciani S, Hagiwara S 1986 Effects of the external pH on Ca channels: Experimental studies and theoretical considerations using a two-site, two-ion model. Proc Natl Acad Sci USA 83:654–658
Johnson RG, Scarpa A 1976 Ion permeabiity of isolated chromaffin granules. J Gen Physiol 68:601–631
Kline R, Morad M 1978 Potassium efflux in heart muscle during activity: Extracellular accumulation and its implications. J Physiol (Lond) 280:537–558
Kraig RP, Ferreira-Filho CR, Nicholson C 1983 Alkaline and acid transients in cerebellar microenvironment. J Neurophysiol (Bethesda) 49:831–850
Konnerth A, Lux HD, Morad M 1987 Proton-induced transformation of calcium channel in chick dorsal root ganglion cells. J Physiol (Lond) 386:603–633
Krishtal OA, Pidoplichko VI 1980 A receptor for protons in the nerve cell membrane. Neuroscience 5:2325–2327
Krishtal OA, Pidoplichko VI 1981 A receptor for protons in the membrane of sensory neurons may participate in nociception. Neuroscience 6:2599–2601

DISCUSSION

Thomas: It's nice to see protons doing something useful on the outside of your cells, though it's too early to say what they are doing in snail neurons.

Oberleithner: In the experiment where you applied K^+ to a single cell, I assume you depolarized the cell. What's the conclusion from these experiments?

Morad: The idea was that increasing the K^+ concentration (to 30 mM) depolarizes the cluster of bovine chromaffin cells, resulting in the secretion of adrenaline. Since the vesicles contain high proton concentrations, the release of their contents into the extracellular space would cause local acidification, which, in turn, would inactivate the Ca^{2+} current. In single, isolated cells, in the absence of any confined extracellular space, the likelihood of an effect on local pH is very small. Consistent with this idea, we found that in single cells, the Ca^{2+} current was not altered by step elevation of K^+ or by nicotine.

Thomas: So the cell you are patch-clamping is not itself releasing noradrenaline?

Morad: We don't know that exactly. It doesn't really matter because all the other cells in the cluster secrete their vesicular contents in response to an increase in KCl.

Thomas: So there is a cloud of noradrenaline and acid products in the vicinity of the cell.

Morad: Right. The vesicles release adrenaline, H^+, ATP, neuropeptides and other agents, so you have to do the control experiments. Nicotine, adrenaline (10^{-5} M) and ATP (10 mM) had no effect on the calcium current. But we found that increasing the H^+ concentration to pH 6.5 had a marked suppressive effect on the Ca^{2+} current.

Vaughan-Jones: Aren't you simply showing that Ca^{2+} currents in these cells are very susceptible to proton block? You don't have to transform them through this high Na^+ conductance state to achieve proton block.

Morad: It may be, but I showed the same thing in neurons. There is no necessity for Ca^{2+} channels to have been transformed under these particular conditions.

Vaughan-Jones: Chromaffin cells may just be more sensitive to proton block than neurons.

Morad: I don't think so. Our experiments showed that the steady-state sensitivity of chromaffin cell Ca^{2+} channels to external H^+ concentrations was not significantly different from that of neuronal cells.

Vaughan-Jones: I don't know how the Ca^{2+} current in the chromaffin cell inactivates, if at all. If it doesn't inactivate significantly, then you would expect secretion to begin again after an external proton load as the Ca^{2+}-conducting state is resumed; the transformation into a Na^+-conducting state is only transient.

Morad: We assume that in the chromaffin cells Ca^{2+} channels are normally shut off by the proton release.

Syková: Martin, you showed that change of extracellular pH by only several tenths of a pH unit can transform these channels. Do you think that such a small pH change is sufficient to induce this transformation? In other words, can the small changes which occur regularly as a result of neuronal activity under physiological conditions also have this effect?

Morad: I think very small changes in extracellular pH are sufficient to activate $I_{Na(H)}$. In outside-out patches, most of the experiments were done going from pH 7.9 to 7.3 or 7.4, so that only a few channels were activated. Naturally, we tried to maximize the activation of $I_{Na(H)}$ in order to study the current in detail. In fact, I feel that very small changes in the local concentration of extracellular H^+ can transform the Ca^{2+} channel. Such changes in pH are known to occur in the extracellular space.

Thomas: Can you see any advantage to the cell in having its Ca^{2+} channels sometimes switched to acting as Na^+ channels? Are there situations where a Na^+ channel might be more useful?

Morad: I can only speculate! Under conditions where repetitive neuronal discharges are known to occur, in order to prevent Ca^{2+} overload of the cells, Ca^{2+} channels could transport Na^+ to prevent Ca^{2+}-induced cytotoxicity.

Vaughan-Jones: Are these the N-type channels, or L or T-type? It is notable that they are insensitive to dihydropyridines.

Morad: I don't think that these are N-type channels. We find only two types of Ca^{2+} channels in DRG neurons, low threshold and high threshold. Since these channels are not blocked by the Ca^{2+} channel blocker, nifedipine, just as reported for the N-channel, one may be tempted to relate $I_{Na(H)}$ to the N-type channel current but there is no real evidence.

Gillespie: I wasn't quite clear what the evidence was for the inactivated state going to the H^+ open state. Did you completely inactivate the Ca^{2+} channels with prolonged depolarizing voltages?

Morad: The way to do this is to look at the voltage dependence of $I_{Na(H)}$. The membrane potential is held at -80 mV, +50 mV or at zero for 10–15 seconds. Then we activate $I_{Na(H)}$ by increasing the extracellular proton concentration. We know that when the membrane is depolarized at potentials positive with respect to -20 mV, I_{Ca} is inactivated; on the other hand, $I_{Na(H)}$ can still be activated. Thus we concluded that transformation of the Ca^{2+} channel can occur from the voltage-inactivated state of I_{Ca}.

Gillespie: To what extent?

Morad: Fully—at -20 mV the voltage-gated channel is completely inactivated after a couple of seconds. The IV relationship for $I_{Na(H)}$, however, remains totally intact and unaltered.

Thomas: So the protons can override everything?

Morad: Proton-induced transformation is the dominant state. But it seems to need a recognition site. The site exists in neurons, but in heart I have not, as yet, been able to activate $I_{Na(H)}$. We could speculate that $I_{Na(H)}$ is related to a subtype of Ca^{2+} channel.

Thomas: Is the Na^+ channel blocked by amiloride?

Morad: No, the TTX-sensitive Na^+ channel is not blocked by amiloride in neuronal cells, but curiously, both the voltage-gated T-type Ca^{2+} channels and $I_{Na(H)}$ are selectively blocked by this drug. Benzyamyl is an even more potent analogue of amiloride, it completely blocks $I_{Na(H)}$ and T-type current at 0.1 μM. Amiloride, however, does not block the L channel.

Harvey: There may be another explanation for the absence of amiloride effect at positive membrane potentials. In tight epithelia, amiloride binding to the Na^+ channel and amiloride block of this Na^+ current are reduced at positive membrane potentials.

Morad: I believe you. There is no way that I could tell that from our data on voltage-gated Ca^{2+} currents, but with respect to $I_{Na(H)}$ your comment is consistent with our experimental findings.

Harvey: The proton-gated effect, is it voltage dependent?

Morad: No, both activation and inactivation of $I_{Na(H)}$ are very voltage insensitive.

Harvey: That means your proton-gated effect doesn't see the intramembrane electrical field, therefore it has to be on the outside.

Morad: Right. In one experiment, I dialysed the cell with high proton concentrations, pH 5, and the activation of $I_{Na(H)}$ remained unaffected. So I think that the proton binding site is at the outer mouth of the Ca^{2+} channel.

Harvey: The pK is 6.8. Have you looked at the effects of molecules which may bind to residues with such a pK?

Morad: No, this is a very difficult problem.

Grinstein: In the experiments with the clusters where you are assuming that it's the pH that's inhibiting the channel, can you repeat that inhibition by increasing the buffering power in the medium?

Morad: We have started doing these experiments. Using 10–20 mM Hepes in the media we think that we have a strong buffering power, nevertheless, we get good effects. We need to do more experiments on this issue.

Grinstein: There are several Ca^{2+} channels which become Na^+ selective when you remove extracellular Ca^{2+}. Is the effect described in the literature transient like you see in your pH-induced conductance changes?

Morad: No. We published a paper describing such an effect (Carbone & Lux 1987). The voltage-gated Ca^{2+} channel can transport Na^+ when the extracellular Ca^{2+} concentration is less than 10^{-7} M. Such channels remain voltage gated, and the response is not transient. However, note that external Ca^{2+} blocks such channels with a K_D of 0.1 μM. The block of $I_{Na(H)}$ by external Ca^{2+} had a K_D of about 5 mM.

References

Carbone E, Lux HD 1987 Single low-voltage-activated calcium channels in chick and rat sensory neurones. J Physiol (Lond) 386:571–601

Aldosterone-controlled linkage between Na^+/H^+ exchange and K^+ channels in fused renal epithelial cells

Hans Oberleithner, Ulrich Kersting and Birgit Gaßner

Department of Physiology, University of Würzburg, Röntgenring 9, 8700 Würzburg, Federal Republic of Germany

Abstract. Aldosterone maintains acid-base balance and K^+ homeostasis by controlling H^+ and K^+ secretion in renal epithelial cells. We have shown recently in the amphibian distal nephron that aldosterone activates a Na^+/H^+ exchange system in the luminal cell membrane, leading to transepithelial H^+ secretion and cytoplasmic alkalinization. Since H^+ secretory fluxes are paralleled by K^+ secretion, it was postulated that the hormone-induced increase of intracellular pH activates the luminally located K^+ channels. In 'giant' cells fused from individual cells of the distal nephron, we measured simultaneously cytoplasmic pH and cell membrane K^+ conductance during acidification of the cell cytoplasm. The experiments demonstrate that cell membrane K^+ conductance is half-maximal at an intracellular pH of 7.42, and that a positive cooperative interaction exists between K^+ channel proteins and H^+ ions (Hill coefficient = 6.5). Moreover, the cellular K^+ conductance is most sensitive to cytoplasmic pH in the range modified by aldosterone. This supports the hypothesis that intracellular H^+ activity, regulated by the Na^+/H^+ exchanger, serves as the signal to couple aldosterone-induced K^+ secretory flux to H^+ secretion in renal tubules.

1988 Proton passage across cell membranes. Wiley, Chichester (Ciba Foundation Symposium 139) p 201–219

In the vertebrate kidney, aldosterone stimulates Na^+ retention, and H^+ and K^+ secretion. The classical view of the cellular mechanisms underlying transport of these three ion species is that aldosterone stimulates Na^+ entry across the luminal cell membrane (Schwartz & Burg 1978), thereby depolarizing the membrane potential (Koeppen et al 1983). This depolarization facilitates cell-to-lumen K^+ secretion (Field et al 1984) via hormone-activated (Wiederholt et al 1973) K^+-selective channels and H^+ extrusion via a rheogenic H^+ pump (Steinmetz 1974). From experiments in the amphibian distal nephron, we have evidence that transepithelial fluxes of Na^+, K^+ and H^+ ions are functionally linked, but that the mechanisms involved deviate from the previous view (Oberleithner et al 1987). We propose a model in which hormone-induced stimulation of the Na^+/H^+ exchanger is the primary event leading to cytoplasmic alkalinization and subsequent activation of the pH-sensitive K^+ channels. If the mechanism whereby the hormone-induced interaction between a counter-transport mechanism (Na^+/H^+ exchanger) and an ionic channel (K^+ channel) is to be physiologically relevant, the K^+ channels must be sensitive to intracellular pH (pH_i) within the range encompassed during exposure to the hormone (7.45–7.73). To test this hypothesis, we have simultaneously measured the intracellular pH and K^+ conductance of fused distal tubules from frog kidney during gradual acidification of the cytoplasm.

Methods

The tubule fusion technique has been described elsewhere (Oberleithner et al 1987). The individual steps of tubule fusion are summarized in Fig. 1. Shortly, kidneys of *Rana pipiens* were isolated and perfused with amphibian Ringer solution composed of: 97 mM NaCl, 3 mM KCl, 5 mM Hepes, 1 mM $CaCl_2$, 1 mM $MgCl_2$, 5 mM glucose. After the blood was rinsed off, 3 ml of the same solution containing 0.1% (w/v) collagenase was injected. The ventral surface of both kidneys containing primarily diluting segments was removed in small pieces and incubated for 50 min in the collagenase-containing solution. Then the pieces were gently sucked up and down with a 200 μl Eppendorf pipette until a suspension of small tubule fragments was obtained. The tubules were centrifuged (400***g***, 2 min) and the supernatant was removed. 1 ml of the fusion medium (30% polyethylene glycol, M_r 4000, dissolved in Leibovitz-15 medium, diluted to 170 mosm/l, pH 8.2) was slowly added and the cells were again centrifuged (400 ***g***, 2 min). Finally, they were suspended in Leibovitz-15 medium (200 mosm/l, pH 7.8) and transferred onto a thin microscope coverslip pre-treated with poly-L-lysine (0.1 g/l; Serva, Heidelberg, FRG) and incubated for 24 to 48 h at 6 °C. Over this time period membrane fusion proceeded and tubules were ready for microelectrode measurements.

Fused tubules were superfused with high perfusion rates and extracellular solutions were exchanged within seconds. The superfusion solutions were composed similarly to the kidney perfusate except that NaCl was substituted for by K gluconate. Since Ca^{2+} ions bind to gluconate, $CaCl_2$ (about 8 mM) was added under the control of a Ca^{2+}-sensitive macroelectrode to obtain the same Ca^{2+} activity as in the gluconate-free perfusion solution. KOH (0.1 M) was used to titrate the solutions to the appropriate pH (between 6.5 and 8.0). $BaCl_2$ (3 mM) was added for certain experiments. Fused cells were penetrated with three microelectrodes. One conventional microelectrode (filled with 1 M KCl) was used to inject negative current pulses ($1–5 \times 10^{-9}$ A, 200 ms duration), another was used to monitor the cell membrane potential

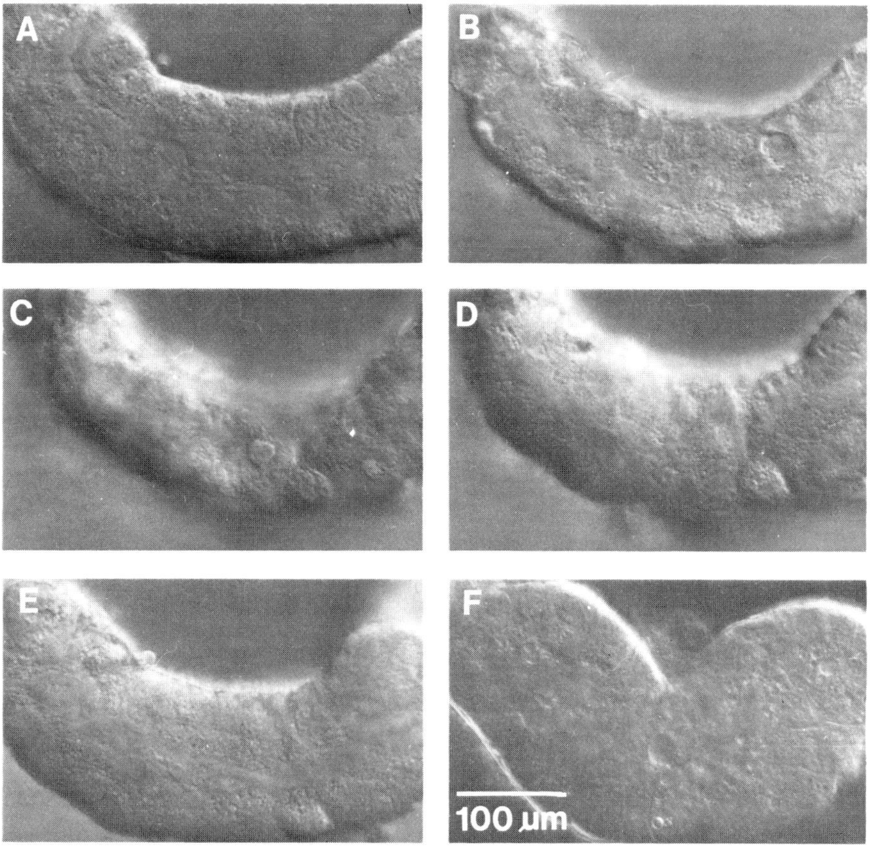

FIG. 1. The various steps of cell-cell fusion within an individual tubule segment. (A) Tubule in culture medium (Leibovitz-15) after isolation from the collagenase-treated frog kidney. (B) Superfusion with fusion medium containing 30% polyethylene glycol (PEG). (C) Dehydrated tubule 3 min after exposure to PEG. (D) Omission of PEG and rehydration. (E) Tubule a few hours after PEG treatment. (F) Fused tubule after PEG treatment kept in culture medium at 6 °C for 48 hrs. Note that the tubule approaches a spherical shape.

(V_m) and a third, a pH-sensitive liquid ion-exchange microelectrode, measured intracellular pH (pH_i) when the cell cytoplasm was gradually acidified. Manufacture, calibration and properties of the H^+-selective electrodes have been described (Ammann et al 1981). pH_i was calculated from the equation: $pH_i = pH_o - (V_m{}^H - V_m)/S$. pH_o is the extracellular pH, $V_m{}^H$ is the cell membrane H^+ electrochemical potential difference, the symbol S is the electrode slope, measured in calibration solutions that mimicked the cytosolic composition. The cell input resistance was calculated according to Ohm's law: $R_m = \Delta V_m/I$, where ΔV_m is the cell membrane potential deflection in response to the injected current pulse (I). Then R_m was related to the surface of the cell (F), according to the equation: R_m (Ωcm^2) = R_m (Ω) $\times$ F (cm^2). Since fused tubules were either spherical or cylindrical, F was calculated as the surface of a sphere ($F = 4\pi r^2$) or of a cylinder ($F = 2r\pi h$), where r (measured optically) is the radius of the sphere or cylinder and h is the height of the cylinder.

The passive membrane conductance of early distal tubule cells is determined by Cl^- and K^+ channels (Oberleithner et al 1982). Since we wanted to study the whole cell K^+ conductance, Cl^- ions were largely removed from the perfusate by substitution with impermeant anions, and Na^+ ions were replaced by K^+ ions (*vide supra*).

Results and discussion

The cell membrane resistance of the fused tubule increases gradually to a new steady-state value when the superfusion medium is acidified by one pH unit (Fig. 2). Addition of Ba^{2+} ions (3 mM) to the superfusion medium leads to no further significant increase of the cell membrane resistance. Obviously, extracellular acidification has rendered pH_i acidic and thus has inactivated this portion of the cell membrane conductance (the reverse of R_m) which is Ba^{2+} inhibitable and most probably is the K^+ conductance. In this paper we have related the pH_i-induced conductance changes to the Ba^{2+}-inhibitable conductance and termed it the Ba^{2+}-sensitive K^+ conductance. Fig. 3 demonstrates the relationship between pH_i and G_K, where G_K is the percentage inhibition of the Ba^{2+}-sensitive K^+ conductance. Apparently, G_K is highly sensitive to H^+ in a rather narrow range of pH. The sigmoidal shape of the two curves indicates that there is a positive cooperative interaction between the K^+ channel proteins and H^+. Binding of one H^+ to the channel causes a conformational alteration such that the affinity for the next H^+ to bind is increased. Furthermore, the curves suggest that multiple H^+ bind simultaneously to a single K^+ channel protein or to separate proteins within a microdomain. This can be derived from the slope (i.e. Hill coefficient) of the non-linear regression between intracellular pH and the corresponding G_K values. A Hill coefficient of approximately 6.5 is calculated. Recently, it has been postulated (Hunter & Giebisch 1987) that the K^+ channel in the same nephron portion of the kidney has a multi-barrelled structure, including a main gate which controls the overall probability of the channel being open and at least four individual channel subunits. It is tempting to speculate that cytosolic H^+ may interact with the individual subunit gates in a positive cooperative manner. This would represent a most suitable, fine-tuning mechanism over a wide K^+ conductance range, while maintaining the high selectivity of the K^+ channels.

The fused cell preparation allows simultaneous measurements of intracellu-

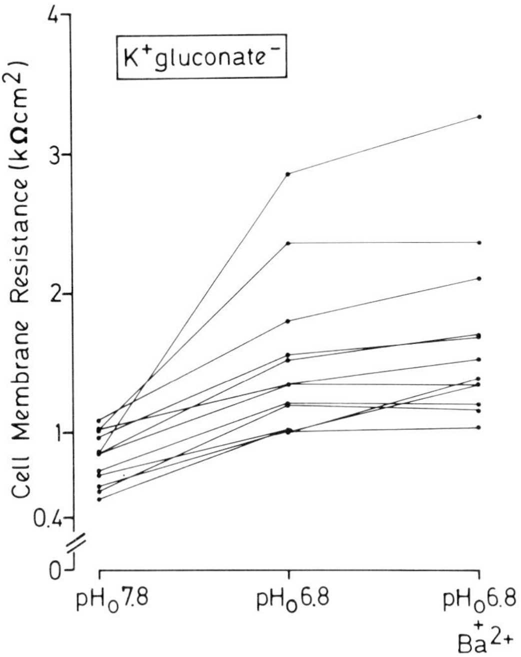

FIG. 2. Cell membrane resistance of fused tubules exposed to control pH (pH_o = 7.8), to acidic pH (pH_o = 6.8) and to acidic pH plus 3 mM $BaCl_2$ (pH_o = 6.8 + Ba^{2+}). Note that there is virtually no further increase in cell membrane resistance in response to Ba^{2+} after the cells have been acidified by the superfusion medium.

lar pH and cell membrane G_K in renal epithelium. Although we cannot distinguish between apical and basolateral K^+ permeabilities, we assume that quantitatively in this preparation the apical G_K dominates over basolateral G_K. This assumption is based on observations in both amphibian (Oberleithner et al 1982) and mammalian (Greger 1985) preparations that the major basolateral membrane conductance is for chloride ions. Since we deleted chloride ions from the perfusates, apical G_K was the major cell membrane conductance in this study.

Free H^+ and Ca^{2+} activities are well balanced by cytoplasmic buffers and may both serve as intracellular messengers to control epithelial transport at the molecular level. Such a regulatory system in which H^+ compete with Ca^{2+} for specific binding sites has been recently proposed (Garty et al 1987). Steroids, e.g. progesterone, induce meiotic division in amphibian oocytes by increasing cytoplasmic calcium (Baulieu et al 1978) activity whilst decreasing intracellular H^+ activity. Transient intracellular alkalinization is a characteristic feature of the post-progesterone period (Rodeau & Vilain 1987) and may be explained by the activation of quiescent Na^+/H^+ exchangers in the cell

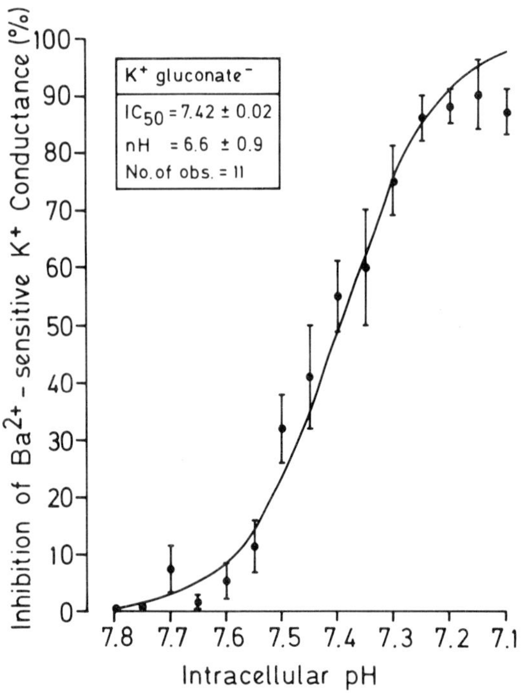

FIG. 3. Relationship between Ba^{2+}-sensitive K^+ conductance and intracellular pH measured in fused tubules. The IC_{50} value indicates half-maximal inhibition of the K^+ conductance, whereas the nH value expresses the Hill coefficient. Both parameters were taken from a graphical plot in which the individual pH_i values (x axis) were related to the individual, fractional G_K values (y axis: $\log[(1\text{-}G_K)/G_K]$). The sigmoidal curve was calculated by applying the IC_{50} and nH values of the individual tubules in a modified Michaelis–Menten kinetics equation:

$$G_K\ (\%) = 100 \times [H^+]^{nH}/(IC_{50}{}^{nH} + [H^+]^{nH}).$$

In this equation, $[H^+]$ corresponds to the intracellular H^+ activity, G_K is the percentage inhibition of the Ba^{2+}-sensitive K^+ conductance.

membrane. We have reported cytoplasmic alkalinization by 0.28 pH units in renal epithelial cells induced by the steroid hormone aldosterone (Oberleithner et al 1987). The present study indicates that the G_K of the diluting segment cell is most sensitive to cytoplasmic pH well within the range of physiological fluctuations. Aldosterone increases cytoplasmic pH from 7.45 to 7.73, which, according to Fig. 3, decreases the percentage inhibition of the K^+ conductance by more than 100-fold. Indeed, K^+ secretory fluxes are induced by aldosterone (Oberleithner et al 1987). It is well accepted that aldosterone is the key hormone involved in the regulation of Na^+, H^+ and K^+ homeostasis. In the mammalian distal nephron, aldosterone induces K^+ and

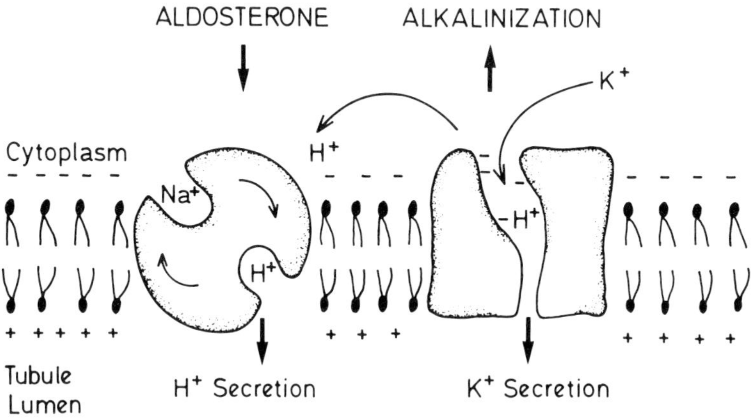

FIG. 4. Action of aldosterone in target cells of the amphibian kidney. Aldosterone activates Na^+/H^+ exchange in the lumen cell membrane. H^+ are extruded and the cell cytoplasm becomes more alkaline. The neighbouring K^+ channel proteins are deprotonated. This improves the conductance properties and facilitates K^+ secretion. (From Oberleithner et al 1987, with permission.)

H^+ net secretion, while Na^+ is retained (Fanestil & Park 1981). Our previous experiments indicate that the transport of these three ions is linked. After a lag period of 30 to 60 min, aldosterone activates the apical Na^+/H^+ exchanger leading to transepithelial H^+ net flux (Weigt et al 1987) and sustained cytoplasmic alkalinization (Oberleithner et al 1987). We can envisage that protons are released from their binding sites on the K^+ channels, which are thereby activated in a positive cooperative fashion. In this model, an ion-carrier (Na^+/H^+ exchanger) and an ion channel (K^+ channel) are functionally linked, while cytoplasmic H^+ serves as the intracellular messenger in regulating the conductance properties of the K^+ channel (Fig. 4). It is not yet known whether aldosterone, in a manner similar to the proposed action of progesterone in oocytes, regulates the H^+ extrusion mechanism and the K^+ conductance by releasing Ca^{2+} from binding sites along the inner face of the cell membrane. The Na^+/H^+ exchanger can indeed be activated by Ca^{2+} (Grinstein & Cohen 1987) and the conductance of a Ca^{2+}-activated and H^+-inactivated K^+ channel will most likely depend on the ratio of cytoplasmic Ca^{2+}: H^+.

Acknowledgements

This work was supported by Deutsche Forschungsgemeinschaft, SFB 176–A6. We thank Drs S. Silbernagl and J. Schwegler for helpful suggestions, Mrs M. Schulze for preparing the figures and Mrs I. Schönberger for typing the manuscript.

References

Ammann D, Canter F, Steiner RA, Schulthess P, Shijo Y, Simon W 1981 Neutral carrier based hydrogen ion selective microelectrode for extra- and intracellular studies. Anal Chem 53:2267–2269

Baulieu EE, Godeau F, Schroderet M, Schroderet-Slatkine S 1978 Steroid-induced meiotic division in xenopus laevis oocytes: surface and calcium. Nature (Lond) 275:593–598

Fanestil DD, Park CS 1981 Steroid hormones and the kidney. Annu Rev Physiol 43:637–649

Field MJ, Stanton BA, Giebisch GH 1984 Differential acute effects of aldosterone, dexamethasone, and hyperkalemia on distal tubular potassium secretion in the rat kidney. J Clin Invest 74:1792–1802

Garty H, Asher C, Yeger OJ 1987 Direct inhibition of epithelial Na^+ channels by a pH-dependent interaction with calcium, and by other divalent ions. Membr Biol 95:151–162

Greger R 1985 Ion transport mechanisms in thick ascending limb of Henle's loop of mammalian nephron. Physiol Rev 65:761–797

Grinstein S, Cohen SJ 1987 Cytoplasmic (Ca^{2+}) and intracellular pH in lymphocytes. Role of membrane potential and volume-activated Na^+/H^+ exchange. Gen Physiol 89:185–213

Hunter M, Giebisch G 1987 Multi-barreled K channels in renal tubules. Nature (Lond) 327:522–524

Hunter M, Oberleithner H, Henderson RM, Giebisch G 1988 Whole-cell potassium currents in single early distal tubule cells. Am J Physiol, in press

Koeppen BM, Biagi BA, Giebisch G 1983 Intracellular microelectrode characterization of the rabbit cortical collecting duct. Am J Physiol 244:F35–F47

Oberleithner H, Guggino W, Giebisch G 1982 Mechanism of distal tubular chloride transport in Amphiuma kidney. Am J Physiol 242:F331–F339

Oberleithner H, Weigt M, Westphale HJ, Wang W 1987 Aldosterone activates Na^+/H^+ exchange and raises cytoplasmic pH in target cells of the amphibian kidney. Proc Natl Acad Sci USA 84:1464–1468

Rodeau JL, Vilain JP 1987 Changes in membrane potential, membrane resistance, and intracellular H^+, K^+, Na^+, and Cl^- activities during the progesterone-induced maturation of urodele amphibian oocytes. Dev Biol 120:481–493

Schwartz GJ, Burg MB 1978 Mineralocorticoid effects on cation transport by cortical collecting tubules. Am J Physiol 235:F576–F585

Steinmetz PR 1974 Cellular mechanisms of urinary acidification. Physiol Rev 54:890–956

Weigt M, Dietl P, Silbernagl S, Oberleithner H 1987 Activation of luminal Na^+/H^+ exchange in distal nephron of frog kidney. An early response to aldosterone. Pflügers Arch Eur J Physiol 408:609–614

Wiederholt M, Schoormans W, Fischer F, Behn C 1973 Mechanism of action of aldosterone on potassium transfer in the rat kidney. Pflügers Arch Eur J Physiol 345:159–178

DISCUSSION

Thomas: Do these cells fire spontaneous action potentials?

Oberleithner: No, they don't.

Thomas: So they are definitely not nerve cells. Yours is a very nice technique. Are you fairly certain that aldosterone works via stimulation of a Na^+/H^+ exchanger?

Oberleithner: Yes.

Steels: When you added aldosterone the membrane potential decreased but in this condition there is a higher conductance of K^+. How do you reconcile these observations?

Oberleithner: This is one thing which I cannot explain yet. If we use aldosterone, at the same time as the cytoplasmic alkalinization there is a significant decrease in the membrane potential of about 4 mV. According to the increase in K^+ conductance, the membrane potential would be expected to rise. We have not further investigated this but we know that in early distal tubules there are two major ion conductances in the presence of non-bicarbonate buffers—one is for K^+ and one is for Cl^-. If, and we cannot exclude this, a basolateral Cl^- conductance increases in response to aldosterone, then the membrane would depolarize. Furthermore, there are some indications that a Cl^-/HCO_3^- exchanger may be involved in the action of aldosterone.

Steels: But you did your experiments in HCO_3^--free media?

Oberleithner: You are right, it doesn't fit—a significant response would be expected only with a HCO_3^--containing solution in the bath.

Malnic: The conventional explanation for the action of aldosterone involves the insertion of new Na^+ channels into the apical membrane. Could you observe the action of aldosterone in your preparation by clamping the internal pH, for example by increasing the buffer capacity, to see if the mechanism which you showed is the only one or if some other mechanism of insertion may be occurring as well?

Oberleithner: We have no data to distinguish whether there is insertion of Na^+ channels. However, Wenhui Wang in Gerhard Giebisch's group at Yale clamped intracellular pH in aldosterone-treated cells and showed that K^+ channels are inserted in response to the hormone. They measured overall K^+ conductance before and after aldosterone at constant pH and showed that the K^+ conductance does increase. They propose that this mechanism can be inhibited by amiloride.

Morad: You showed that when you give aldosterone the steady-state pH of the cell increased to pH 7.7. Could that have shifted the Na^+/H^+ curve in a parallel way to the left or to the right, thereby shifting the set-point?

Thomas: Why was the pH_i increasing if you didn't stimulate Na^+/H^+ exchange?

Oberleithner: Because the set-point of the Na^+/H^+ exchanger was shifted.

Morad: The base line proton concentration had gone down, which means that the internal Na^+ concentration must have been high.

Oberleithner: It does not necessarily mean that the Na^+ concentration is high because the Na^+/K^+ pump could dissipate the increased Na^+ load. We measured the intracellular Na^+ concentration years ago and found no increase.

Morad: My explanation accounts for why the membrane didn't hyperpolarize.

Oberleithner: There is no significant Na^+ permeability in early distal tubule cells.

Aickin: You suggest that aldosterone is changing the set-point for Na^+/H^+ exchange rather than stimulating it. It struck me that the recovery from the NH_4^+ rebound was very much faster after aldosterone than before. If it was simply a changed set-point, I don't think you would expect to see that.

Vaughan-Jones: I agree, you seem to have a change in slope and a change in set-point, if you compare the rate of pH_i recovery as a function of pH_i before and after addition of aldosterone.

Moolenaar: What is the time course of the aldosterone effect?

Oberleithner: It starts about 25 minutes after application of the hormone and levels off after about an hour. We have demonstrated this time course in the cell and we have shown transepithelial H^+ secretion over the same period.

Moolenaar: From the kinetics of the pH recovery, can you tell whether it is an effect on V_{max} or a shift in the set point?

Oberleithner: We haven't calculated that yet. The problem is that we have to go through the same intracellular pH range before and after adding aldosterone. After we have treated the cell with aldosterone, with the same pre-pulse method, the same exposure and everything else kept comparable to the control experiment, the intracellular pH range is different.

Machen: You have an increase in pH that you think is due to the Na^+/H^+ exchanger. You start at pH 7.4 and go to pH 7.7. In many cell types the 'set-point' of the Na^+/H^+ exchanger is pH 7.1 or 7.2, so it should be turned off. I have never seen one that has a set-point of pH 7.7. Is it possible that there are different set-points for Na^+/H^+ exchangers in different cells?

Oberleithner: I am aware of this obvious discrepancy. There may be a difference between amphibian and mammalian kidneys.

Vaughan-Jones: It is documented that external pH appears to alter the set-point of Na^+/H^+ exchange. Whether it's a real alteration or a kinetic effect is not known (Jean et al 1985). There seems to be an interdependence between the internal and external pH sensitivity of Na^+/H^+ exchange in brain synaptosomes. The effect has not yet been explained adequately.

Machen: So it increases the set-point, if you increase extracellular pH?

Vaughan-Jones: Yes, Na^+/H^+ exchange seems to 'switch off' at a more alkaline pH_i when external pH is raised.

Grinstein: Wouter Moolenaar showed some data where he alkalinized the medium and looked at the effect of growth factors or TPA and cells in alkaline medium wouldn't react.

Boron: People often look at a pH recovery from an acid load and where the pH plateaus they call that plateau pH the 'set-point'. That's not necessarily the set-point of any acid extrusion mechanism. One must demonstrate that, in the steady state, blocking an acid extruder with a drug such as SITS or amiloride has no effect on pH_i. If addition of the blocker produces a decline in pH_i, then one can conclude that the transporter was active during the steady state. Thus, if one adds a growth factor and observes an increase in steady-state pH_i, one cannot automatically conclude that the 'set-point' for the Na^+/H^+ exchanger has been altered. At least in principle, the increase in pH_i could have been caused by a fall in the rate of intracellular acid loading, e.g. by metabolic acid production.

Moolenaar: Do these fused cells respond to activators of protein kinase C by a shift in pH?

Oberleithner: We are planning to do such experiments.

Steels: In your fused cells, do you know anything about their polarity?

Oberleithner: No, that's one thing we lose by this process. The cells are losing membranes; we don't know yet which membrane but I think they lose apical as well as basolateral membranes. However, we don't mind that there is no longer any polarity because we don't study transepithelial transport; rather we are looking at intracellular events.

Frömter: Before fusion these cells were probably involved in transepithelial H^+ transport; as a result, cell pH might have shifted to a steady-state value different from the steady-state pH of the non-polarized fused cells. This would indicate a change in set-point; in other words, small changes in pH might affect K^+ conductance *in vivo* less strongly than under your experimental conditions.

Oberleithner: I have never really compared intracellular pH between the various studies but, if I recall correctly, it is approximately the same. We have an extracellular pH of 7.8 and an intracellular pH of about 7.4-7.5.

Stone: You said that you have applied this to mammalian systems: what characteristics can be carried over to this system, where the H^+ and K^+ secretory cells are different.

Oberleithner: Dr Ulrich Kersting in our laboratory has done some experiments recently, trying to fuse cultured dog kidney cells (MDCK cells). In a fused MDCK cell a few hours after fusion—we used acridine orange to make the nuclei visible—we can see more than 40 nuclei inside an individual fused cell. We observed such a cell over two weeks. One day after fusion some of the nuclei have been metabolized. After about a week the remaining nuclei start to fuse: the nuclei become large and sit within a large cell of about 100 μm in diameter. We looked for the aldosterone response in this epithelium. When cells are cultured in a Petri dish, after a few days the monolayer starts to form domes. This dome formation is dependent on aldosterone. In a dome the basolateral side faces the Petri dish, whereas the luminal side is exposed to the bath solution. In the presence of aldosterone and with a bath pH of 7.4, the pH inside the dome reaches 8.4. This extremely alkaline pH can be reduced by amiloride in the bath, consistent with inhibition of Na^+/H^+ exchange. So we have good evidence that the action of aldosterone in the amphibian system is comparable to its action in the mammalian preparation.

Vaughan-Jones: Do you get a unique dependence of conductance on internal pH when you change pH by other methods? You have only shown the relationship when pH_i is becoming more acid after external acidosis. If you make the external pH alkaline and plot the curve again, do you get the same answer?

Oberleithner: Yes, we have done the opposite and it looks very similar.

Vaughan-Jones: And what about doing an ammonia pre-pulse? Under these conditions you would not be changing external bulk pH.

Oberleithner: We showed that when the cell faces an external acid pH, the cytoplasm gradually acidifies. If the external medium is rapidly returned to pH 8, the cell interior remains acid, i.e. intracellular pH is acidic in an alkaline extracellular pH.

Malcolm Hunter has done another series of experiments in isolated cells of amphibian kidney. Measuring whole cell currents he set a constant pH on the inside and changed the external pH. There was no change in the overall cell membrane conductance.

Vaughan-Jones: But does the pH_i--conductance relationship hysterese? If it did, it might indicate that protons are not the blockers, they are simply changing something else which then blocks.

Oberleithner: I have to re-check that.

Harvey: In short circuit conditions in frog skin, we see the same type of pH dependence of the K^+ conductance of the basolateral membranes. Except that in our case the K^+ channels seem to be less sensitive to intracellular protons because the Hill coefficient is 2.

Oberleithner: In our cells the Hill coefficient is about 6.

Harvey: You are changing extracellular pH to bring about corresponding changes in the intracellular pH and then you look at the conductance. Isn't the K^+ conductance also sensitive to extracellular pH in the kidney (Kubota et al 1983)?

Oberleithner: No, that's what I wanted to point out. We measured the K^+ conductance in dialysed cells where extracellular pH can be clamped and intracellular pH can be kept constant. There was no change in the conductance. In another set of experiments, we acidified the intact cytoplasm by an extracellular acid load and then went back to the control pH of 7.8 on the outside. Under these conditions, the overall K^+ conductance is sharply reduced.

Russell: You showed some data in which you plotted the salt conductance versus the pH (Fig. 3). That was done with ammonium. The relationship did not look steep nor did it appear to be sigmoid. What's the difference?

Oberleithner: This plot had a large scatter. These were whole cell current measurements in dialysed cells of the amphibian kidney. The problem was that we were not able to change the pipette pH step-wise in an individual cell. So we dialysed the cell with medium at pH 7.8 while we determined the current–voltage relationship. Then we filled another pipette with a medium of pH 7.6 and tried another cell. This resulted in the large scatter.

Vaughan-Jones: Can you do the experiment backwards?

Oberleithner: Yes, we can acidify the cytoplasm and thereby block the K^+ channels, and we can re-alkalinize the cytoplasm, thereby removing the block, with comparable kinetics.

Russell: As I understand it, the distal part of some kinds of kidney tubules can not only secrete K^+ but under appropriate conditions they can absorb K^+ (Wright & Giebisch 1985). Is the ability to do one or the other somehow connected with intracellular pH?

Oberleithner: K^+ filtered in the kidney is completely reabsorbed along the nephron and all the K^+ appearing in the urine is secreted in the distal nephron. If you treat an animal with K^+-free media, you can cut down K^+ secretion more or less completely but you cannot induce significant net reabsorption.

Malnic: Along the collecting duct you may have reabsorption. This was shown along the medullary collecting duct by Wingo (1987).

Steinmetz: Is this effect inhibited by spironolactone (an aldosterone antagonist)? Are there receptors?

Oberleithner: Spironolactone binds to the intracellular receptors for aldosterone. If you pre-perfuse the cell with this drug and then do the aldosterone experiment, the effect of aldosterone is completely inhibited.

Steels: Dr Oberleithner, you did your experiments in the presence of Hepes. When you measure the intracellular pH in the presence of HCO_3^-, what happens then?

Oberleithner: We have done experiments in the presence of Hepes and in CO_2/HCO_3^-. Intracellular pH is more acidic, by about 0.1 pH units, in the latter conditions.

Moolenaar: In fibroblasts it is the opposite.

Russell: Is not the difference due to this outwardly directed Na^+-$nHCO_3^-$ co-transport system, so when you give CO_2 this co-transporter acts to shunt the alkalinity of your cell, i.e. to make it relatively acidic?

Oberleithner: Yes, I agree; if we inhibit the Na^+-HCO_3^- co-transport system, by DIDS for example, we alkalinize the cell.

Moolenaar: Concerning inhibition of anion exchangers, there has been a report by Madshus & Olsnes (1987) that you can distinguish pharmacologically the Na^+-dependent anion exchangers from the Na^+-independent one. Na^+-dependent exchange is inhibited by probenecid and Na^+-independent exchange by ethacrynic acid. This was in *VERO* epithelial cells. Does anybody have experience with those inhibitors?

Grinstein: Probenecid inhibits the red cell anion exchanger.

Sabolic: The ammonia pre-pulse technique works quite well with whole cells but I tried it in red blood cell membrane vesicles and renal brush border vesicles and it didn't work at all.

Thomas: It doesn't work well on *all* whole cells: it needs a high membrane permeability to NH_4^+.

Sabolic: I just wonder whether NH_4^+, when it enters the cell, binds to intracellular membranes and other proteins. This would mean that the amount of ammonia in the cell could be much bigger than we assume, so that in an ammonia-free buffer the cell can develop a much bigger ΔpH. This could not happen in plasma membrane vesicles.

Vaughan-Jones: Are you saying that you can't change internal pH by adding NH_4Cl or that you can't get an acid load?

Sabolic: I can't get an acid load.

Vaughan-Jones: That means you can't get NH_4^+ ions across the membrane.

Sabolic: But in whole red blood cells or proximal tubule cells one can get an acid load by this technique.

Grinstein: Yes, but the cells have a membrane potential and a Na^+/K^+ pump and they accumulate a lot more NH_4^+.

Thomas: Snail neurons accumulate NH_4^+ rather slowly so it's not a very good way of acidifying them.

Oberleithner: It also depends on the volume:surface area ratio of the cells. If we apply the NH_4^+ pre-pulse technique in a single cell in the epithelium, the pH response from the same load for the same period of time is completely different to that in fused cells. My experience is that the bigger the cell, the better the method works.

Frömter: Could you please specify what you observe in small cells and what in big cells?

Oberleithner: In a big cell, there is a larger acidification and a slower recovery which you can easily observe. In a small cell, the recovery might be so fast that you miss it.

Harvey: Doesn't it also depend on the buffering power? Because if you use permeable buffers, this will reduce the acidification effect of NH_4^+.

Sabolic: We used 10 mM Hepes, which is not very permeable.

Thomas: Richard Vaughan-Jones showed that in heart muscle you don't get very good NH_4^+ loading, if you block Cl^-/HCO_3^- exchange. In this case, it's not the entry of NH_4^+ so much as the cell trying to restore its intracellular pH. NH_4^+ entry in other systems may be poor but if there is pH regulation from the alkaline state, that would produce NH_4^+ from the NH_3 entering the cell. If there is no regulation from the alkaline state, then the cell relies on NH_4^+ going in through K^+ channels or possibly on the Na^+ pump. If that isn't very much then you get a lousy acid loading at the end of it all.

Aickin: This is exactly what happens in dog fish and crab muscle.

Frömter: What do you use to acid load crab muscle?

Aickin: Very long exposure to NH_4^+ or, preferably, addition of CO_2.

Vairo: What about the nigericin method, would that work?

Aickin: I suppose so, but I don't know.

Thomas: Claire (Aickin) had a beautiful effect once when she had a mouse muscle in K^+-free Ringer solution for a long time, then put NH_4^+ on. From the pH_i changes, it looked as though the NH_4^+ was entering via the Na^+ pump. I tried to reproduce that on snail neurons but it didn't work.

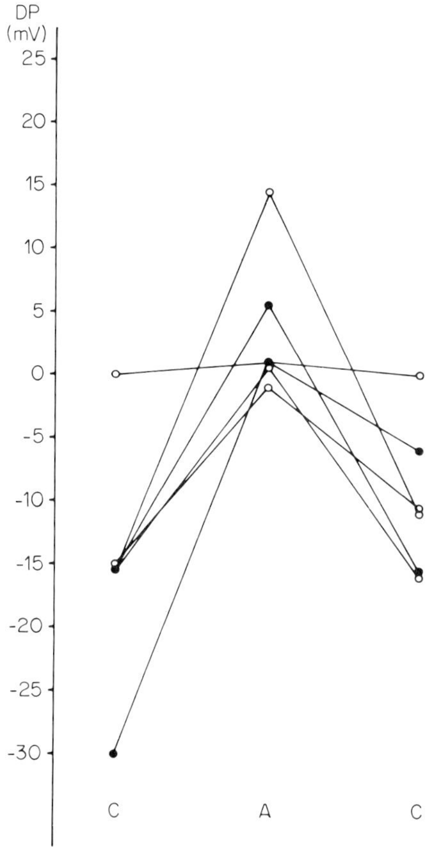

Fig. 1. (*Malnic*). Transepithelial potential difference in early distal tubule of rat kidney during luminal perfusion with N-methyl glucosamine Ringer's solution. C, control; A, 10^{-3} M amiloride; DP, transepithelial potential difference.

Boron: In our lab, we have observed that NH_3 pulses produce large acid loads in the presence of Na^+. However, in the absence of Na^+, the plateau phase acidifications are often very slow.

Vaughan-Jones: But for many Na^+/K^+ pumps, the K_m for external K^+ is about 1 mM. The affinity of the Na^+/K^+ pump in many systems for $NH_4{}^+$ is substantially lower than that for K^+. So if the pump is working close to saturation at normal concentrations of extracellular K^+, the amount of $NH_4{}^+$ loading you get through the pump will be negligible.

Malnic: Hans Oberleithner has shown that in the early distal tubule of amphibian kidney there is H^+ secretion by Na^+/H^+ exchange (Oberleithner et al 1984). In our laboratory, Miguel Lopes has tried to determine the mechanism of H^+ secretion in the cortical distal tubule of the mammalian kidney. This was done by stopped-flow microperfusion studies using ion-exchange resin pH electrodes.

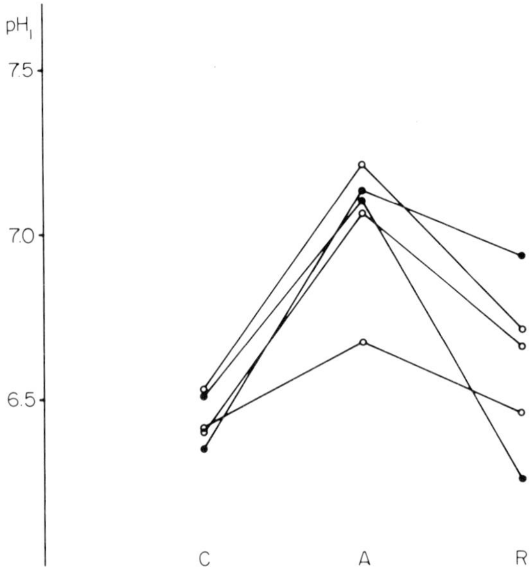

Fig.2. Luminal pH in early distal tubule during stationary microperfusion with N-methyl glucosamine Ringer's solution. C, control; A, 10^{-3}M amiloride; R, recovery; pH$_l$, luminal pH.

Distal tubules were perfused with a solution containing 10^{-3} M amiloride at low Na^+ concentration. We obtained the characteristic reversal of the transepithelial potential (Fig.1).

We also measured pH: there was a marked alkalinization of the stationary pH in these early distal tubules (Fig.2).

That suggests a relationship between the capacity to establish pH gradients and the presence of Na^+. In another experiment we used a technique first used by Hans Oberleithner, in which we perfused the peritubular capillaries of our system with a low Na^+ solution (Na^+ substituted for by N-methyl glucosamine), while the lumen was perfused with a solution containing normal concentrations of Na^+. In this condition luminal Na^+ falls progressively towards zero (Oberleithner et al 1984). In the amphibian diluting segment, it had been shown that the luminal pH gets more alkaline under these conditions and we saw the same thing in the early distal tubules of rat kidney (Fig.3).

The tubules are in a HCO_3^--free system: both the capillaries and the lumen are perfused with phosphate Ringer's solution. These results suggest that the early distal tubule of the mammalian kidney, which is not exactly equivalent to the early distal diluting segment of the amphibian kidney, may have also a Na^+-dependent acidification mechanism.

Harvey: Have you tried inhibitors of the proton ATPase?

Malnic: No, we haven't, we have only used amiloride.

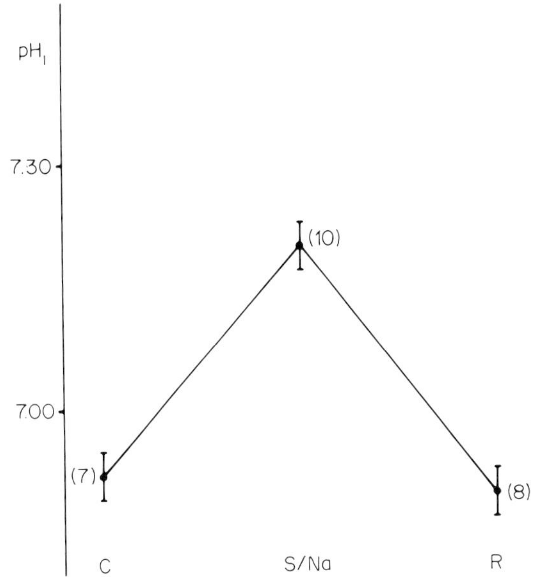

Fig.3. Luminal pH in early distal tubule of rat kidney during stationary microperfusion with sodium Ringer's solution and capillary perfusion with sodium-free (N-methyl glucosamine) Ringer's solution. C, control; S/Na, sodium-free capillary perfusion; R, recovery. pH_l, luminal pH. Mean ± SE. (*n*), number of perfusions.

Harvey: Do you think that there is a Na^+ channel in these membranes?

Malnic: Since the effect of amiloride on the transepithelial potential difference is quite clear it is very possible that there is a channel.

Harvey: One possible interpretation, besides a Na^+/H^+ exchanger, is that there is an electrogenic proton pump which is dependent on membrane potential. If you block Na^+ uptake through the channel with amiloride, you get a negative serosal or blood-side potential. This would retard proton secretion via an electrogenic proton pump and give the result that you saw.

Thomas: I thought you said that the electrogenic proton pump wasn't very sensitive to membrane potential?

Harvey: It's extremely sensitive to membrane potential once you go below -50 mV. It reverses at -100 mV (Ehrenfeld et al 1985).

Malnic: This is a possibility. We would have to do some specific studies to investigate this.

Harvey: You found a non-specific effect of amiloride at concentrations above 10^{-4} M, we have shown that this concentration of amiloride inhibits H^+-ATPase.

Malnic: AC Cassola and Nancy Reboucas in our laboratory have studied the transepithelial potential difference along the distal tubule. They blocked Na^+ channels by amiloride and K^+ channels by barium. In a number of experiments they obtained a lumen positive potential. They tried to see if this was due to H^+ secretion. Transepithelial potential difference in the late distal tubule is normally lumen negative. Amiloride reverses the potential difference; when you remove amiloride the potential difference returns to its normal value. The same effect had been found in the cortical collecting duct, where acetazolamide reduces the potential difference to zero (Koeppen & Helman 1982). So it is possible that in the late distal segment there may be an H^+-ATPase like that found in the collecting duct. This behaviour, however, was found only very rarely in distal tubule of rat kidney and only in very late distal tubules. In most of the tubules, a lumen positive potential was found after treatment with amiloride but no effect of acetazolamide was seen. Their current hypothesis is that this positive potential is somehow related to other ion transport, perhaps K^+, but not to H^+, at least in the early distal tubule. There may be some initial collecting duct segments which are found in the late distal tubule and these may behave in the same way as the collecting duct. However, this is not the case for most distal tubules.

Oberleithner: There is a transepithelial potential in the kidney distal tubule which is negative and which can be eliminated by amiloride. My first thought would be of a Na^+ permeability, which is blocked by amiloride. Then, from what I know from tight epithelia and from the late portions of the nephron, I would place the rheogenic H^+ pump in conjunction with the Na^+ permeability. Thus the amiloride-induced luminal pH change can be interpreted as an inhibition of the H^+ pump by the hyperpolarized luminal cell membrane potential rather than by direct amiloride-induced inhibition of the Na^+/H^+ exchanger.

Malnic: If you compare the change in potential with the change in pH in steady-state conditions, they are not equivalent. The change in potential corresponds to about half or less of the pH change.

Steinmetz: The time course of the aldosterone effect is very different in these systems. In your preparation it takes 30 minutes to show an effect, in the kidney or turtle bladder, it takes two to three hours.

Thomas: You think this is suspiciously fast?

Steinmetz: No, it's interesting that it's fast. I wonder whether there are several aldosterone effects to be observed in different segments of the nephron. This may explain the early kaliuresis after administration of aldosterone. What Dr Oberleithner observes may be the major aldosterone effect but there are others. There is an aldosterone effect in the outer medullary collecting duct that stimulates acidification. There is an effect in the cortical collecting duct that increases Na^+ absorption. These are all fairly slow effects, they take 2–3 hours to reach their maximum.

References

Ehrenfeld J, Garcia-Romeu F, Harvey BJ 1985 Electrogenic active proton pump in Rana esculenta skin and its role in sodium ion transport. J Physiol (Lond) 359:331

Jean T, Frelin C, Vigne P, Barby P, Lazdunski M 1985 Biochemical properties of the Na^+/H^+ exchange system in rat brain synaptosomes. Interdependence of internal and external pH control of the exchange activity. J Biol Chem 260:9678–9684

Koeppen BM, Helman SI 1982 Acidification of luminal fluid by the rabbit cortical collecting tubule perfused 'in vitro'. Am J Physiol 242:F521–531

Kubota T, Biagi BA, Giebisch G 1983 Effects of acid base disturbances on basolateral membrane potential and intracellular potassium activity in the proximal tubule of Necturus. J Membr Biol 73:61–68

Madshus IH, Olsnes S 1987 Selective inhibition of sodium linked and sodium independent bicarbonate chloride antiport in Vero cells. J Biol Chem 262:7486–7491

Oberleithner H, Lang F, Messner G, Wang W 1984 Mechanism of hydrogen ion transport in the diluting segment of frog kidney. Pfluegers Arch Eur J Physiol 402:272–280

Wingo CS 1987 Potassium transport by medullary collecting tubule of rabbit: effects of variation in K intake. Am J Physiol 253:F1136–1141

Wright FS, Giebisch G 1985 Regulation of potassium excretion. In: Seldin DW, Giebisch G (eds) The Kidney: Physiology and Pathophysiology, vol 2. Raven Press, New York

Extracellular pH and stimulated neurons

E. Syková, J. Svoboda, A. Chvátal and P. Jendelová

Institute of Physiological Regulations, Czechoslovak Academy of Sciences, Bulovka, Pavilon 11, 180 85 Prague 8, Czechoslovakia

Abstract. The use of ion-sensitive microelectrodes enabled us to follow the dynamic changes in extracellular pH (pH_e) together with those in the extracellular concentration of some biologically important ions, particularly K^+ and Ca^{2+}. Activity-related changes in pH_e were studied in isolated spinal cords of frogs and in spinal cords of rats *in vivo*. Repetitive electrical stimulation of an afferent input led either to triphasic alkaline-acid-acid changes (90% of frogs) or to triphasic alkaline-acid-alkaline changes (10% of frogs and rats) with the greatest changes in the lower dorsal horns. The transient acid shift by as much as 0.15–0.25 pH units is dominant and builds up during the stimulation. The changes in pH_e were also found in response to various adequate stimuli applied to the skin on the hind limb. Using specific inhibitors of Na^+/H^+ exchange, K^+-Cl^- co-transport, Cl^-/HCO_3^- exchange, the Na^+/K^+ pump and carbonic anhydrase, we found pH_e homeostasis to be impaired and stimulation-induced changes in pH_e decreased. We conclude that the pH_e changes evoked by electrical or adequate stimulation of an afferent input are not determined by changes in extracellular strong ion concentration differences due to accumulation of lactate, since we found no effect of NaF, a metabolic blocker of lactate production. However, lactate accumulation has been demonstrated during seizures, spreading depression and anoxia. Recently, it has been recognized that the observed pH_e changes can affect permeability of membrane ionic channels, neuronal excitability and glial cell function.

1988 Proton passage across cell membranes. Wiley, Chichester (Ciba Foundation Symposium 139) p 220–235

The depolarization of neurons is accompanied by translocation of ions, particularly of H^+, Na^+, K^+, Cl^-, Ca^{2+}, HCO_3^-, and of other molecules across their membranes, which modifies the extracellular ionic composition and extracellular space (ECS) volume (for review see Nicholson 1980, Syková 1983, 1986, Thomas 1984). Recently, it has been recognized that changes in pH_e by only several tenths of pH units evoke marked effects on neuronal ionic channels (Krishtal & Pidoplichko 1980, Iljima et al 1986, Konnert et al 1987), affect neuronal excitability (Gruol et al 1980, Balestrino & Somjen 1988) and glial cell function (Deitmer & Schlue 1987, Walz & Hinks 1987).

The use of ion-sensitive microelectrodes enabled us to follow the dynamic ionic changes, including pH, in ECS of the central nervous system (CNS) as a result of neuronal activity. Until recently, the dynamic changes in the extracellular pH (pH_e) have been recorded under a variety of pathological experimental conditions. Transient changes in pH_e by several tenths of pH units with dominating acid shifts of 0.1–0.5 pH units were found in the rat cortex and cerebellum during local surface electrical stimulation (Urbanics et al 1978, Kreig et al 1983), seizures (Siesjö et al 1985), spreading depression, anoxia and ischaemia (Lehmenkühler et al 1981, Kreig et al 1983, Harris & Symon 1984, Mutch & Hansen 1984). Under more 'physiological' conditions, the changes in pH_e by as much as 0.25 pH units were found during repetitive electrical stimulation in the isolated rat optic nerve or vagus nerve (Ransom et al 1985, Endres et al 1986) and in the spinal cord of rats and frogs during stimulations of peripheral nerves (E. Syková and J. Svoboda, unpublished work 1987, Chvátal et al 1988).

The mechanism of activity-evoked transient changes in pH_e and the mechanisms by which pH_e is stabilized are far from clear and probably not the same under physiological and pathophysiological conditions. According to Stewart (1981), the pH_e in body tissues is determined by three independent variables: 1) extracellular strong ion difference ($[SID]_e$) (strong ions means completely dissociated ions); 2) extracellular total weak acid concentration ($[A_{tot}]_e$); and 3) partial pressure of CO_2 (p_{CO2}). These three independent variables completely define pH_e and other dependent variables such as $[H^+]$, $[HCO_3^-]$, $[OH^-]_e$ and $[CO_3^{2-}]_e$. Out of these three independent variables, changes in $[SID]_e$ represented by the value of strong base cations minus strong acid anions are likely to be the most important mechanism involved in the activity-related pH_e changes in the CNS, since substantial changes in $[A_{tot}]_e$ and p_{CO2} are unlikely (Stewart 1981, Kreig et al 1983). The extracellular changes in the concentration of strong ions, particularly Na^+, K^+, Ca^{2+}, Cl^- and possibly lactate, occur during neuronal activity and may therefore considerably affect pH_e. It may be useful to state here Stewart's (1981) important conclusions, so simple and so different from conventional thinking:

. . . '1. Given the assumption that $[A_{tot}]$ stays constant and that p_{CO2} is not affected by membranes because they are so CO_2 permeable, it then follows that body fluids can only interact across membranes by exchanging strong ions so as to alter their [SID] values.

2. Only [SID] changes can bring about changes in $[H^+]$, $[HCO_3^-]$, or any other dependent variable, assuming constant p_{CO2} and $[A_{tot}]$.

3. Transport of weak ions, or dependent variable ions, such as H^+, HCO_3^-, and OH^-, across membranes separating body fluids cannot result in changes of $[H^+]$, $[HCO_3^-]$, etc, unless these ions are accompanied by strong ions, and then it is the effect of strong ions on [SID] that determines the $[H^+]$, $[HCO_3^-]$, etc, not the transport of the H^+ etc. Dependent ion transport is essentially

irrelevant to the determination of [H^+], [HCO_3^-], etc. . . .'

In our studies we examined transient changes in pH_e in the spinal dorsal horns of rats and in isolated spinal cords of frogs evoked by electrical as well as by adequate stimulation of the afferent input, i.e. by relatively 'physiological' stimulation. In all experiments *in vivo* as well as *in vitro*, triphasic changes in pH_e, alkaline-acid-alkaline, were found, and they exhibited a typical depth profile in the spinal dorsal horns which corresponded to stimulation-evoked neuronal excitation. We explored the origin of these changes and the mechanisms of post-stimulation pH_e homeostasis. We provide an interpretation of the observed activity-related pH_e changes and an insight into the pH_e regulatory mechanisms operating in the CNS.

Methods

Experiments were performed either on rats anaesthetized by Nembutal (pentobarbitone) or on isolated spinal cords of frogs. Rats were immobilized and ventilated with air enriched with a gas mixture of 5% CO_2/95% O_2. Arterial blood pressure was monitored and rectal temperature was maintained at 37 °C. Inhibitors of enzymic activity and of ionic transports were applied in an artificial cerebrospinal fluid which was pre-warmed to 37 °C and bubbled with 5% CO_2/95% O_2. Dissection of the isolated spinal cord was described in detail by Czéh et al (1981). Bicarbonate-buffered Ringer solution at 18–20 °C, pH 7.30–7.35 either of standard composition or with added inhibitors was used for spinal cord superfusion.

Double-barrelled ion-sensitive microelectrodes were used to measure dynamic changes in extracellular ionic activity. The ion-sensitive barrel contained either a pH-sensitive cocktail (Fluka) or K^+-sensitive Corning 477317 or a Ca^{2+}-sensitive cocktail (Fluka). The second barrel served as the reference electrode. For further details see Kříž et al (1975) and Chvátal et al (1988).

Results and discussion

Activity-dependent changes in pH_e

High-frequency repetitive electrical stimulation (10–100 Hz) of a peripheral nerve evoked transient extracellular acidification in the dorsal horns of rats and frogs by as much as 0.15–0.25 pH units. The acid phase reached a plateau (a so-called 'ceiling' level) after about 15–20 s of stimulation, i.e. it had a slower rise time than the simultaneously recorded increase in $[K^+]_e$ which reached a plateau after 6–9 s. The acid phase had a rise time almost as fast as the decrease in $[Ca^{2+}]_e$ (Fig. 1). When stimulation continued for more than 20–30 s, the pH_e began to return to the base line, as did the $[K^+]_e$ and $[Ca^{2+}]_e$ (Chvátal et al 1988). The acid shift was preceded by a very small initial

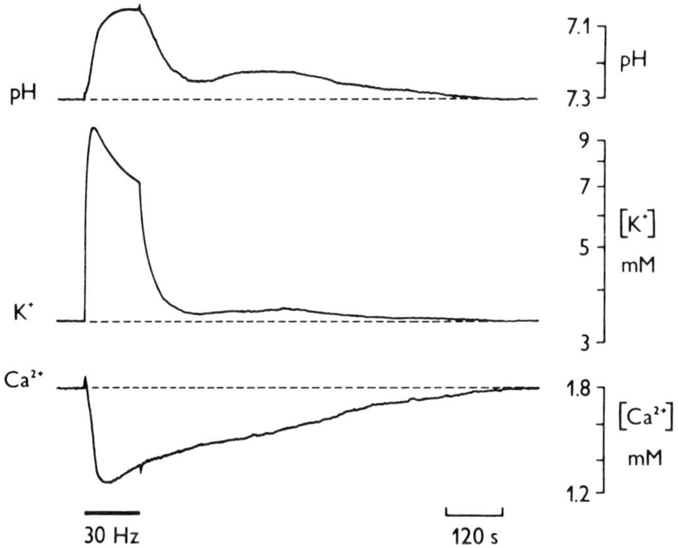

FIG. 1. Simultaneous records of changes in extracellular pH, $[K^+]$ and $[Ca^{2+}]$ in the 8th spinal cord segment of the isolated frog spinal cord. Changes produced by stimulation of the 8th dorsal root at a frequency of 30 Hz applied for 120 s were recorded in the lower dorsal horn, at a depth of 350 μm from the dorsal spinal surface. Tips of the three double-barrel ion-sensitive microelectrodes were about 100 μm apart.

alkaline shift (less than 0.01 pH units) lasting 1–2 s, clearly visible during only some experiments (Fig. 2). The initial alkaline shift was less pronounced in the dorsal horns than the previously described alkaline shift in the rat cortex during direct tissue stimulation (Kreig et al 1983). The latency of the initial alkaline shift corresponded to the rise in $[K^+]_e$.

A typical depth profile of the stimulation-evoked acid shifts was found in the spinal cord of rats and frogs, with the greatest changes in the lower dorsal horn (Fig. 3). The depth profile of pH_e changes corresponded well to the maximum of focal potentials elicited by single electrical stimuli, to the depth profiles of stimulation-evoked increases in $[K^+]_e$ (Czéh et al 1981) and to decreases in $[Ca^{2+}]_e$ (Chvátal et al 1988).

When stimulation was discontinued, the pH_e returned to the base line with an approximately similar time course as the simultaneously recorded K^+ transients. The post-stimulation K^+ undershoot recorded in the rat spinal cord *in vivo* was associated with an alkaline shift below the original pH_e base line by 0.02–0.03 pH units (Fig. 2). While the post-stimulation K^+ undershoot is present regularly in the mammalian spinal cord, it is present only rarely in the frog spinal cord (Syková 1983). The recovery in the isolated frog spinal cord was frequently not smooth but a second spontaneous acid shift occurred

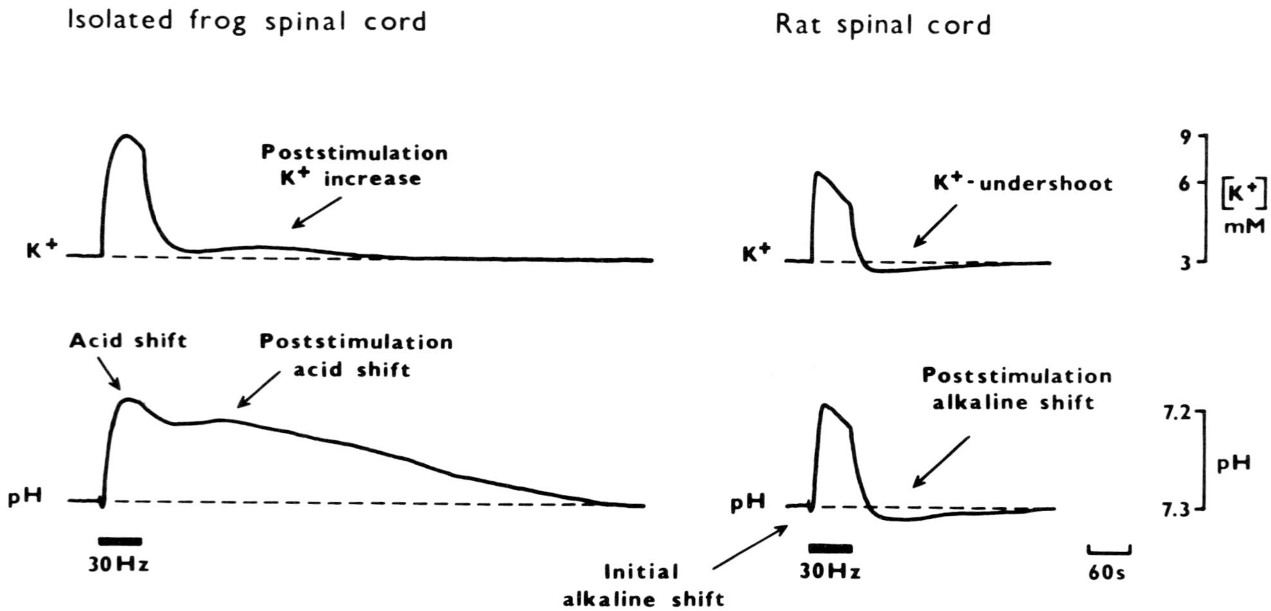

FIG. 2. Typical records of activity-related changes in extracellular pH and $[K^+]$ in the isolated frog spinal cord and in the rat spinal cord *in vivo*. Changes in pH_e and $[K^+]_e$ were produced in dorsal horns by electrical nerve stimulation at a frequency of 30 Hz applied for 60 s. Note the characteristic triphasic, alkaline-acid-acid changes in pH_e in the isolated frog spinal cord and alkaline-acid-alkaline changes in the rat spinal cord and their relation to the simultaneously recorded changes in $[K^+]_e$.

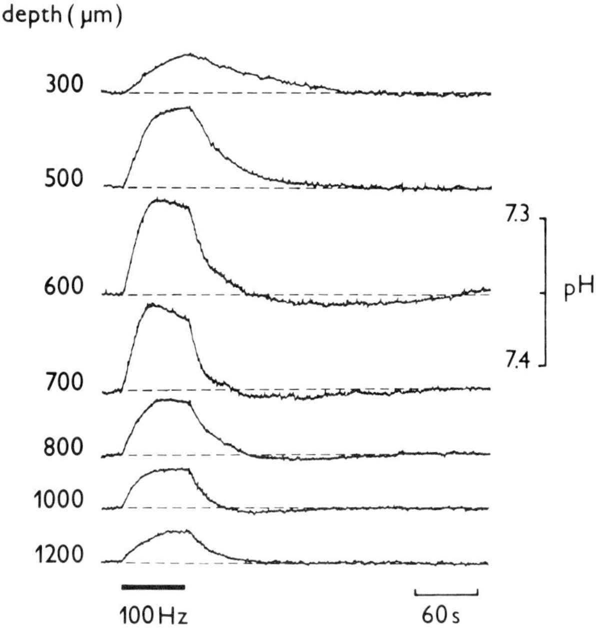

FIG. 3. Distribution of changes in extracellular pH in the L_5 spinal cord segment of a rat. Changes in pH_e in response to repetitive electrical stimulation at a frequency of 100 Hz applied for 60 s were recorded at various depths in the spinal cord, as measured from the dorsal spinal surface. Stimulation was delivered through a pair of needles inserted intramuscularly into the planta.

during post-stimulation recovery to the original pH_e base line, which was associated with a spontaneous increase in $[K^+]_e$ with a similar time course (Figs. 1 and 2). All four stimulation-evoked pH_e changes, 1) the initial alkaline shift, 2) the stimulation-related acid shift, 3) the post-stimulation alkaline shift ('pH_e undershoot') and 4) the post-stimulation acid shift, were found to increase with the stimulation frequency and intensity in a manner similar to that previously described for K^+ and Ca^{2+} transients (Czéh et al 1981, Chvátal et al 1988).

Transient acid shifts by 0.02–0.06 pH units were also found in the corresponding lumbar spinal segments when adequate stimuli (pinching of the toes with a pair of forceps, pinching of the skin, heat application) were applied to the skin of the ipsilateral hind limb (E. Syková & J. Svoboda, unpublished work 1987).

Spontaneous elevations in $[K^+]_e$ due to spontaneous neuronal firing were often seen in the frog spinal cord, especially when the cord was superfused with elevated $[K^+]$ or ouabain. The spontaneous elevation in $[K^+]_e$ was always accompanied by acid shifts of a similar time course.

TABLE 1 Effects of inhibitors on activity-related pH_e changes

	NaF	*Acetazolamide*	*Amiloride*		*Ouabain*	*Furosemide*	*SITS*	*Elevated* $[K]_e$
Effect of inhibitor	lactate production	carbonic anhydrase	high affinity Na^+ channel	low affinity Na^+/H^+ exchange	Na^+/K^+ pump	K^+-Cl^- co-transport	Cl^-/HCO_3^- exchange	Na^+/K^+ pump stimulation, membrane depolarization
Shift in resting level of pH_e (pH units)	alkaline (0.05)	alkaline (0.05)	0	alkaline (0.05–0.1)	acid (0.05)	alkaline (0.1–0.15)	alkaline (0.1)	acid (0.2)
Initial alkaline shift	0	0	↓	↑	0	↓↑	↓↓	↑↑
Activity-related acid shift	0	↓↓	↑	↓↓	↓↓↓ (slower)	↓↓↓	↓↓↓	↓↓
Post-stimulation alkaline shift	0	↑↑	0	0	↓↓↓	↓↓↓	↓	↑↑
Post-stimulation acid shift	↑↑/↓	↓↓	↑	↓↓	↓↓ (slower)	0	↓↓↓	↓↓

0, no effect; ↓–↓↓↓, decrease in amplitude; ↑–↑↑↑, increase in amplitude.

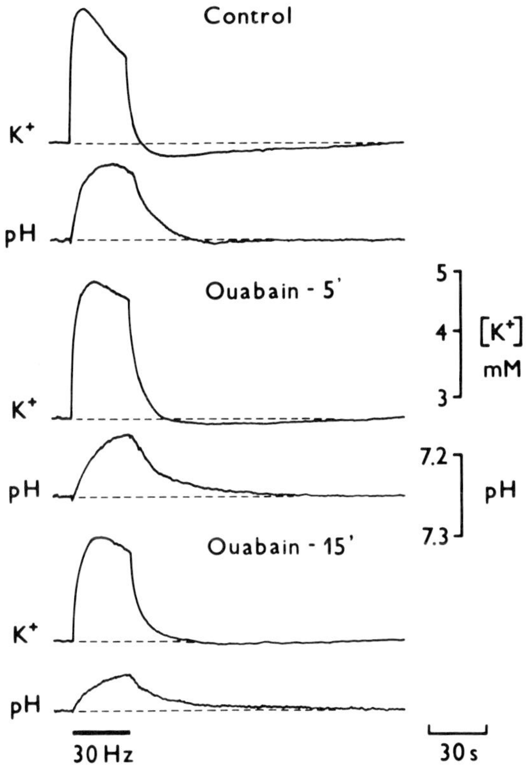

FIG. 4. Effects of ouabain applied to the spinal surface of the rat spinal cord. Simultaneous records of changes in pH_e and $[K^+]_e$ in the L_5 spinal cord segment at a depth of about 700 μm were evoked by repetitive electrical stimulation at a frequency of 30 Hz applied for 30 s through a pair of needles inserted into the planta. Note that the rise time of acid shift and its recovery to the base line are slowed down and the post-stimulation alkaline shift disappeared 5 and 15 minutes after application of 10^{-5} M ouabain.

To summarize, stimulation-evoked neuronal activity leads to rather complicated pH_e changes, triphasic alkaline-acid-alkaline shifts. The most prominent is the transient acidification which builds up during the stimulation. The overall activity-related change of acid-base balance does not last longer than the changes in $[K^+]_e$ and $[Ca^{2+}]_e$ and presumably $[Cl^-]_e$ and $[Na^+]_e$, i.e. the changes in $[SID]_e$.

pH_e homeostasis: application of inhibitors

Various inhibitors of enzymic activity and of ion transport were used to study pH_e homeostasis. We studied their effects on base line pH_e, initial alkaline shift, acid shift which developed during the stimulation, and post-stimulation alkaline or acid shifts. The results are summarized in Table 1.

Alkaline shifts in pH_e. The origin of the initial alkaline shift remains obscure. One of the reasons is that the alkaline shift is very small, inconstant and short-lasting, nevertheless apparently not an artifact. In fact, an electrode artifact imitating an initial acid shift and resulting from measurement of pH as a potential difference between the very high resistance barrel containing the ion-exchanger and the low-resistance reference barrel, was seen in several of our experiments (see Fig. 2). We offer several hypotheses concerning the origin of the initial alkaline shifts and post-stimulation alkaline shifts which are based on our data as well as those of others, being aware of the fact that all the described alkaline shifts do not necessarily have the same mechanism.

Since it has been recognized that changes in pH_e are the consequence of alterations in $[SID]_e$, we can hypothesize about the initial and post-stimulation increase in $[SID]_e$. An increase of only 3 mM in $[SID]_e$ would account for a 0.05 pH unit alkaline change. The source of an increase in $[SID]_e$ might be due to K^+ which accumulate in ECS, the movement of Cl^- into glial cells (K^+-Cl^- co-transport), etcetera. The post-stimulation alkaline shift was found to be effectively blocked by furosemide, suggesting that K^+-Cl^- co-transport may play an important role in pH_e homeostasis. Another interesting possibility is that the increase in $[SID]_e$ results from the reduction of the volume of ECS (Syková 1987), without a comparable reduction in impermeable $[HCO_3^-]_e$. However, the $[HCO_3^-]_e$ could double as a result of extracellular space shrinkage. We cannot exclude the possibility that the initial alkaline shift is due to flow of $[HCO_3^-]$ down its electrochemical gradient, which in neurons results in efflux.

The uptake process, represented by the K^+ undershoot and almost certainly involving the Na^+/K^+ pump, seemed to have the same time course as the post-stimulation alkaline shift. Both were found to be selectively blocked by ouabain. At all concentrations tested (from 10^{-7} M to 10^{-4} M), ouabain prolonged the post-stimulation recovery of $[K^+]_e$ and pH_e to their base lines (Fig. 4). Furthermore, the superfusion of the spinal cord with ouabain increased the resting level of $[K^+]_e$ by as much as 1.1 mM and evoked an acid shift in pH_e base line by 0.07 pH units. These data demonstrate that both post-stimulation pH_e and K^+ homeostasis are related to the Na^+/K^+ pump activity.

Acid shifts in pH_e related to stimulation. With respect to Stewart's ideas, it is unlikely that the observed acid shifts result from a direct movement of H^+ through the neuronal membrane, even when changes in surface $[H^+]$ may occur through Na^+/H^+ exchange or the H^+ channel (Meech & Thomas 1987, Thomas 1984). The observed acid shifts of 0.15–0.25 pH units could be either

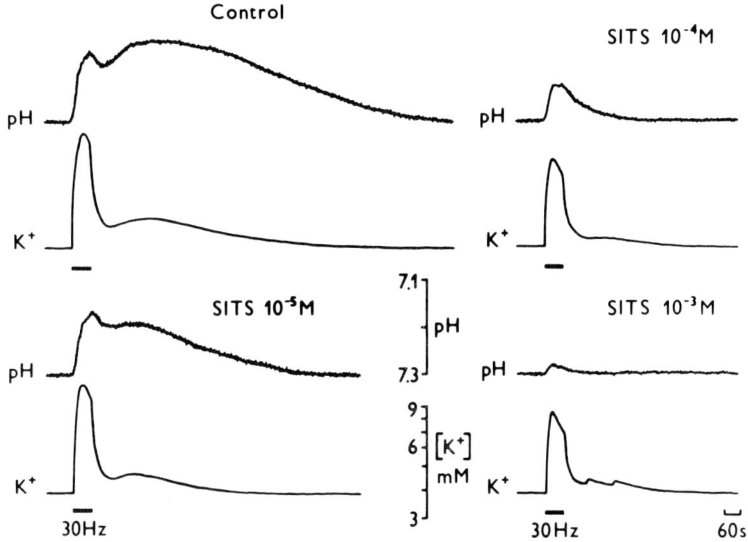

FIG. 5. Depression of simultaneously recorded changes in pH_e and $[K^+]_e$ in the lower dorsal horn of the isolated frog spinal cord by superfusion of the cord with SITS (10^{-5}, 10^{-4} and 10^{-3} M). Changes were evoked by tetanic stimulation of the 8th dorsal root at a frequency of 30 Hz applied for 60 s.

a consequence of ion movements leading to alterations in $[SID]_e$ of about 8–12 mM (for calculations see Kreig et al 1983) or due to the metabolic production of acid.

In our experiments, the observed attenuation of acid shifts due to amiloride was about 20–40%, to furosemide 60–80% and to SITS 80–90% (Fig. 5). The fact that NaF, which blocks the glycolytic portion of metabolic processes and lactate production, is almost without effect suggests that under the relatively 'physiological' conditions in our experiments the dominating mechanism is not alteration in $[SID]_e$ due to lactate production. Ion extrusion mechanisms across cell membranes (Na^+/H^+ or Cl^-/HCO_3^- exchange or K^+-Cl^- cotransport) play an important role in pH_e homeostasis. A decrease in $[SID]_e$ is probably due to extrusion of Cl^- or HCO_3^-, since production of strong acid anions is unlikely. Lactate may be produced under extreme conditions, e.g. seizure, spreading depression or anoxia. Siesjö et al (1985) found that lactic acid values during seizures were not significantly altered during the first 30–60 s. They suggest that the Na^+/H^+ exchange across cell membranes is enhanced due to increased Na^+ permeability. The resulting influx of Na^+ and efflux of H^+ would lead to extracellular acidification. Another argument against the role of lactate production in stimulation-evoked acid shifts is the fact that normalization of extracellular lactate content is slow, lasting about 15 minutes after the seizures, while the pH recovery observed in our experiments (especially *in vivo*) was almost as fast as the build up and never longer than the recorded changes in extracellular concentration of K^+ and Ca^{2+}. There is other evidence against extracellular lactate accumulation. When the peripheral nerve was stimulated for five or more minutes the pH_e changes (similarly, the changes in $[K^+]_e$ and $[Ca^{2+}]_e$) were not enhanced; on the contrary, once they had attained a plateau, the acidification (and the elevated $[K^+]_e$ and $[Ca^{2+}]_e$ gradually decreased to the original base line (E. Syková et al, unpublished work 1987). The situation in the spinal cord may, however, be different from that in the cerebellum and cortex where long-term direct tissue stimulation may lead to accumulation of lactate (Kreig et al 1983).

Carbonic anhydrase in the brain is found predominantly in the glia and myelin (see Kreig et al 1983). Since the fluid in the extracellular space is buffered by an HCO_3^--CO_2 system, inhibition of carbonic anhydrase by application of acetazolamide (4.0 or 8.0 mM) could result in changes of pH_e. Indeed, acetazolamide evoked an alkaline shift of the pH_e base line by about 0.05 pH units. The stimulation-evoked acidification was depressed and the post-stimulation alkaline shift was enhanced. Our findings, as well as those of Kreig et al (1983), provide evidence that carbonic anhydrase and glia play a significant role in pH_e homeostasis.

Post-stimulation or delayed acid shifts were always accompanied by a spontaneous rise in $[K^+]_e$ and were selectively blocked by 4 mM Mn^{2+} while the acid shifts which build up during the stimulation were depressed by only 60–80% (E. Syková et al, unpublished work 1987). This suggests that post-stimulation acid shift is postsynaptic in nature, while the pH_e changes during the stimulation arise from activity in neurons, as well as in pre-synaptic afferent fibres themselves (see Syková 1983).

Conclusions

We conclude that acid-base changes in neuronal microenvironment occur under physiological as well as pathological conditions and may significantly affect CNS function.

It is apparent that the Na^+/H^+ exchange, K^+-Cl^- co-transport, Na^+/K^+ pump activity, Cl^-/HCO_3^- exchange (or Na^+-H^+-HCO_3^-/Cl^- exchange), carbonic anhydrase and, presumably, the extracellular space shrinkage due to neuronal and/or glial cells swelling play an important role in pH_e homeostasis. One of the pertinent questions at the present time is to what extent the changes in pH_e are due to ionic changes in neurons or in glial cells or due to changes in extracellular space volume. We need to study and clarify the role of glia in activity-related pH_e homeostasis. Recently described HCO_3^- conductance in the glial cell membrane (Astion et al 1987) points to the mechanism that will tend to buffer temporary changes in pH_e at the expense of glial

pH_i. To elucidate further the mechanisms of acid-base changes in the extracellular space may require the use of methods other than those available at present. However, it is evident that the described pH_e changes are of physiological significance and as such support the concept of the extracellular space as a dynamic structure maintaining the brain cell microenvironment and communication channel between cells.

References

Astion ML, Coles JA, Orkand RK 1987 Effects of bicarbonate on glial cell membrane potential in Necturus optic nerve. Neurosci Lett 76:47–52

Balestrino M, Somjen GG 1988 Concentration of carbon dioxide, interstitial pH and synaptic transmission in hippocampal formation of the rat. J Physiol (Lond) 396:247–266

Chvátal A, Jendelová P, Kříž N, Syková E 1988 Stimulation-evoked changes in extracellular pH, calcium and potassium activity in the frog spinal cord. Physiol Bohemoslov 37:203–212

Czéh G, Kříž N, Syková E 1981 Extracellular potassium accumulation in the frog spinal cord induced by stimulation of the skin and ventrolateral columns. J Physiol 320:57–72

Deitmer JW, Schlue W-R 1987 The regulation of intracellular pH by identified glial cells and neurones in the central nervous system of the leech. J Physiol 388:261–283

Endres W, Grafe P, Bostock H, ten Bruggencate G 1986 Changes in extracellular pH during electrical stimulation of isolated rat vagus nerve. Neurosci Lett 64:201–205

Gruol DL, Barker JL, Huang LM, McDonald JF, Smith Jr TG 1980 Hydrogen ions have multiple effects on the excitability of cultured mammalian neurons. Brain Res 183:247–252

Harris RH, Symon L 1984 Extracellular pH, potassium and calcium activities in progressive ischaemia of rat cortex. J Cereb Blood Flow Metab 4:178–186

Iljima T, Ciani S, Hagiwara S 1986 Effects of the external pH on Ca channels: Experimental studies and theoretical considerations using a two-site, two-ion model. Proc Natl Acad Sci USA 83:654–658

Konnert A, Lux HD, Morad M 1987 Proton-induced transformation of calcium channel in chick dorsal root ganglion cells. J Physiol (Lond) 386:603–633

Kreig RP, Ferreira-Filho CR, Nicholson C 1983 Alkaline and acid transients in cerebellar microenvironment. J Neurophysiol 49:831–850

Krishtal OA, Pidoplichko VI 1980 A receptor for protons in the nerve cell membrane. Neuroscience 5:2325–2327

Kříž N, Syková E, Vyklický L 1975 Extracellular potassium changes in the spinal cord of the cat and their relation to slow potentials, active transport and impulse transmission. J Physiol (Lond) 249:167–182

Lehmenkühler A, Zidek W, Staschen M, Caspers H 1981 Cortical pH and pCa in relation to DC potential shifts during spreading depression and asphyxiation. In: Syková E et al (eds) Ion-selective microelectrodes and their use in excitable tissues. Plenum Press, New York, p 225–229

Meech RW, Thomas RC 1987 Voltage-dependent intracellular pH in Helix aspersa neurones. J Physiol (Lond) 390:433–452

Mutch WAC, Hansen AJ 1984 Extracellular pH changes during spreading depression and cerebral ischemia: Mechanisms of brain pH regulation. Cereb Blood Flow Metab 4:17–27

Nicholson C 1980 Dynamics of the brain cell microenvironment. Neuroscience Res Prog Bull 18:177–322
Ransom BR, Carlini WG, Connors BW 1985 Brain extracellular space: Developmental studies in rat optical nerve. Ann New York Acad Sci 481:87–105
Siesjö BK, von Hanwehr R, Nergelius G, Nevander G, Ingvar M 1985 Extra- and intracellular pH in the brain during seizures and in the recovery period following the arrest of the seizure activity. J Cereb Blood Flow Metab 5:47–57
Stewart PA 1981 How to understand acid-base. Elsevier, New York
Syková E 1983 Extracellular K^+ accumulation in the central nervous system. Prog Biophys Mol Biol 42:135–189
Syková E 1986 Activity-related fluctuations in extracellular ion concentrations in the central nervous system. NIPS 1:57–61
Syková E 1987 Modulation of spinal cord transmission by changes in extracellular K^+ activity and extracellular volume. Can J Physiol Pharmacol 65:1058–1066
Thomas RC 1984 Experimental displacement of intracellular pH and the mechanism of its subsequent recovery. J Physiol 354:3–22
Urbanics R, Leniger-Follert E, Lübbers DW 1978 Time course of changes of extracellular H^+ and K^+ activities during and after direct electrical stimulation of the brain cortex. Pflügers Arch Eur J Physiol 378:47–53
Walz W, Hinks EC 1987 Extracellular hydrogen ions influence channel-mediated and carrier-mediated K^+ fluxes in cultured mouse astrocytes. Neuroscience 20:341–346

DISCUSSION

Frömter: I should like to add to your list of possible mechanisms that might explain these observations: rheogenic Na^+-HCO_3^- co-transport. When the cell depolarizes on stimulation, HCO_3^- could enter the cell, leaving the extracellular fluid more acidic. The blocking effect of SITS supports this interpretation.

Oberleithner: This is indicated by your work with furosemide and SITS. Rheogenic Na^+-HCO_3^- co-transport depends on the cell membrane potential. We have done similar experiments in frog distal cells, where this system is present. HCO_3^- is moved across the basolateral cell membrane because of changes in membrane potential. I think that you could explain elegantly most of the your findings, if you postulate rheogenic Na^+-HCO_3^- co-transport.

Syková: I agree, that is possible.

Schlue: Is the pH in the extracellular spaces of the spinal cord different from the pH in the bathing medium?

Syková: The resting pH level *in vitro* was essentially the same as that in Ringer solution, i.e. pH 7.3-7.35. In rat spinal cord the extracellular pH (pH_e) is slightly more acidic (7.1–7.25) than the pH of blood. We also measured the extracellular p_{CO2} and p_{O2}; these were within the physiological range.

We measured pH_e within the spinal cord *in vivo* and found that it varies at different depths from 7.1 to 7.25. In dorsal horn we described the various K^+ concentrations which are related to the spontaneous activity of the interneurons (Syková et al 1983). Many neurons in the dorsal horn fire spontaneously. This spontaneous firing is accompanied by permanent elevation of the extracellular K^+ concentration in the lower dorsal horn, and may also be accompanied by a decrease in pH_e.

Morad: In the experiments where you increase the K^+ concentration in the bath, H^+ was secreted into the extracellular space. In those experiments, did you add SITS to see whether you could dissociate the efflux of K^+ from depolarization?

Syková: Not yet.

Morad: It's curious that K^+ always seems to be coming out at the same time that H^+ is coming out.

Syková: In the spinal cord there are spontaneous changes of K^+. Sometimes there is a burst of firing, especially in isolated spinal cord, which is associated with the K^+ increase, and with a spontaneous acid shift that follows a similar time course.

Morad: But when you added SITS, did the K^+ efflux decrease?

Syková: Yes, but the depression of the K^+ transients with SITS was only about 20–30%, while there was nearly complete inhibition of the pH change.

Grinstein: You showed that *in vivo* there was a much greater inhibition of the K^+ flux than *in vitro*.

Syková: The effect of SITS is generally the same in the spinal cord of the rat *in vivo* and the spinal cord of the frog *in vitro*. I showed the experiment *in vivo*, in which the inhibition of K^+ accumulation 15 minutes after the application of 3mM SITS was greater, but less than the inhibition of the acid shift. *In vivo* we applied SITS to the surface of the spinal cord, and therefore used a higher concentration than that used *in vitro* (see Fig.5).

Grinstein: You don't have any explanation for why the K^+ flux is depressed by SITS?

Morad: The issue is whether the neurons are firing. In the presence of SITS, you can say that you are blocking some excitatory process so the neurons aren't all firing, resulting in accumulation or depletion of ions. To investigate this, I would simply increase the K^+ concentration, as you did. You can then examine whether H^+ accumulates and how it is affected by SITS. In this way, you can decide whether this is a secondary or primary phenomenon.

Syková: Yes, SITS can partially block excitation. However, when we applied SITS, we observed an almost immediate, rapid effect on pH, while the K^+ accumulation was not substantially depressed. The effect on K^+ transients was slower, i.e. it required a longer application of SITS.

Thomas: You are expecting SITS to diffuse a long way. This rat presumably still has an intact blood supply.

Syková: That is completely intact. You have to apply SITS in higher concentrations. *In vivo* we recorded pH and K^+ changes at a depth of 700 μm, applied 3 mM SITS to the spinal surface, then recorded five and fifteen minutes later.

Thomas: Whereas the frog spinal cord doesn't have a blood supply and so it is more accessible.

Schlue: You stated that part of the changes in external pH are due to shrinkage of the extracellular space. Is it possible to exclude this contribution in the same experiment by blocking KCl uptake, for example, with furosemide?

Syková: The inhibition of transient extracellular pH changes with furosemide varies from 40–80%. One question is what is the source of that shrinkage of the extracellular space? We don't know whether only the neurons or the glial cells or both are swelling. A possible relationship between swelling of the cells and pH_e changes is under investigation in our laboratory.

Kaila: I wonder whether KCl co-transport has to be implied in the increase in cell volume. If the product of the extracellular concentrations of K^+ and Cl^- increases, that would account for cell swelling caused by an influx of ions through plasmalemmal channels.

Syková: Several mechanisms have been proposed for the activity-related shrinkage of the extracellular space, including transport of KCl. We can't tell at the moment what is the main mechanism of that shrinkage, or to what extent the glial cells, neurons and/or fibres are swelling.

Thomas: The HCO_3^- efflux that you would predict, Kai, to occur through GABA-activated channels would cause an extracellular alkalinization rather than an acidification.

Kaila: Yes, a GABA-activated bicarbonate conductance should produce an initial alkalosis at the extracellular surface.

Syková: There is an initial alkalinization.

Kaila: It looks very fast. Do you think it could be truncated by the slow response time of the H^+-selective microelectrode?

Thomas: It might, in addition, be swamped by the subsequent acidification.

Syková: That's exactly what happens. If you block the acidification or slow it down, the initial alkaline shift becomes larger. These are two antagonistic processes. It was shown in cerebellum by Kraig et al (1983) that the size of the initial alkaline shift depends on the frequency of stimulation. The alkaline shift seems to be related to synaptic activity.

Thomas: Eva, you are not sure where the pH changes are, presumably your recording site is chosen simply as an area of maximal changes. You don't know how close you are to the main front. We are all interested in H^+ fluxes over some distance. In snail neurons the effect was magnified by reducing the buffering power. Have you looked at extracellular buffering power?

Syková: We haven't, but I assume that the buffering power is about 30–34 mM.

Thomas: Your solutions contain HCO_3^- but that was a lousy buffer for the space outside the snail neuron. The buffering power could be much lower than you think. Would it be possible to add 20 mM Hepes to your solution?

Syková: It would be possible in experiments on isolated frog spinal cord, but essentially we obtained the same results *in vivo* and *in vitro*.

Thomas: In the frog spinal cord it might be very interesting if you were to do the experiment with a higher concentration of Hepes.

References

Kraig RP, Ferreira-Filho CR, Nicholson C 1983 Alkaline and acid transients in cerebellar microenvironment. J Neurophysiol 49:831–850

Syková E, Kříž N, Preis P 1983 Elevated extracellular potassium concentration in unstimulated spinal dorsal horns of frogs. Neurosci Lett 43:293–298

General discussion IV

Experimentally induced intracellular pH gradients

de Hemptinne: We did some experiments in which we induced a pH gradient within a single elongated cell (skeletal muscle) or a strand of electrically coupled cells (cardiac Purkinje cells). The purpose of these experiments was to check whether an experimentally induced intracellular pH gradient can be seen to spread out in the longitudinal direction within a single cell or within cells that are electrically coupled.

The muscle preparation was placed in a perfusion chamber which was separated into two compartments by a thin rubber membrane. After making a small hole in the membrane, one extremity of the muscle was carefully pulled through this hole, to give a roughly equal mass of tissue in each compartment. The membrane did not affect the physical continuity of the skeletal muscle cell or the intercellular connections of the Purkinje cells.

In one compartment, the conditioning compartment (cc), intracellular acidification was induced by applying then removing NH_4Cl from the superfusion medium in the presence of amiloride. Intracellular pH was measured simultaneously in the test compartment (tc) at different distances (0.25 mm and 0.5 mm) from the rubber membrane. We observed almost no pH change within the cells in the tc.

When NH_4Cl was applied to the cc, the intracellular pH of the cells in that compartment became more alkaline.

Intracellular pH in the tc in the vicinity of the rubber membrane (0.25 mm) slowly became more acidic.

This observation suggests that intracellular NH_4^+ diffuses down the core of the fibres from cc to the tc and leaves the cell as NH_3 in the tc, thereby causing a local acidification.

Thomas: What was the diameter of the preparation?

de Hemptinne: The sheep Purkinje fibres were about 1 mm in diameter. Similar experiments on mouse soleus muscle gave comparable results.

Morad: Are these simultaneous measurements?

de Hemptinne: No, intracellular pH changes were induced by modifying the superfusion solution in the conditioning compartment. pH measurements were recorded in the test compartment.

Thomas: So when NH_4Cl is present in the conditioning compartment, the pH change parallels the rise of internal NH_4^+?

de Hemptinne: Yes, internal NH_4^+ accumulating in the cc diffuses down the core of the fibres or from cell to cell in the cardiac muscle, towards the tc where it can readily leave the cells as uncharged NH_3.

Gillespie: Given the diffusion of H^+ in free solution, would you expect to see an acid shift in your test compartment as a result of acidifying the conditioning compartment?

de Hemptinne: I didn't expect to see a large effect. The buffer capacity of the cells is about 10–30 mM/pH unit, so that a proton load reaching the millimolar range is required to cause a discernible pH change.

Boron: I'm trying to understand the behaviour of pH_i on the *trans* side during the NH_4^+ application. Presumably, NH_3 and NH_4^+ diffuse down the cells, with only a minor difference in their relative mobilities. I would think that NH_3 entering the *trans* side would have no net effect on pH_i, since it would leave the fibre as NH_3. NH_4^+ diffusion to the *trans* side might cause a decrease in pH_i, as NH_4^+ dissociates to form NH_3, which would leave the cell, and H^+, which would be left behind.

de Hemptinne: I agree with that.

Boron: Could you go back, model the system and predict the rate of change of pH_i, given the concentration of mobile buffers? Would there be virtually no pH change?

de Hemptinne: The concentration of H^+ is ten times greater in the conditioning compartment (pH 6.2) than in the test compartment (pH 7.2), so one would expect net diffusion of H^+ down that concentration gradient. However, the intracellular activity of the protons in the cc is in the micromolar range, so the proton load caused by diffusion would be negligible in comparison with the intracellular buffering capacity.

Thomas: I have a student (Alastair Pickering) now injecting HCl into muscle fibres and trying to record the pH changes at various distances. The pH change that he records initially goes away during the injection, presumably as H^+ diffuses from the injection site.

de Hemptinne: In epithelial cells where protons can be secreted to the luminal side and HCO_3^- to the serosal side, this might be able to generate a pH gradient within the cell.

Al-Awqati: How can you have intracellular pH gradients when there is such a high buffering power within the cell? People keep talking about pH gradients inside the cell but nobody has ever demonstrated such a gradient. Where is the evidence?

de Hemptinne: In the experiment I have described, an intracellular proton gradient is generated.

Al-Awqati: Yes, but you have made it by injection, what is the evidence that proton gradients exist *in vivo*?

Gillespie: If you image with pH-dependent fluorescent dyes, you see regional variations in signal which could quite easily be differences in H^+ concentrations. Taking into account possible dye binding and compartmentalization, differences are still seen. These gradations in fluorescence can also be modified by NH_4^+ pulses, Cl^- removal or agonists. They may therefore be real differences in pH within the cell cytoplasm.

Al-Awqati: One has to be careful about artifacts when interpreting these sorts of results.

Thomas: Well you certainly can produce pH gradients artificially, whether they occur naturally is a very different question.

Structural properties of the proton translocating complex of the clathrin-coated vesicle

Dennis K. Stone, Shi-Zhang Sun and Xiao-Song Xie

Departments of Physiology and Internal Medicine, University of Texas Southwestern Medical Center, 5323 Harry Hines Boulevard, Dallas, Texas 75235—9030, USA

Abstract. The clathrin-coated vesicle proton pump is a representative member of the new class of endomembrane proton ATPases that share an inhibitor profile which distinguishes them from classic F_1F_0 and E_1E_2-type proton pumps. The coated vesicle proton pump is a large (530 kDa) heteroligomer composed of eight polypeptides with molecular masses of 116, 70, 58, 40, 38, 34, 33 and 17 kDa. The 200-fold purified enzyme catalyses ATP-generated proton pumping when reconstituted in liposomes composed of pure lipids. Subunit function has been determined by partial reaction analysis of subunit and subcomplex activities. The isolated 17 kDa subunit, when co-reconstituted with bacteriorhodopsin, forms a dicyclohexylcarbodiimide-inhibitable proton channel. Selective removal of the 116 kDa subunit transforms the proton ATPase from a Mg^{2+}-activatable to a Ca^{2+}-activatable ATPase. Subsequent dissociation and reconstitution of subunits reveals that the 70, 58, 40 and 33 kDa components are required, in composite, to form a functional ATP-hydrolytic core, and that no single subunit or subcomplex deficient in these subunits can catalyse ATP hydrolysis.

1988 Proton passage across cell membranes. Wiley, Chichester (Ciba Foundation Symposium 139) p 238–251

Proton-translocating ATPases have recently been identified in an extraordinarily wide array of intracellular organelles, including clathrin-coated vesicles, endosomes, lysosomes, endoplasmic reticulum and Golgi membranes. In addition, proton pumps have been identified as mediators of acidification of specialized organelles, such as chromaffin granules, synaptic vesicles and tertiary granules of leucocytes. These proton pumps share a characteristic inhibitor profile — sensitivity to N-ethylmaleimide (NEM) and resistance to vanadate and azide — which serves to distinguish them from E_1E_2-type ATPases and classic F_1F_0-type ATPases. Proton ATPases sharing these characteristics have also been found in plant and fungal vacuolar membranes (Stone & Xie 1988).

Functionally, these pumps conduct a diverse set of activities, ranging from pH-dependent processing of endocytosed ligand-receptor complexes (endosomes) to energization of substrate storage (chromaffin granules). It is notable that these pumps are largely responsible for endomembrane acidification. One definite exception is the proton ATPase, which is the final effector of distal nephron acidification; the proton pump of renal medulla and turtle urinary bladder is, by its inhibitor profile, a *bona fide* member of this new class of proton pumps (Gluck et al 1982, Stone et al 1984b). Less clear is the nature of the electrogenic proton pump of Ehrlich ascites tumour cell plasma membrane (Heinz et al 1981). Although this pump can regulate intracellular pH, it does so only under highly artificial conditions in which an intracellular acid load is caused by NH_4Cl, and not the physiological metabolite, lactic acid (Stone et al 1984b). In this regard, there is no convincing evidence that these proton ATPases serve as important cellular pH defence mechanisms in any eukaryotic system. Although a number of endomembrane proton ATPases have been solubilized and partially purified, only the coated vesicle (Xie & Stone 1986a) and chromaffin granule ATPases (Moriyama & Nelson 1987) have been shown to catalyse active proton pumping after reconstitution of the purified enzyme.

We have focused on a structural and biochemical analysis of one of these proton pumps, the bovine brain clathrin-coated vesicle proton translocating complex. This ATPase has been solubilized with the non-ionic detergent $C_{12}E_9$ and purified 200-fold. Table 1 summarizes important properties of this enzyme and those of the bovine adrenal chromaffin granule proton pump. As is evident, these pumps are large heteroligomers which share a number of

TABLE 1 Properties of the purified clathrin-coated vesicle and bovine chromaffin granule proton pumps

	Clathrin-coated vesicle	*Chromaffin granule*
Molecular mass (kDa)	530	
Constituent polypeptides (kDa)	116	115
	70	70
	58	58
	40	40
	38	
	34	
	33	
	17[a]	17[a]
Specific activity (μmol P_i mg protein^{-1} min^{-1})	17	8

[a] Dicyclohexylcarbodiimide binding.

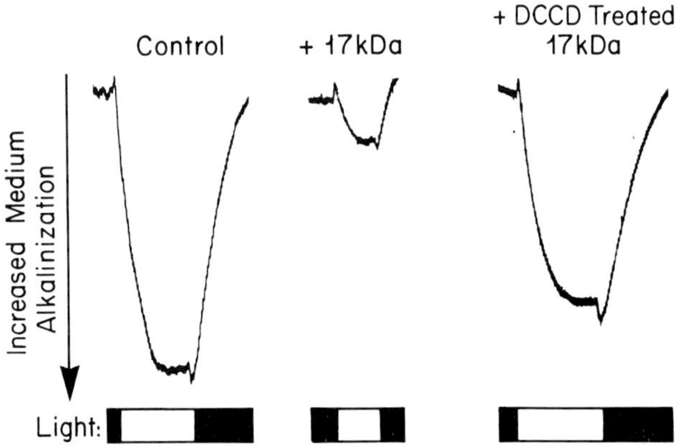

FIG. 1. Co-reconstitution of the 17 kDa polypeptide with bacteriorhodopsin. For all experiments, bacteriorhodopsin (45 μg) was reconstituted with 625 μg of lipids and assayed as described in Sun et al 1987. Control, bacteriorhodopsin alone. The 17 kDa polypeptide (0.8 μg) was co-reconstituted with bacteriorhodopsin for the trace designated + 17 kDa. In the final trace, 0.8 μg of the 17 kDa subunit was preincubated with 0.7 μmol DCCD/mg protein before co-reconstitution with bacteriorhodopsin.

polypeptides of similar molecular mass. In this regard, these proton ATPases resemble classic F_1F_0-type ATPases, such as the mitochondrial ATPase. Particularly notable is labelling of the 17 kDa components of these pumps with [^{14}C]dicyclohexylcarbodiimide (DCCD), which is highly reminiscent of the 8 kDa DCCD binding protein of mitochondrial F_0. To determine the function of these potential subunits, we have undertaken the dissociation and molecular reconstitution of the component polypeptides of the clathrin-coated vesicle proton ATPase.

Methods

Purified clathrin-coated vesicle proton translocating ATPase (Xie & Stone 1986b) was used as the starting material for all described studies. Preparation of the 17 kDa proteolipid and its reconstitution with bacteriorhodopsin have been described elsewhere (Sun et al 1987). Dissociation of the catalytic sector of the proton pump was achieved by treatment of the enzyme with 3 M urea; resolution of the individual subunits and subcomplexes was achieved by glycerol gradient centrifugation, performed essentially as described in Xie & Stone 1986a. ATPase activity was assessed by the liberation of $^{32}P_i$ from [γ^{32}P]ATP (Stone et al 1984a).

Results and discussion

The inhibition of the clathrin-coated vesicle proton pump by DCCD and the exclusive labelling of the 17 kDa component with [^{14}C]DCCD suggested that the low molecular weight component of the pump might act as a proton channel which lowers the activation energy of transmembranous proton movement, analagous to mitochondrial F_0. To test this hypothesis, the 17 kDa component was isolated from the intact holoenzyme by toluene extraction. Polyacrylamide gel electrophoresis and autoradiography using [^{14}C]DCCD revealed the presence of the isolated 17 kDa component in the toluene extract. To assess function of this apparent proteolipid, the toluene residue was evaporated and the resultant protein was resuspended in 0.1% $C_{12}E_9$. As an assay of proton conductivity, the proteolipid was co-reconstituted with the light-energized proton pump, bacteriorhodopsin, by the octylglucoside dilution procedure (Sun et al 1987). Proteoliposome acidification was assessed by measuring the pH of the medium. Shown in Fig. 1 is the result of this experiment. The first trace (control) is bacteriorhodopsin alone. As shown, energization of the system by light results in proteoliposome acidification with a resultant alkalinization of the medium. The trace '+17 kDa' illustrates the results of co-reconstitution of the proteolipid with bacteriorhodopsin. As shown, proteoliposome acidification is markedly attenuated, a result consistent with the proteolipid having induced a proton leak in the system. Finally, DCCD treatment of the proteolipid (third trace) results in restoration of the acidification signal. In separate experiments, it was determined that DCCD had no effect on bacteriorhodopsin alone and that the proteolipid was sensitive to trypsin. Close examination of Fig. 1 reveals a somewhat surprising result: the release of the proton gradient after the light is extinguished is not accelerated in the presence of the proteolipid. If indeed the proteolipid functioned as a proton pore, an enhanced release rate might be anticipated. Although further experiments are required to clarify this issue, a likely explanation lies in the rate of proton conductivity facilitated by the proton channel. While precise determinations have not been made, a reasonable estimate of proton conductivity through chloroplast F_0 is about 7000 protons/sec (Nelson 1980). Given that bacteriorhodopsin has a rate-limiting step of >1 msec, this would indicate that the attenuation of proteoliposome acidification provoked by the proteolipid is due to the complete prevention of gradient formation in proteolipid-containing proteoliposomes, while normal acidification occurs in a reduced subpopulation of liposomes containing bacteriorhodopsin alone. In such an event, the proton release rate would not be expected to differ from that in the control setting.

Taken together, the results of this experiment demonstrate the function of one pump subunit, the 17 kDa proteolipid. The most notable implication of this finding is the structural resemblance of this endomembrane pump to classic F_1F_0-type ATPases.

To determine the molecular composition of the catalytic sector, dissociation of the ATPase under non-denaturing conditions has been necessary. We have tested a number of empirical manoeuvres in an attempt to achieve this. The basic approach has been to subject the purified holoenzyme to a given condition and then perform glycerol gradient centrifugation to determine if any subunit dissociation and resolution has occurred. If component polypeptides were resolved, then each gradient fraction was analysed for its capacity to catalyse ATP hydrolysis and/or proton pumping after reconstitution. Several observations have been made using this approach.

Before describing these results, certain features of the clathrin-coated vesicle proton ATPase require clarification. First, the enzyme has a specific ATPase activity of 15 μmol P_i mg protein^{-1} min^{-1} when assayed in the presence of 3 mM Mg^{2+}, and 6 μmol P_i mg protein^{-1} min^{-1} when assayed in the presence of 3 mM Ca^{2+}. Both are activated by chloride and phosphatidylserine, and inhibited by NEM and NO_3^-. In addition, both activities are equally inhibited by a polyclonal antibody we have generated against the enzyme. Second, Mg^{2+} will support ATP-energized acidification by the reconstituted enzyme, whereas Ca^{2+} will not. Thus Ca^{2+} prevents coupling of ATP hydrolysis to proton pumping, and Ca^{2+}-activated ATP hydrolysis serves as a partial reaction useful in determining the polypeptide requirements for ATP hydrolysis, independent of the substructure required for coupling and vectoral proton movement.

It was determined that treatment of the purified holoenzyme with nonylglucoside and/or an alkaline pH (pH = 10.5), followed by gradient centrifugation, resulted in fractions which were depleted of the 116 and 38 kDa polypeptides. Under these conditions, the resultant subcomplex had a complete loss of Mg^{2+}-activated ATPase activity and an absolute increase in Ca^{2+}-activated activity. The preparation could not be reconstituted to yield ATP-driven proteoliposome acidification. Attempts at restoring Mg^{2+}-activated ATPase activity by readdition of the 116 and 38 kDa subunits to the polypeptide-depleted subcomplex were unsuccessful. At the present time we cannot determine whether loss of these two polypeptides from the complex is causally related to the metal activation transition, or whether the loss of these polypeptides is simply a coincidental event accompanying the detergent/pH-provoked shift.

Subsequently, a variety of manoeuvres were tested in an attempt to achieve dissociation. Of these, it was found that treatment of the holoenzyme with 3M urea, followed by gradient centrifugation, resulted in the generation of a diverse group of subcomplexes and individual subunits. None of these subcomplexes would support either Mg^{2+}-activated ATP hydrolysis or ATP-generated proteoliposome acidification after reconstitution. Most informative was the isolation of a subcomplex (fraction I) which contained only the 70 and 58 kDa polypeptides. This preparation had no Ca^{2+}-activated ATP hydrolytic activity. This observation assumes importance in that both of these subunits have been identified previously as nucleotide-binding subunits and have been

TABLE 2 Structural and functional properties of urea-generated subcomplexes

Fraction	*Polypeptides*	*ATPase activity (nmol P_i released/20 min)*
I	70, 58	0
II	70, 58, 40»33	2.3
III	33	0
I + II	70, 58, 40»33	21.5
I + III	70, 58, 33	3.6
I + II + III	70, 58, 40, 33	59.2

ATPase assays were performed using 5 mM $CaCl_2$, 40 mM Tris/2-[N-morpholino]ethane sulphonic acid pH 7.0, 30 mM KCl, and 3 mM [γ^{32}P]ATP (600 cpm/nmol) (Stone et al 1984a). Amounts of protein of the fractions used in the assay were: I, 1.2 μg; II, 1.8 μg; III, 1.0 μg. Individual or mixed fractions were incubated at 25 °C for 20 min in the presence of 2.5 μg of phosphatidylserine prior to assay.

proposed to constitute the catalytic sector of the enzyme (Stone & Xie 1988). Our observation, although not incompatible with this view, clearly indicates that other factors are required for catalysis. These factors may themselves constitute a portion of the molecular catalytic centre, or they may be required to induce conformational changes in the 70–58 kDa subcomplex needed to render the catalytic site operational.

What is the nature of these additional factors? Although resolution is only partial at present, it appears that the 40 kDa and 33 kDa components of the pump are necessary for Ca^{2+}-activated ATP hydrolysis. This view stems from a series of add-back experiments performed with other fractions obtained after urea treatment and glycerol gradient centrifugation. These particular fractions, designated as II and III, had trivial Ca^{2+}-activated ATP hydrolysis when tested alone, but when combined with fraction I gave a dramatic increase in Ca^{2+}-activated hydrolysis. Biochemically, fraction II contains near equimolar amounts of the 70, 58 and 40 kDa polypeptides, and a smaller amount of the 33 kDa component. Fraction III contains predominantly the 33 kDa polypeptide. The results are summarized in Table 2. That these add-back experiments are a reconstitution of subunits is suggested by further studies in which it was determined that activation of Ca^{2+}-activated ATP hydrolysis by subunit combination was more than a simple first order process.

We thus conclude, pending further resolution, that four polypeptides (70, 58, 40 and 33 kDa) are required to constitute a functional catalytic core. Removal of the 116 kDa polypeptide may play a role in the transition from Mg^{2+} to Ca^{2+} activation. Taken together, these experiments strongly support the notion that the clathrin-coated vesicle proton pump, and by inference all endomembrane pumps, are F_1F_0-type ATPases. Determination of the molecular events responsible for the transition from Mg^{2+} to Ca^{+} activation, we believe, will allow definition of the coupling mechanism which coordinates ATP hydrolysis with the vectoral transmission of protons through the 17 kDa proton channel.

Acknowledgements

This work was supported by grants from the National Institutes of Health (DK33627) and by the Perot Family Foundation. The authors would like to thank Ms Dedrian Copeland and Ms Linda Lee for their expert secretarial assistance.

References

Gluck S, Kelly S, Al-Awqati Q 1982 The proton translocating ATPase responsible for urinary acidification. J Biol Chem 257:9230–9233

Heinz A, Sachs G, Schafer JA 1981 Evidence for activation of an electrogenic proton pump in Ehrlich Ascites tumor cells during glycolysis. J Membr Biol 61:143–154

Moriyama Y, Nelson N 1987 The purified ATPase from chromaffin granule membranes is an anion dependent proton pump. J Biol Chem 262:9175–9180

Nelson N 1980 Proton channels in chloroplast membranes. Ann NY Acad Sci 358:25–35

Stone DK, Xie XS 1988 Proton translocating ATPases: Issues in structure and function. Kidney Int 33:767–774

Stone DK, Xie XS, Racker E 1984a Inhibition of clathrin coated vesicle acidification by duramycin. J Biol Chem 259:2701–2703

Stone DK, Xie XS, Wu LT, Racker E 1984b Proton translocating ATPases from clathrin coated vesicles, renal medulla vesicles and Ehrlich ascites tumor cells. In: Forte J et al (eds) Proton Transport in Epithelia. Wiley, New York p 219–231

Sun SZ, Xie XS, Stone DK 1987 Isolation and reconstitution of the dicyclohexylcarbodiimide-sensitive proton pore of the clathrin coated vesicle proton translocating complex. J Biol Chem 262:14790–14794

Xie XS, Stone DK 1986a Isolation and reconstitution of the clathrin-coated vesicle proton translocating complex. J Biol Chem 261:2492–2495

Xie XS, Stone DK 1986b Lipid requirements for reconstitution of the proton translocating complex of clathrin coated vesicles. Proc Natl Acad Sci USA 83:8913–8917

DISCUSSION

Thomas: Can you tell us why these vesicles need to be acid inside?

Stone: With the coated vesicle it is a little controversial; with endosomes it is clear why this happens, at least to some extent. The best example is probably iron-transferrin, which is taken up as a complex by the transferrin receptors. The release of iron from the transferrin requires exposure to an acid environment.

Thomas: So it makes an acid microenvironment inside.

Stone: Exactly, there are two different pI's for the dissociation, and with acidification the iron is released within the endosome.

Thomas: Lysosomes are a sub-class of these sorts of vesicles and you use them primarily because they are easy to get hold of and relatively easy to purify.

Stone: We calculated that we have used over 4000 lbs of cow brain, so these vesicles are not abundant.

Sabolic: Forgac & Berne (1986) described isolation of the proton pump by solubilization from the same source, i.e. calf brain clathrin-coated vesicles. They used three different detergents for solubilizing. The solubilized proteins were reconstituted into phospholipid vesicles and exhibited proton pumping similar to yours. But the molecular weight of their proton pumps was between 200 000 and 247 000 Da.

Stone: Forgac recently confirmed our results (Arai et al 1987), and other laboratories have as well (Xie & Stone 1986).

Steinmetz: It seems that you have a fairly homogeneous population of pumps. There are eight polypeptides, of which six are needed for pumping. The molecular weight of the ATPase complex is about 530 kDa. The question is can you see it by any technique?

Stone: If you look at the purified enzyme by negative staining, you see large structures that are calculated to be around 500–600 kDa. We haven't focused much on the immunocytochemistry or transmission EM work.

Grinstein: Did I understand correctly that Cl^- is necessary for ATPase activity in the uncoupled system?

Stone: It's not absolutely required, but it will stimulate the activity of the solubilized form of the enzyme about twofold. Within the native vesicle, Cl^- is certainly required for dissipation of the electrical gradient generated by the pump. So there are at least two effects of Cl^-.

Thomas: How do you distinguish that from just inhibition by glucuronate or whatever you used to substitute for Cl^-?

Stone: We compared about 15 different ions. Of these, no anion had any effect, except Cl^-, which was stimulatory, and NO_3^-, which was inhibitory.

Thomas: But the levels of Cl^- in your solutions are probably ten times higher than those found *in vivo*, aren't they?

Stone: 10 mM. The K_m for the activation is 7.2 mM. Hans Oberleithner determined the Cl^- concentration in *Amphiuma* distal neuron cells to be 10–20 mM (Oberleithner et al 1982).

Boron: Before the clathrin-coated vesicles form, this membrane is presumably exposed to the extracellular fluid. Are these proton pumps active at that time? If not what keeps them turned off and then turns them on?

Stone: You should ask Qais Al-Awqati, because he has become a fan of the Cl^- channel being the regulator and he is probably right. That touches on a real issue: in a cell, certainly there is a range of pH within acidic compartments. The question is, are there active proton pumps on the cell surface or not? Heinz et al (1981) found an electrogenic proton pump on Ehrlich ascites tumour cells. In plasma membrane preparations of these cells, we have found an electrogenic proton pump that is sensitive to N-ethyl maleimide. However, if you look at the rate of correction of an acidosis imposed by NH_4Cl, there is a slow recovery. If you use a physiological acid, such as lactic acid by supplying glucose, to produce an equivalent acid load, then the lactate-H^+ symporter described by Spencer and Lehninger (1976) will correct that acidosis very rapidly. If you block that symporter with the specific inhibitor lactic acid anhydride, then there is no recovery. This means that the dominant regulator of the mechanism of pH correction is the nature of the accompanying anion.

Apart from kidney cells and possibly osteoclasts, I don't think that there is any good evidence that these pumps play a physiological role in intracellular pH regulation. Is the pump turning over? Could it be a back- up mechanism? The contrary evidence which is always cited is that coated vesicles in proximity to the plasma membrane are not acidic, therefore they don't have a proton pump, therefore the plasma membrane doesn't have a pump. That would mean that the coated vesicles that we isolate must come from the peri-Golgi region. I am not sure this is right. Clearly they are not acidic, Dick Anderson has shown that *in situ*. It does not mean that they are not acidifying for the following reasons: 1) coated vesicles only last 30 seconds as a coated vesicle after endocytosis, then they are stripped by uncoating enzyme. 2) The vesicles have just endocytosed 25 mM HCO_3^- which has to be consumed before you can detect an acidification signal. 3) These coated vesicles also have Na^+/K^+ ATPase, an electrogenic pump which produces an interior positive potential that is counter-productive to acidification since it is in parallel with the electrogenic proton pump. (You can acidify coated vesicles but you get greater acidification if you treat with vanadate or ouabain to inhibit the Na^+/K^+ ATPase.) I am not convinced that inward-bound coated vesicles which fuse with the acidic endosome within several minutes don't have proton pumps. It may be true that the only source of those pumps is the Golgi, future experiments are required to determine that.

Russell: The Cl^- channel that is used to shunt the voltage, is that in the native plasma membrane before it endocytoses to form the clathrin-coated vesicle?

Stone: If you can't answer the question of whether the proton pump is in those vesicles, you can't say where the Cl^- channel is. Professor Al-Awqati might want to comment on that because he is doing beautiful work on that channel.

Al-Awqati: All the vesicles that both Dennis (Stone) and I have worked on that have a proton ATPase always have the Cl^- channel. We found recently that it is very heavily regulated. It looks like it can be closed and opened by what we think is a membrane-associated kinase and a phosphatase. It has been shown that the pH in endocytic vesicles or intracellular compartments which have the proton ATPase is very variable. Some, such as the Golgi or parts of the Golgi, are not acid at all, while the lysosomes can be very acidic. What we think is producing these changes is the degree of openness of the Cl^- channels which is regulated by these membrane-associated kinases and phosphatases.

Russell: So you visualize receptors on these vesicles that are reacting to some intracytoplasmic function of pH?

Al-Awqati: We can only say right now, based on physiological studies, what is happening and what effect ATP has on the conductance of the channel. We have recently purified the channel, so I hope that we will be able to work out the regulatory events, and investigate whether any receptors participate in the activation. The aim of all this work is there may be a receptor which is important for protein transport: a protein comes in from the endoplasmic reticulum and binds to a receptor in the Golgi which activates a kinase or a phosphatase that changes the pH or the membrane potential. That could then act as a targeting signal.

Sabolic: We measured the specificity of the Cl^- channel for various anions. In endosomes from kidney cells and from liver cells these anion channels are not the same. For example, the channel in kidney endosomes, besides Cl^-, also takes up bromide, iodide and thiocyanate but the one in the liver cells does not take thiocyanate. I think that nobody has measured this with thiocyanate before. I know that there are many isolations of endosomal or other intracellular membranes with proton pumps showing very similar characteristics but we now see some differences there.

Stone: The earliest work on that came from Nathan Nelson who compared in parallel the activities of synaptic vesicles, lyosomes and chromaffin granules (Cidon et al 1983). He found differences in the conductivities of the Cl^- channels that were associated with the pumps. All of these compartments within the cell have a different pH—the lysosome is 5, the endosome 6.2 and the Golgi is barely below neutrality. The leading view is that regulation occurs at the level of the Cl^- channel. The question is, is it strictly phosphorylation events or are there intrinsic differences in the Cl^- channels or indeed the pumps that account for the regional differences in pH gradients.

Thomas: Are these Cl^- channels inhibited by SITS?

Stone: Yes, but DIDS is more effective.

Grinstein: If the Cl^- channels are really what makes the difference then in some cases there should be a Δ pH and in other cases a Δ membrane potential, do they really match?

Stone: You mean do they match within the organelle? It's very difficult to measure intraorganelle membrane potential.

Grinstein: Has anybody imaged a cell with cyanines and seen whether the clathrin-coated vesicles have a high potential?

Stone: You can't do it with coated vesicles. You could for endosomes. Coated vesicles are only 1000 Å. Dick Anderson approached this problem by using his DAMP compound. Essentially what he has is a weak permeant base and an antibody that is directed against it. He can use that and look immunocytochemically by electronmicroscopy to determine acidic compartments.

Grinstein: Does anybody know the stoichiometry of the pump? If it were 1:1, it wouldn't be making much sense, pumping protons across the plasma membrane and producing one proton per molecule of ATP hydrolysed.

Stone: I don't know, it is between 2–4, I would imagine, from attempts to measure hydrolysis versus proton uptake.

Machen: So the reason that you need valinomycin in your experiments is that you are reconstituting only the ATPase, and the Cl^- channel is not there?

Stone: That has been lost earlier in the purification.

Sabolic: The Ca^{2+} stimulation of this ATPase is interesting—is there any Ca^{2+} pumping?

Stone: None, it uncouples the enzyme.

Sabolic: It seems that it is easy to show the presence of H^+ ATPase in intracellular organelles but not so easy in the plasma membrane. We (I. Sabolic, Z. Zimolo and G. Burckhardt) did some experiments in the kidney. There is an old story that proximal tubule cells contain not only a Na^+/H^+ exchanger for regulation of intracellular pH and secretion of protons but also an ATP-dependent proton pump. These would both be located on the luminal side of the cell, which means that this proton pump should, according to the data of Schwartz & Al-Awqati (1985), in some way be connected with endosomes by exo- and endocytosis. In an attempt to resolve this problem, we isolated endosomes from rat renal cortex and characterized this ATPase. We then looked for this ATPase in cortical brush border membranes. The problem is that nobody has shown directly that brush border membranes do have a proton pump, because the isolated brush border membranes are right side out and when ATP is added outside it simply cannot get inside the vesicle. We characterized the ATPase activities in closed and in open membrane vesicles,

TABLE 1 (*Sabolic*) Effect of various inhibitors on ATPase activity in closed and opened endocytotic vesicles and brush border membrane vesicles (BBMV). ATPase activity was measured in intact (−OG) and opened (+OG) vesicles by assaying the release of inorganic phosphate (Sabolic & Burckhardt 1988) in the presence or absence of inhibitors. Vesicles were opened by 0.5% octylglucoside for 15 min at room temperature. Results shown are means +/− SEM obtained with 5–6 membrane preparations. Statistically significant differences from control preparations are indicated by the asterisk.

	ATPase activity (μmol Pi/15 min/mg protein)			
	Endocytotic vesicles		*BBMV*	
Inhibitor	*−OG*	*+OG*	*−OG*	*+OG*
None	2.95±0.18	4.38±0.61*	12.13±1.31	8.43±0.35*
NEM	2.37±0.14	2.13±0.39	12.17±1.48	5.87±0.59*
DCCD	2.54±0.24	2.33±0.41	6.74±0.50	2.99±0.82
DES	3.39±0.25	2.74±0.37	5.59±0.80	1.88±0.80

i.e. in the absence and presence of detergent. In the absence of detergent (Table 1), the endosomes contain an ATPase activity which is slightly sensitive to N-ethyl maleimide (NEM), DCCD and not to diethylstilbestrol (DES). After opening the vesicles with detergent this ATPase is strongly stimulated. The stimulation can be attributed to the H^+ ATPase present in these membranes (Sabolic & Burckhardt 1988). This stimulation is completely inhibited by NEM, DCCD and DES.

We then looked in brush border membranes. In closed brush border membranes, which are right side out, there is a much greater ATPase activity than in endosomes. However, this ATPase was not affected by NEM. It is inhibited by DCCD and DES. After opening these membranes some of the activity is lost and there is additional inhibition by NEM, as well as the inhibition by DCCD and DES. We concluded that renal brush border membranes have a DCCD-

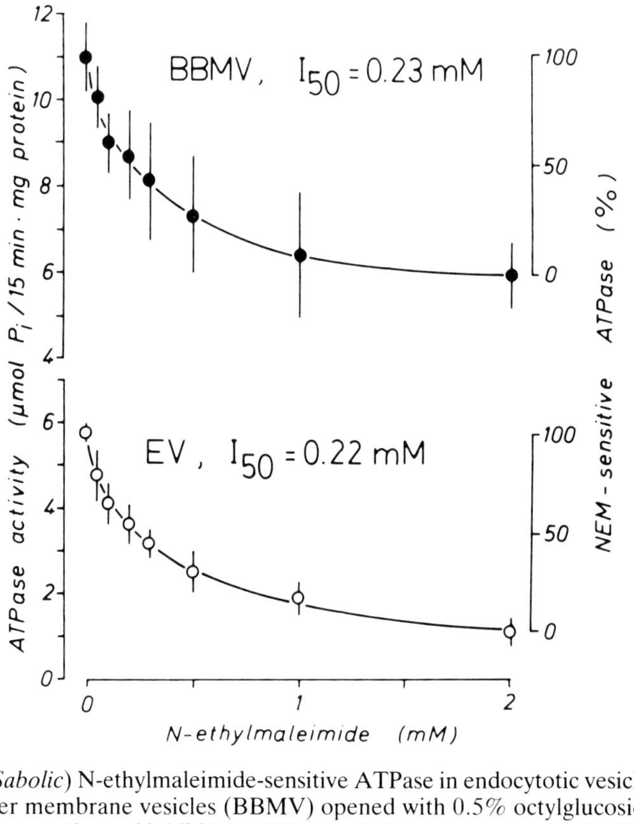

FIG. 1. (*Sabolic*) N-ethylmaleimide-sensitive ATPase in endocytotic vesicles (EV) and brush border membrane vesicles (BBMV) opened with 0.5% octylglucoside; effect of various concentrations of inhibitor. ATPase activity was measured as in Table 1, using 50 µg and 25 µg protein of EV and BBMV, respectively. Each point is the mean +/− SEM obtained from four vesicle preparations.

sensitive and DES-sensitive but NEM-insensitive ecto-ATPase and an NEM-sensitive ATPase present on the internal side of the vesicle membrane.

We then compared the effects of various concentrations of NEM on the NEM-sensitive ATPase in open endosomes and open brush border membranes. In endosomes this ATPase is nearly completely inhibited by 2 mM NEM (Fig. 1). The I_{50} for this inhibition is 0.22 mM. In open brush border membranes, of the total ATPase activity about 40% is sensitive to NEM; the I_{50} for NEM is 0.23 mM. Therefore, at least according to NEM sensitivity, brush border membranes when open have the same ATPase as endosomes.

Thomas: That ecto ATPase, where does it get its ATP from in a physiological situation? Is there any ATP in the tubule fluid?

Sabolic: I don't know, there is some in blood.

Frömter: What is the possible function of an ecto ATPase on other cell membranes? Have they been found on other cells?

Sabolic: I think they have been identified in mouse macrophages (Gordon & Cohn 1970) and some other cell types but nobody talks about their function. One possible function could be that it just splits ATP; it has been shown that ATP in high concentrations or even physiological concentrations, 0.5–1.0 mM, is dangerous for the cell. It makes some cells leaky for various molecules, including nucleotides (Rozengurt et al 1977), monovalent and divalent cations (Sung et al 1985) or even proteins (Otero & Carrasco 1986).

Grinstein: But the leakiness is induced only in the absence of divalent cations.

Sabolic: Not necessarily.

Morad: There have been reports that addition of ATP outside smooth muscle and the heart activates monovalent cation channels in the 0.5–1 mM range.

References

Arai H, Berne M, Forgac M 1987 Inhibition of the coated vesicle proton pump and labeling of a 17,000 Dalton polypeptide by N,N′-dicyclohexylcarbodiimide. J Biol Chem 262:11006–11011

Cidon S, Ben-David H, Nelson N 1983 ATP-driven proton fluxes across membranes of secretory organelles. J Biol Chem 258:11684–11688

Forgac M, Berne M 1986 Structural characterisation of the ATP-hydrolyzing portion of the coated vesicle proton pump. Biochemistry 25:4275–4280

Gordon S, Cohn Z 1970 Macrophage-melanocyte heterokaryons I. Preparation and properties. J Exp Med 131:981–1003

Heinz A, Sachs G, Schafer JA 1981 Evidence for activation of an electrogenic proton pump in Ehrlich Ascites tumor cells during glycolysis. J Membr Biol 61:143–154

Oberleithner H, Guggino W, Giebisch G 1982 Mechanism of distal tubular chloride transport in amphiuma kidney. Am J Physiol 242:F331–339

Otero MJ, Carrasco L 1986 External ATP permeabilizes transformed cells to macromolecules. Biochem Biophys Res Commun 134:453–460

Rozengurt E, Heppel LA, Friedberg I 1977 Effect of exogenous ATP on the permeability properties of transformed cultures of mouse cell lines. J Biol Chem 252:4584–4590
Sabolic I, Burckhardt G 1988 Proton ATPase in rat renal cortical endocytotic vesicles. Biochim Biophys Acta 937:398–410
Schwartz GJ, Al-Awqati Q 1985 Carbon dioxide causes exocytosis of vesicles containing H^+ pumps in isolated perfused proximal and collecting tubules. J Clin Invest 75:1638–1644
Spencer TL, Lehninger AL 1976 L-Lactate transport in Ehrlich ascites tumour cells. Biochem J 154:405–414
Sung SSJ, Young JD, Orgilio AM, Heiple JM, Kaback HR, Silverstein SC 1985 Extracellular ATP perturbs transmembrane ion fluxes, elevates cytosolic Ca^{2+}, and inhibits phagocytosis in mouse macrophages. J Biol Chem 260:13442–13449
Xie X-S, Stone DK 1986 Isolation and reconstitution of the clathrin-coated vesicle proton translocating complex. J Biol Chem 261:2492-2495

General discussion V

Base-induced hyperpolarization of the basolateral membrane potential of renal proximal tubule cells

Steels: Dr M Granitzer and I wanted to investigate further previous observations on the influence of pH on K^+ conductance in *Necturus* renal proximal tubule cells (Steels & Boulpaep 1987). Therefore the kidney was artificially perfused with a HCO_3^- free Hepes-buffered solution on both sides of the tubular wall: the peritubular (P) and luminal (L) compartments. The solutions contained lactate 3.6 mM, butyrate 3 mM, glutamine 0.5 mM, glutamate 0.05 mM, alanine 0.5 mM, lysine 0.2 mM. As shown in Fig.1, a short-term peritubular change in pH from 7.5 to 8.5, in the presence of organic acids on both sides of the tubular wall elicits a sustained hyperpolarization of the basolateral membrane potential, V_1, of about 10 mV.

In previous experiments (Steels & Boulpaep 1987), we assessed the interference of pH with Na^+, Cl^- or K^+ conductance. Any appreciable role for Na^+ or Cl^- was excluded, since depletion of Na^+ or Cl^- from the perfusion fluid did not diminish the base-induced hyperpolarization (BIH) significantly, whereas the K^+ transference number (t_{K+}) of the peritubular membrane increased in alkaline pH. Therefore we attributed the BIH to an increase in K^+ conductance. However, in recent experiments we found that the BIH in the presence of Ba^{2+} at the peritubular and/or luminal side was not significantly reduced, whereas t_{K+} measured in the presence of Ba^{2+} at both pH values was not significantly different from zero, indicating that the K^+ channels were blocked. It is therefore possible that the increase in apparent t_{K+} observed in alkaline pH is not primarily due to a direct action of pH on K^+ channels, but is secondary to a hyperpolarization elicited by the alkaline pH through a mechanism that is independent of the K^+ conductance. It has been shown that K^+ channels are voltage gated (Sackin & Palmer 1987).

In order to evaluate a possible role of the organic acids in determining the BIH, the peritubular pH change from 7.5 to 8.5 was carried out when organic acids were suddenly withdrawn from the perfusion solution either at the peritubular side only, or at both sides. As shown in Fig.1, there is a significant reduction in the pH response of the peritubular membrane potential in the absence of organic acids.

Thomas: There is no electrogenic transport system involving those acids, is there?

Frömter: Yes, there is a Na^+ glutamate co-transporter in the basolateral cell

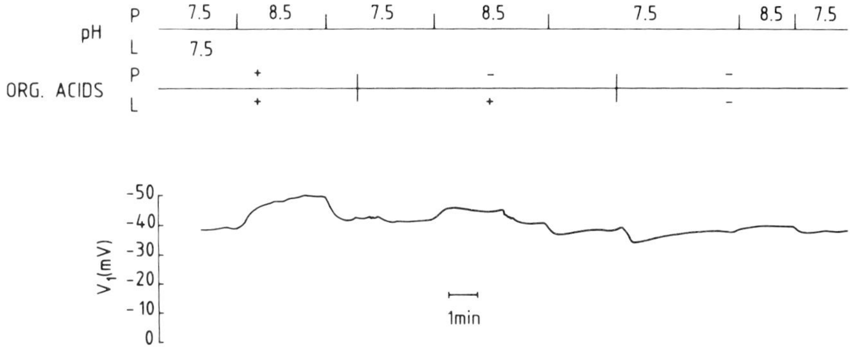

FIG. 1. (Steels) Typical recording of the basolateral membrane potential difference (V_1) of Necturus renal proximal tubule cell, when the peritubular solution (P) was changed from pH 7.5 to pH 8.5 in the presence (+) or absence (-) of organic acids in the artificial perfusion fluid at the luminal (L) and/or peritubular side (P).

membrane, at least in the rat (Samarzija & Frömter 1982). Whether it is also present in *Necturus*, I don't know.

Boron: What's the stoichiometry?

Frömter: The stoichiometry is probably 2:1 Na^+:glutamate.

Thomas: Right, but you don't think it's the pH effect at all.

Steels: On the K^+ channel, I don't think it is an exclusive primary effect. Anagnostopoulos & Planelles (1979) observed an electrical response of the peritubular membrane of *Necturus* proximal tubule to lactate but not to glutamate. Addition of alanine and lysine to the peritubular fluid did not cause any immediate change in peritubular membrane potential of *Triturus* proximal tubule (Hoshi et al 1976). We (Steels & Boulpaep 1987) ranked transport numbers of buffer anions for the basolateral membrane of *Necturus* proximal tubule cell as follows: $t_{HCO3} < t_{butyrate} < t_{phosphate}$.

References

Anagnostopoulus T, Planelles G 1979 Organic anion permeation at the proximal tubule of Necturus. An electrophysiological study of the peritubular membrane. Pfluegers Arch Eur J Physiol 381:231–239

Hoshi T, Sudo K, Suzuki Y 1976 Characteristics of changes in the intracellular potential associated with transport of neutral, dibasic and acidic amino-acids in Triturus proximal tubule. Biochim Biophys Acta 448:492-504

Sackin H, Palmer LG 1987 Basolateral potassium channels in renal proximal tubule. Am J Physiol 253:F476–F487

Samarzija I, Frömter E 1982 Electrophysiological analysis of rat renal sugar and amino acid transport. V Acidic amino acids. Pfluegers Arch Eur J Physiol 393:215–221

Steels PS, Boulpaep EL 1987 pH dependent electrical properties and buffer permeability of Necturus proximal tubule cell. J Membr Biol 100:165–182

Summary

Thomas: We have heard a great deal about a number of different H^+ transport mechanisms and ways of regulating intracellular pH. I would like to summarize these, together with their respective inhibitors and other characteristics (Table 1); additional information would be very welcome. I shall start with electroneutral mechanisms.

Perhaps the most widespread of the mechanisms involving a carrier is the Na^+-dependent H^+ extruder (number 2 in the Table). This is blocked by amiloride in different doses, although in crayfish neurons Bill Moody found that amiloride was ineffective.

TABLE 1 ***(Thomas)*** **How H^+ (and/or equivalent) ions cross cell membranes**

Name (Inhibitor)	*Model*	*Comments*
(a) Electroneutral processes		
1. Diffusion of uncharged weak acids and base molecules (Impossible to inhibit)	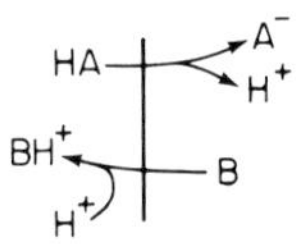	Depends on lipid solubility Physiologically, CO_2 the most important acid
2. Na^+/H^+ exchange (Amiloride)	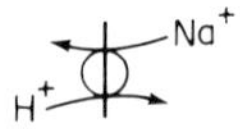	Very widespread Regulates pH_i in many cells
3. Na^+-dependent Cl^-/HCO_3^- exchange (SITS, DIDS)	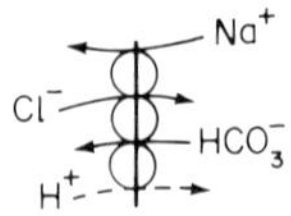	pH_i regulation in many invertebrate cells, some vertebrate Role of H^+ uncertain
4. Cl^-/HCO_3^- exchange (SITS, DIDS)	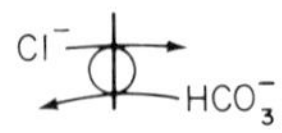	Long known in red blood cells, role in other cells not clear

5. H^+/K^+-ATPase	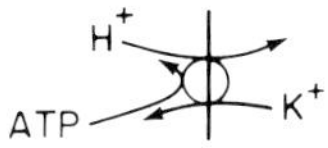	Acid secretion into stomach
6. H^+-lactate co-transport (α-cyano-4-hydroxycinnamate)		Muscle recovery from anaerobic glycolysis
7. Cl^-/Formate exchange		Kidney only?
8. Na^+-lactate co-transport	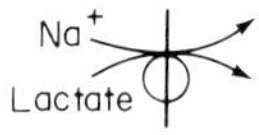	Kidney only?

(b) Electrogenic processes

9. Na^+-HCO_3^- co-transport (SITS etc)	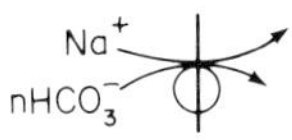	Epithelia and smooth muscle Mechanism unclear, carbonate may participate
10. Electrogenic H^+ pump (ethoxzolamide)	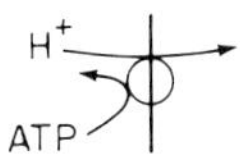	Epithelia only?
11. H^+ channel (Zn^{2+}, Cd^{2+})	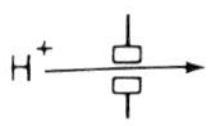	Snail neurons and some eggs when depolarized
12. Inhibitory anion channel	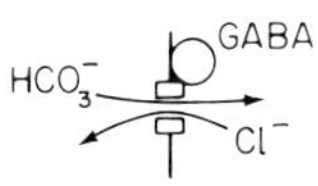	Bicarbonate often higher inside cells than Cl^-

Aickin: In pH_i recovery in mouse skeletal muscle, Na^+/H^+ exchange is slowed by only a very small reduction in external Na^+. This is probably why amiloride works in the presence of Na^+ at a relatively low dose in this preparation.

Frömter: Harmaline is supposed to inhibit a number of Na^+ co-transporters—it acts at Na^+ binding sites.

Thomas: Mechanism number 3 is the Na^+-dependent Cl^-/HCO_3^- exchanger.

Inhibitors of this also inhibit the simple Cl^-/HCO_3^- exchanger (number 4), the only difference being that one is Na^+ dependent. There is some suggestion that probenecid may be specific for the Na^+-dependent mechanism; picryl sulphonic acid may be as well.

Harvey: Meclofenamate inhibits the simple Cl^-/HCO_3^- exchanger in frog skin and in red blood cells (Duranti et al 1986, Cousin & Motais 1979).

Steinmetz: Isn't that an indirect effect through Ca^{2+}-dependent proteins, such as calmodulin?

Frömter: That may not be acting as a specific molecular inhibitor, it could be intracellular down-regulation.

Thomas: The fifth process on my list is the H^+/K^+-ATPase, responsible for acid secretion against an enormous gradient in the stomach.

Machen: There are two inhibitors: SCH 28080 (made by Schering) binds to the K^+ site with high affinity and seems to be specific. Omeprazole (made by Hassle) seems to act primarily after activation in an acidic environment, where it is converted to a compound that binds to sulphydryl groups on the ATPase. Omeprazole appears to be less specific than SCH 28080.

Thomas: Mechanism number 6 is the lactate transporter found in skeletal muscle. Numbers 7 and 8 transport acid only very indirectly, since pH will not be changed until the weak acid anion picks up a proton and crosses the membrane by non-ionic diffusion (mechanism number 1).

Coming to the electrogenic processes, or more accurately rheogenic (current-generating), the first (number 9) was discovered in kidney tubules but may be quite widespread. It is not clear whether it transports bicarbonate, carbonate or both.

Boron: Evidence from Krapf and co-workers at the University of California at San Francisco indicates that the transporter works at ambient p_{CO_2}, in nominally HCO_3^--free solutions. In addition, the work of Warnock's and Aronson's groups suggests that the transporter is partially supported by sulphite.

Thomas: The second of these mechanisms is the electrogenic H^+ pump, which in frog skin is indirectly inhibited by carbonic anhydrase inhibitors.

Boron: DCCD inhibits these pumps in vesicle preparations.

Thomas: Finally two channels: one for H^+ and one for Cl^-, which will, however, let HCO_3^- through. (By a quirk of fate, I started my Ph.D. working on the ionic selectivity of the latter in snail neurons and am now, 27 years later, working on the former, still on snail neurons!)

To bring this symposium to a close, I should just like to say that this field of pH regulation has been a popular subject for investigation for only the past 15 years or so. It is now progressing rapidly as the techniques become more refined, for example the development of fluorescent dyes and ligand-type microelectrodes, enabling people to make measurements on a much greater variety of cells. The importance of pH and the need for it to be strictly regulated by powerful systems are only now being appreciated. Future developments require the synthesis or isolation of more potent and specific inhibitors, particularly some to distinguish between the various mechanisms inhibited by SITS. The role of pH regulation in the action of growth factors, helping the cell to meet the needs imposed by their stimulatory effects, means that this subject will become influential in a wider field.

References

Cousin JL, Motais R 1979 Inhibition of anion permeability by amphiphilic compounds in human red cell: evidence for an interaction of niflumic acid with the band 3 protein. J Membr Biol 46:125–153

Duranti E, Ehrenfeld J, Harvey BJ 1986 Acid secretion through the *Rana esculenta* skin: involvement of an anion-exchange mechanism at the basolateral membrane. J Physiol (Lond) 378:195–211

Index of contributors

Non-participating co-authors are indicated by asterisks. Entries in bold type indicate papers; other entries refer to discussion contributions

Indexes compiled by Liza Weinkove

Subject index